W9-AQE-336

ALGEBRA

THIRD EDITION

Douglas Downing, Ph.D.
Seattle Pacific University

ST. PHILIP'S COLLEGE LIBRARY

QA
152.2
.D69
1996

© Copyright 1996 by Barron's Educational Series, Inc.

Prior editions copyright © 1989, 1983 by Barron's Educational Series, Inc.

All rights reserved.
No part of this book may be reproduced in any form, by photostat, microfilm, xerography, or any other means, or incorporated into any information retrieval system, electronic or mechanical, without the written permission of the copyright owner.

All inquiries should be addressed to:
Barron's Educational Series, Inc.
250 Wireless Boulevard
Hauppauge, New York 11788

Library of Congress Catalog Card No. 96-15413
International Standard Book No. 0-8120-9393-3

Library of Congress Cataloging-in-Publication Data
Downing, Douglas.
Algebra the easy way / Douglas Downing. ; [illustrations by Susan Detrich]. — 3rd ed.
p. cm.
At head of title: Barron's
Includes index.
ISBN 0-8120-9393-3
1. Algebra. I. Title
QA152.2.D69 1996
512 — dc19

96-15413
CIP

PRINTED IN THE UNITED STATES OF AMERICA
9

Illustrations by Susan Detrich

00/17.89

This book is for Charlie Roberts

Acknowledgments

I would like to thank Clint Charlson, Linda Roberts, and my parents Robert Downing and Peggy Downing for their help. Dr. Jeffrey Clark provided important inspiration and Dr. Mary Budinger provided creative assistance. Special thanks also to Susan Detrich for the marvelous illustrations.

Contents

Introduction

This book tells of the adventures that took place in the faraway land of Carmorra. During the course of the adventures, we discovered algebra. This book covers the topics that are covered in high school algebra courses. However, it is not written as a conventional mathematics book. It is written as an adventure novel. None of the characters in the story know algebra at the beginning of the book. However, like you, they will learn it.

Why should you study algebra? For many of you, the answer might be obvious: because you have to, whether you like it or not. Here are some other reasons why you might be interested in learning algebra. First, there are many practical problems that you can solve once you know algebra. You will use algebra constantly if you study engineering, computers, physics, chemistry, economics, or business. Second, you may become interested in algebra because of the way that its logical structure is put together. I hope that you will begin to appreciate the way a logical structure is developed starting from a few basic terms and axioms. You may learn to like other things about algebra. For example, the character Marcus Recordis in the book likes the sense of power he feels when he is able to choose whatever letter he wants for a variable name. And you will find that most new aspects of algebraic notation that seem complicated at first are introduced to reduce the amount of writing that you need to do. If you like lots of tiresome writing, you will not appreciate the ability to write algebraic expressions in a concise manner. However, if you don't, then you can gain a great deal from learning algebra.

Why is algebra so hard? For many people, algebra is admittedly a difficult subject to master. One reason is that some algebraic derivations are very tedious and require a lot of writing. (The binomial formula is a good example.) However, other derivations are very short and elegant, such as the derivation of the equation for a circle. Tedium cannot be the only reason algebra is hard. Part of the difficulty is that algebra requires you to think in very general terms. It requires you to become accustomed to making statements about abstract symbols. Chapter 1 introduces the concept of making general statements about the properties of numbers and the use of letters to act as symbols representing numbers. Chapter 2 introduces the basic problem in algebra: finding the solution to an equation.

It is possible to discover many useful mathematical ideas without ever learning to think in general terms—in other words, without ever learning algebra. Examples of people who did this include the ancient Babylonians. The Babylonians discovered many important mathematical results, including an approximation for π and a version of the quadratic formula. However, the Babylonians never learned to write a formula in general terms and then prove it to be true. The ancient Greeks were the first people who developed a sound logical structure for mathematics. The ideas in this book have been developed by many people in the years since then.

To understand this book you will need a background in arithmetic. You should be familiar with common fractions and decimal fractions. These concepts are reviewed in Chapter 4. Chapter 16 (Proofs by Mathematical Induction) includes some material that is more complicated than most of the book; if you find it heavy going you can skip it without losing the flow of the story. (It *is* interesting, though.)

Mastering algebra requires practice, so the book includes exercises to help you develop more skill. The answers to the exercises are included at the back of the book.

There are several tools that can help you learn algebra by enabling you to accomplish the routine tasks more easily:

- A simple arithmetic calculator. It is now very easy to obtain an inexpensive pocket calculator that can perform the four basic arithmetic functions: addition, multiplication, subtraction, and division. This type of calculator gives you all the computational aid you will need for most of the first half of the book.
- A scientific calculator. In addition to performing basic arithmetic, a scientific calculator will perform exponentiation, root extraction, and logarithms. Previous generations of algebra students suffered from the cumbersome labor required to look up logarithms in a table. The ability to summon logarithms at the touch of a button frees your time to enable you to concentrate on learning the concepts. The ability to calculate trigonometric functions with a scientific calculator is also very helpful if you will later study trigonometry.
- A graphing calculator. A newer, more advanced computational aid is a graphing calculator. While still being easily portable, these calculators contain a display screen that makes it possible to visualize the graph of a curve, such as a polynomial. By the time you reach chapters 11 and 13 of the book, it will be very helpful for you to be able to create your own graphs of algebraic curves, either with a graphing calculator or with a computer as described below. That is the best way for you to develop your own feel for how these curves behave.
- A symbolic algebra calculator. Some advanced calculators even allow you to perform symbolic operations. Standard calculators only performed com-

putations on numbers; for example, they could tell you 2 + 3 = 5. Symbolic algebra calculators can operate on algebraic expressions; for example, they could tell you that $(a + b)^2 = a^2 + 2ab + b^2$. Relying on such a calculator is no substitute for learning to perform these operations by yourself, but these tools can be a big help if you have a lot of algebra to do in a short time.

- A business/financial calculator. A business or financial calculator can perform many of the operations you will need in algebra, in addition to the specialty formulas it will calculate automatically.
- A computer spreadsheet. Spreadsheet programs, such as Lotus 1-2-3 or Microsoft Excel, can be used both to perform calculations and to draw graphs. Some of the exercises in this book give suggestions for how to use spreadsheets to calculate points and draw graphs. Some spreadsheet programs also contain built-in routines for the complicated calculations used near the end of the book, such as finding matrix inverses and determinants.
- A computer programming language. Learning to write your own programs is the best way to understand the way that computers solve problems. The exercises give several examples of programs. Most of these are written in BASIC, one of the easiest languages for beginners to learn. One example is in Pascal to show the features of a more powerful language. The basic concepts of these programs can be adapted to other languages.

Algebra is only the beginning of our discoveries in mathematics. If you are interested in further adventures in the land of Carmorra you may read the books *Trigonometry the Easy Way* and *Calculus the Easy Way.*

Good luck. You're now about to set out on the journey of learning algebra.

1 Rules of Behavior for Numbers

The people in the strange, faraway land called Carmorra knew arithmetic quite well. Of all the people in the kingdom, the person who could do arithmetic the fastest was Marcus Recordis, the Royal Keeper of the Records. So it was natural that Mrs. O'Reilly, the owner and manager of the Carmorra Beachfront Hotel, asked Recordis and the other members of the Royal Court to help her keep track of the hotel guests. Everyone felt it was time for a vacation, so we made plans to spend a few weeks at the hotel. I joined Recordis, the king, and Professor Stanislavsky on the journey.

"I know all my addition and multiplication tables by heart," Recordis said proudly. "Well, except for the twelves—I always did have trouble with the twelves. But I know arithmetic so well that I know exactly how numbers behave."

We performed many addition calculations while we were at the hotel. For example, if there were 56 guests in the hotel on Monday night and 6 more guests checked in on Tuesday, then there were $56 + 6 = 62$ guests in the hotel Tuesday night. We also needed to use subtraction. For example, if there were 62 guests Tuesday night and 10 guests checked out on Wednesday, then there were $62 - 10 = 52$ guests Wednesday night.

One day five families each with four people arrived at the front desk at the same time. We used multiplication to calculate that a total of $5 \times 4 = 20$ people had arrived. (We called the answer to a multiplication problem the *product*, so we could say that 20 was the product of 5 and 4.) Later that day, 28 people in a scout troop arrived, and we needed to split them up so that there were four people in a room. This problem called for division. We calculated that it would take $28 \div 4 = 7$ rooms to accommodate the entire troop. We had also decided that we could write division problems as fractions, like this: $28 \div 4 = 28/4 = 7$.

We had learned all of this in arithmetic, so we had no trouble managing the hotel's brisk business. However, occasionally Recordis would become confused. One weekend he got the cards for Saturday and Sunday mixed up. "I know that six guests checked in one day, and two guests checked in the other day, but I can't remember whether it was six on Saturday and two on Sunday or two on Saturday and six on Sunday. So I don't know if the total number of guests that checked in was $2 + 6$ or $6 + 2$. We'll have to call them all to find out when they checked in."

"What difference does it make?" the professor said, looking up from her work with her latest special research project. "When you add two numbers, it doesn't make any difference which order you add them in. We know that $2 + 6 = 8$, and we also know that $6 + 2 = 8$, so we're sure that 8 guests checked in over the weekend."

"That worked out lucky for us this time," Recordis sighed in relief.

"It wasn't luck!" the professor insisted. "If you add *any* two numbers together, the order doesn't matter." She tried some examples:

$$3 + 5 = 8 \qquad 5 + 3 = 8$$

$$16 + 10 = 26 \qquad 10 + 16 = 26$$

$$9 + 12 = 21 \qquad 12 + 9 = 21$$

$$1 + 9 = 10 \qquad 9 + 1 = 10$$

Recordis could quickly see that this rule would save a little bit of work, so he wrote it in the notebook in which he recorded Significant Things:

> If you add any two numbers together, it doesn't matter in which order you add them.

He called this rule the "Order-doesn't-make-a-difference Property of Addition." (The formal name is the *commutative* property.)

The king joined the discussion with a cautionary note. "How do we know that this property is true for *every* pair of numbers?" he asked.

"We've already checked it for five pairs of numbers," Recordis said. "Isn't that enough to satisfy you?" Although Recordis had been skeptical of this new property at first, it only took a few examples to convince him that the property was absolutely true.

"We could check to make sure that it is true for every possible pair of numbers," the professor said. "Then we'd be absolutely sure."

Recordis grumbled that this sounded like a lot of work, but he began to list all of the numbers:

$$1, 2, 3, 4, 5, 6, 7, 8, 9, 10, 11, 12, \ldots$$

"By the way, how many numbers *are* there?" he asked.

"You can't list them all!" the king said. "It would take you forever to make a list of all of the numbers. There must be an infinite number of numbers—meaning that there is no limit to how many there are."

"That's correct," the professor said. "Since you can always add 1 to any number to produce a bigger number, you can never find the biggest possible number." We used a sideways 8, like this ∞, to stand for infinity.

"It woulld be very difficult to show that the order-doesn't-make-a-difference property is true for all possible numbers," the king said.

"This isn't fair!" Recordis protested. "You would need to find only *one* example in which the property does not hold in order to disprove it. If you could find one single situation where

(first number) + (second number)

does *not* equal

(second number) + (first number)

then we would know for sure that the property does not hold for *all* pairs of numbers. On the other hand, it would take me forever to prove to you that it *does* hold true for all pairs of numbers."

We puzzled over this problem for a while. We finally decided to take a break and have lunch at the hotel restaurant, where we happened to be joined by the chief ferry-boat loader from the ferry dock adjacent to the hotel. The ferry-boat loader explained that his job was quite tricky, since it was imperative that the ferry be perfectly balanced, or else it might tip over. "It all depends on whether I have an odd number of cars or an even number of cars," the ferry loader said.

("What are odd and even numbers?" Recordis whispered to the professor. The record keeper tended to forget most things, so the professor quickly explained. "An even number is a number that can be split evenly in half, such as 2, 4, 6, 8, 10, and so on. An odd number is a number that can't be split in half, such as 1, 3, 5, 7, and so on. They're easy to tell apart, because the last digit of an even number is always 2, 4, 6, 8, or 0, and the last digit of an odd number is always 1, 3, 5, 7, or 9.")

"If there is an even number of cars on a particular trip, then I balance the ferry by placing half of the cars on one side and half on the other side," the ferry loader continued. "If there is an odd number of cars, then I need to place one car in the slot in the middle." (See Figure 1-1.)

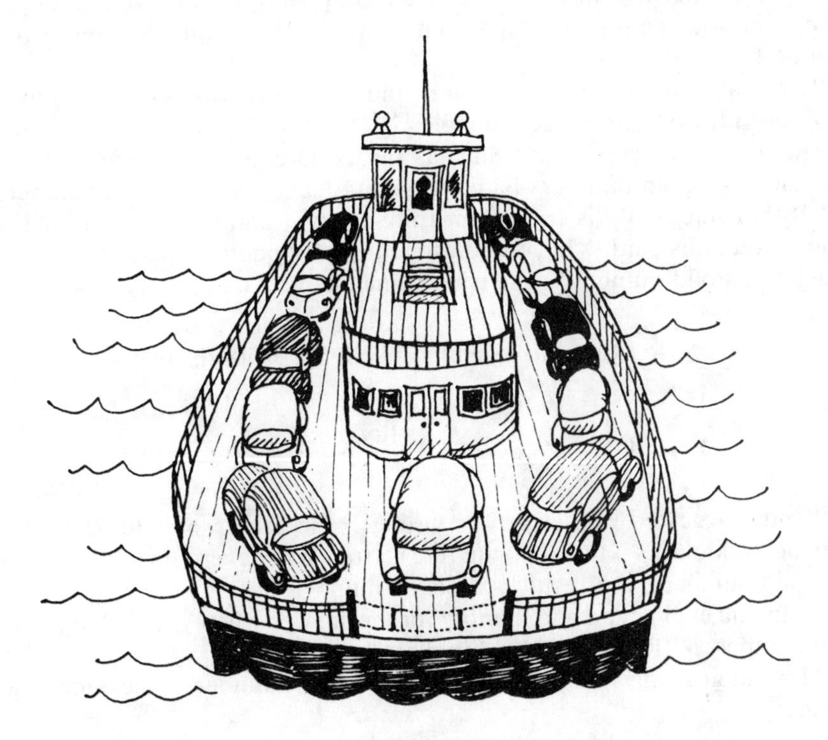

Figure 1–1

ST. PHILIP'S COLLEGE LIBRARY

"The problem is that the last group of cars arrives just before the ferry leaves. I know whether there is an even number of cars on the ferry before the last group arrives, and I can count the number of cars in the last group, so I know whether that number is even or odd. Unfortunately, I don't know whether the total number of cars on the trip is going to be even or odd, so I don't know whether or not to put one car in the middle slot."

He pulled out a napkin and scribbled a formula for us:

(total number of cars on ferry)

= (number of cars on ferry before last group arrives)

+ (number of cars in last group)

"I know whether those two numbers after the equal (=) sign are even or odd, but I need to know whether the total number of cars is even or odd."

"We will need to develop a new behavior rule for numbers," the king said. "We have already found one behavior rule that we are almost certain is true. We need to make up a new rule that tells what happens when you add together odd and even numbers."

We tried some examples to see what happened when two even numbers were added together.

$2 + 4 = 6$ (even)	$10 + 8 = 18$ (even)
$12 + 24 = 36$ (even)	$100 + 48 = 148$ (even)
$2 + 62 = 64$ (even)	$12 + 12 = 24$ (even)

"I see a pattern!" the professor said triumphantly. "It looks as though, whenever you add together two even numbers, the result is another even number!"

"I wouldn't bet the kingdom on it, though," Recordis said. "We may have been lucky with these examples."

The ferry manager agreed that, as far as he could remember, adding together two even numbers had always produced another even number.

"If that rule really is true, then I'm sure that another obvious rule is true," Recordis said. "If you add together two odd numbers, the result must be an odd number." We tried some examples:

$3 + 5 = 8$ (even)	$7 + 11 = 18$ (even)
$15 + 31 = 46$ (even)	$1 + 9 = 10$ (even)
$13 + 13 = 26$ (even)	$1 + 3 = 4$ (even)

"Your suggested rule clearly is wrong!" the professor told Recordis. "In fact, it looks as if the opposite rule is true: Whenever you add together two odd numbers, the result is an even number."

"Can you ever get an odd number for the answer when you add together two numbers?" Recordis asked in puzzlement.

"Try adding one odd number and one even number," the king suggested.

3 + 6 = 9 (odd)	15 + 98 = 113 (odd)
36 + 5 = 41 (odd)	1,983 + 1,984 = 3,967 (odd)
1 + 2 = 3 (odd)	7 + 8 = 15 (odd)

Recordis entered these seemingly correct rules in his notebook:

RULES FOR ADDING ODD AND EVEN NUMBERS

- If you add together two even numbers, the result is an even number.
- If you add together two odd numbers, the result is an even number.
- If you add together one even number and one odd number, the result is an odd number.

"These rules will help a lot!" the grateful ferry loader said when it was time to return to work.

"I am still worried, though," the king said. "We have not been able to *prove* that these rules work. Suppose that they work all the time except for one exception. And if that exceptional situation happens to occur one day, and the ferry loader follows the rules, then the ferry will be unbalanced and it will tip over. That would be disastrous."

"I had thought that we already knew everything that it was possible ever to know about numbers," Recordis said. "After all, we do know our multiplication tables. But I can see how it will help if we can come up with Behavior Rules for numbers."

After lunch Mrs. O'Reilly gave us a very easy problem. She showed us a list of the number of hours that each hotel employee had worked that week, and we had to calculate the payroll. Since each employee was paid $5 per hour (and none of them worked overtime) it was a simple matter of multiplication to calculate the total pay. For example, John Jones had worked 25 hours, and therefore earned $5 \times 25 = 125$ dollars, and Sue Smith had worked 29 hours, earning $5 \times 29 = 145$ dollars.

"Our method of operation in this case is very clear," the professor said. She wrote out a simple rule for how to calculate pay:

> Look at the number of hours on the list. Then multiply by 5.

"That is a very good, easy-to-follow rule," Recordis agreed. "However, there must be a shorter way to write the rule." Since Recordis' job involved a lot of writing, he often suffered from writer's cramp, and was always looking for ways to reduce the amount of writing he needed to do.

The king suggested that we could write the expression for pay as follows:

5 times (number of hours worked)

"We can shorten that further by using the multiplication sign," the professor suggested.

$$5 \times (\text{number of hours worked})$$

"We can just write (hours) instead of (number of hours worked)," the professor added.

$$5 \times \text{(hours)}$$

"Good idea," Recordis agree. "Why didn't I think of that?"

"To make it even shorter, we can just use the letter h to stand for hours," the king suggested.

$$5 \times h$$

"Hold it!" Recordis objected violently. "You can't use a letter to stand for a number! Letters and numbers are fundamentally different entities. They can't be mixed, or we will end up in real trouble."

"But the expression $5 \times h$ is much shorter than $5 \times$ (number of hours worked), even though the two expressions mean exactly the same thing," the king said. "We made the expression shorter just to make things easier for you."

"I have always liked things to be shorter," Recordis agreed, "but not at the expense of utter confusion!"

"What exactly does the expression $5 \times h$ mean?" the professor asked. "I'd be more comfortable about accepting the expression if we clarified that."

"The letter h is a symbol that stands for the number of hours," the king said. "It is a very general symbol, because we can apply it to a wide variety of specific circumstances. In any particular situation, we fill in the appropriate value of h. Once we have filled in the value, the expression $5 \times h$ turns into a regular arithmetic expression, and we know how to calculate those. For example, if h is 20 in a particular case, then the expression $5 \times h$ becomes 5×20, which is 100."

"The letter h acts just like the word 'king'," the professor realized. "The word 'king' by itself is a general term that does not refer to a particular person, just as the letter h is a general symbol that does not refer to a particular number. But in a specific situation you can specify which person you mean by 'king', just as in a specific situation you can specify which number you mean by h."

Recordis still objected to the unfamiliar, uncomfortable concept of using letters to stand for numbers. "How can you tell which letter you should use?" he demanded.

"We can use whatever letter we feel like using," the professor said. (We did later adopt some guidelines to tell us what letter we would normally want to use. We decided that we would use a letter from the beginning of the alphabet, such as a, b, or c, to stand for a general *known* quantity. We decided to use letters from near the end of the alphabet, such as x, y, or z, to stand for an *unknown* quantity that we were trying to find the correct value for. There are also some cases where the letter to use is obvious from what it is supposed to represent; for example, we will often use t to represent time, r to represent radius, w to represent width, and n to represent the number of objects in a particular group.)

Recordis had another objection. "There aren't nearly enough letters," he said. "There are only 26 letters, so after we have used all 26 of them we will be absolutely stuck."

"We can use a letter more than once!" the professor exclaimed. "For example, if we use the letter h to stand for hours today, there is no reason why we can't use the letter *h* to stand for something else, such as height, tomorrow."

"But how will we tell which one it means?" Recordis asked.

"We will have to tell from the context—in other words, we will have to know what problem we are working on," the king said. "And we will have to make sure that we don't use the same letter to stand for two different things in the same problem. But there is no reason why we can't use a letter to stand for one thing when we are working on one problem and another thing when we are working on a different problem."

"I personally know four people named John," the professor said. "And that usually doesn't present a problem—just so long as I make clear at the beginning of a conversation which particular John I happen to be talking about at the moment."

Recordis was still grumbling to himself, but since he could not think of any more objections at the moment the king went on to explain some more advantages of using letters to stand for numbers. (A letter that stands for a number is often called a *variable*.)

"We can write the order-doesn't-make-a-difference property of addition in a much shorter fashion," the king said. "Let's use the letter a to represent one number and the letter b to represent another number. Then we can say that

$$a + b = b + a$$

for any values of a and b.

"Now we have a very useful tool that can allow us to develop lots of general results!" the professor said excitedly.

We were interrupted by a loud cackling laugh echoing across the yard. In the next instant a terrifying figure clad in a black cape stood before us.

"You have no idea what you are getting yourselves into!" the mysterious stranger said wickedly. "You are entering the realm of a subject known as *algebra*. You will find that algebra requires a whole new way of thinking—one that is completely different from the memorization of arithmetic results that you have done up to now. Once you learn how to think of numbers in abstract, general terms, you will be able to develop many powerful new concepts—but you will find them terrifyingly difficult to understand. And," he laughed again, "once you fail at algebra, you will find that the doors to all of the rest of mathematics and science will be closed to you forever! Then I shall become king of Carmorra!"

The rest of us cowered in fear (not noticing the *non sequitur*), but the king stood defiantly and looked at the stranger directly in the eyes. "Never!" he said. "We accept your challenge, and we shall learn this subject that you call algebra. And," he added, "by the way, who are you?"

The stranger passed out copies of his business card. "You may call me the gremlin," he said. "But shall we begin. Before you develop rules of behavior for numbers, you will need to define precisely what a number is."

"Everybody knows what a number is," Recordis said. "We learned that in kindergarten." Recordis made a list of numbers:

$$1, 2, 3, 4, 5, 6, 7, 8, 9, 10, 11, 12, \ldots$$

"We called these numbers the *natural numbers* or the *counting numbers*. We even showed how they could be represented on a diagram we called a *number line*." (See Figure 1-2.)

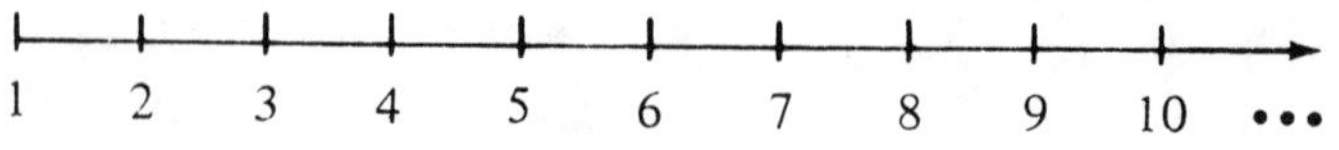

Figure 1–2

However, it turned out to be no easy matter to define the word "number." Recordis suggested that a number was the result you got when you counted something, but then we were unable to define the word "count." We tried other definitions without success.

All this time the gremlin stood laughing at us. He finally posed another question for us. "You have symbolized the operation of addition by a plus sign (+), but what exactly does the operation of addition mean?"

"That's even easier!" Recordis said. "Addition is the operation of adding two numbers together to form a sum." However, we quickly realized that this definition was no good because it used the word "adding" in the definition of the word "addition." Try as we might, we were again unable to come up with a suitable defintion. "Even if I can't define addition, I know what it is when I see it," Recordis muttered.

The gremlin laughed again. "You have hit a dead end even earlier than I had anticipated you would. I shall soon be back to claim the kingdom." He spread his cape and a gust of wind blew him out the window.

Recordis began to moan. "I wonder if people who know algebra get paid more than people like us who don't," he sobbed.

"We will learn algebra, or else we will give our lives in the attempt," the king said. "We will start by giving a precise definition of every term that we use."

We were trying to decide which term to define first when the professor suddenly realized what was wrong. "We can't define *everything!*" she said. "We can only define a word by using other words. Let's say that the word "number" is the first word we want to define. We can't use any words in its definition, because we would not as yet have defined any words."

"The dictionary is able to print a definition for every word," Recordis objected.

"But the dictionary must inevitably have circular definitions!" the professor said. "Suppose we define 'big' to mean 'large,' and 'large' to mean 'big.' That's what I call a circular definition, because it leads you around in circles."

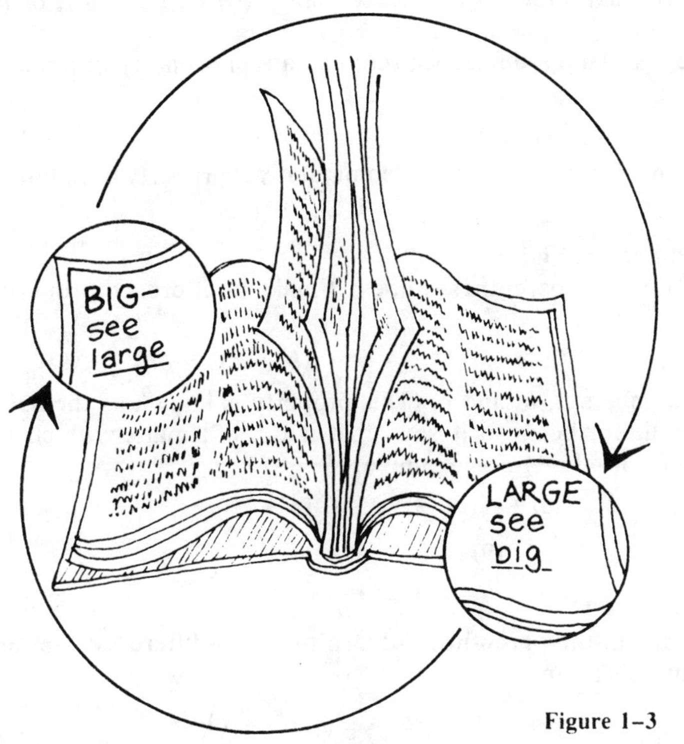

Figure 1–3

Sure enough, the dictionary definition of the word "number" also turned out to be circular. The dictionary defined "number" to mean "a symbol representing how many." When we looked up the definition of the word "many," we found "a large indefinite number."

"What this means," the king said, "is that we have to start with some terms that we will assume we know the meaning of. We can call these the *undefined terms*. Then we will make definitions of the other terms by using these undefined terms."

We decided that we would let *number* and *addition* be undefined terms, since we all had an intuitive idea of what they mean. Then the professor realized that we also needed some behavior rules for these terms. The rules (called *axioms* or *postulates*) would pin down the undefined terms to make sure that they behaved in the way that we intuitively thought they did. We would not prove these postulates. Instead, we would just assume that they were true.

"All right," the king said. "First we will list some properties that we will just assume to be true. Then we can later develop some other properties that we will prove to be true, with no doubt about it."

We spent the next week working on a list of properties that seemed to describe the behavior of numbers. I won't detail all of the arguing that occurred, but we finally came up with the following list:

POSTULATES FOR THE NATURAL NUMBERS

- Every natural number a has a successor $a + 1$. (For example, the successor of 5 is 6, the successor of 10 is 11, and so on.)
- Every natural number a (except 1) has a predecessor $a - 1$. (The predecessor of 4 is 3, the predecessor of 9 is 8, and so on.)
- The natural numbers can be put in order. In other words, if a and b are two natural numbers, then either a is greater than b (written $a > b$), or a is less than b ($a < b$), or a is equal to b ($a = b$.)
- The order-doesn't-make-a-difference (commutative) property of addition:

$$a + b = b + a$$

for any two numbers a and b.
- The order-doesn't-make-a-difference (commutative) property of multiplication:

$$a \times b = b \times a$$

for any two numbers a and b.
- The where-you-put-the-parentheses-doesn't-make-a-difference property of addition:

$$(a + b) + c = a + (b + c)$$

for any three numbers a, b, and c. (This property is known as the *associative* property of addition, because it says that it doesn't matter which numbers associate with each other.) For example,

$$(5 + 4) + 3 = 5 + (4 + 3)$$

$$(9) + 3 = 5 + (7)$$

$$12 = 12$$

- The where-you-put-the-parentheses-doesn't-make-a-difference (associative) property of multiplication:

$$(a \times b) \times c = a \times (b \times c)$$

for any three numbers a, b, and c. For example,

$$(2 \times 6) \times 4 = 2 \times (6 \times 4)$$

$$12 \times 4 = 2 \times 24$$

$$48 = 48$$

The associative properties of addition and multiplication mean that you can leave out the parentheses in expressions such as $(a + b) + c$ and $a \times (b \times c)$ and just write $a + b + c$ and $a \times b \times c$.

We added two more postulates that also seemed useful:

- If you add together two natural numbers, the result will always be a natural number.
- If you multiply together two natural numbers, the result will always be a natural number.

These two properties are called *closure* properties. Imagine that you are fenced in with a set of natural numbers and you are trying to get out. Whenever you add or multiply together two natural numbers, you are still trapped within the set of natural numbers (as if you were closed in). On the other hand, if you divide two natural numbers or subtract two natural numbers, you can sometimes escape outside the set of natural numbers. For example, $6 - 8$ and $4/3$ are not natural numbers. Therefore, the natural numbers do not obey the property of closure with respect to subtraction or division.

We added one more postulate. Mrs. O'Reilly asked us to calculate a holiday bonus for an employee who was to receive $3 for every hour he worked during the two weeks prior to the holiday. The first week he worked 26 hours and the second week he worked 24 hours, so Recordis figured that his bonus was

$$3 \times 26 + 3 \times 24 = 78 + 72 = 150$$

(If more than one operation occurs in a single expression, then the order in which the operations are performed sometimes makes a difference. For example, $2 + 3 \times 4 = 2 + 12 = 14$, whereas $(2 + 3) \times 4 = 5 \times 4 = 20$. In an expression containing both additions and multiplications, it is customary to perform the multiplications first, and then the additions. The only exception to this rule occurs if you are directed otherwise by a set of parentheses, since you always do the operations inside the parentheses first.)

The professor suggested that it would be easier to add the number of hours first and then multiply by 3, like this:

$$3 \times (26 + 24) = 3 \times 50 = 150$$

"Does that always work?" Recordis asked.

We experimented with some numbers and this property seemed to be true, so we added it to our list of postulates. This property has become known as the *distributive* property.

If a, b, and c are any three numbers, then

$$\begin{aligned} a \times (b + c) &= (a \times b) + (a \times c) \\ &= a \times b + a \times c \end{aligned}$$

"We must be careful to avoid ending up with too many postulates," the king said. "We will want to *prove* most of our behavior rules, rather than simply assume them to be true."

"Are we just making these postulates up?" the professor asked philosophically. "Or do they exist already, in which case we are just discovering them?"

"Who cares?" Recordis said. "I think that gremlin was wrong! Algebra will be as easy as pie—none of these properties are hard at all! The only hard part is getting used to the idea of using letters to stand for numbers, and I think I am even beginning to get used to that idea. I remember that at first I didn't like spinach, but once I became accustomed to it I thought it was great."

Next, we developed our first result that we proved, rather than assumed. (A proved result is called a *theorem*.) We set out to prove the addition properties for odd and even numbers. We realized that any even number could be written in the form $2 \times n$, where n is some natural number. An odd number can be written in the form $2 \times n + 1$.

We tried adding together one even number called $2 \times n$ and another even number called $2 \times m$, calling the resulting sum s.

$$s = (2 \times m) + (2 \times n)$$

Using the distributive property,

$$s = 2 \times (m + n)$$

From the closure property, $m + n$ must be a natural number, so s can be written in the form

$$s = 2 \times (\text{some natural number})$$

Therefore, s must be even.

Next, we tried adding together two odd numbers, called $2 \times m + 1$ and $2 \times n + 1$ (again calling the result s).

$$s = (2 \times m + 1) + (2 \times n + 1)$$

$$s = 2 \times m + 2 \times n + 2$$

$$s = 2 \times (m + n + 1)$$

Since $m + n + 1$ must be a natural number, it follows that s must be even.

We had one more combination to do: the sum of an odd number (which we called $2 \times m + 1$) and an even number (which we called $2 \times n$):

$$s = (2 \times n) + (2 \times m + 1)$$

$$s = 2 \times (m + n) + 1$$

Since $2 \times (m + n)$ must be even, it follows that $2 \times (m + n) + 1$ must be odd.

"We did it!" the professor exclaimed in amazement. "We can prove general behavior rules by using symbols to stand for letters! I wasn't even sure that it could be done!

"Now we don't have to worry about the ferry boat tipping over, no matter what the actual number of cars happens to be."

We were all pleased that we had been able to prove our very first theorems in algebra. The professor was excitedly making plans for further exploration of the subject of algebra, but unfortunately our vacation was over and it was time to return to Capital City.

Note to Chapter 1

- There are times when we need more than 26 letters to represent different numbers. In these cases we can put a little number (called a *subscript*) next to the letter, allowing us to use the same letter again. For example, a_1, a_2, a_3, a_{10}, a_{11} are all different variables. Another trick is to use capital letters sometimes. For example, we use A and a as separate variable names if we need to.

Exercises

Verify the associative properties of addition and multiplication in the following cases. To do this, calculate the result of the expression in two different ways.

For example: $3 \times (4 \times 5) = 3 \times 20 = 60$

$(3 \times 4) \times 5 = 12 \times 5 = 60$

1. $12 \times 6 \times 2$
2. $11 \times 5 \times 16$
3. $33 \times 1 \times 5$
4. $2 \times 2 \times 2$
5. $54 + 6 + 20$
6. $23 + 11 + 18$
7. $11 + 12 + 8$
8. $6 + 19 + 4$

Verify the distributive property in these cases.

For example: $6 \times (2 + 10) = 6 \times 12 = 72$

$6 \times 2 + 6 \times 10 = 12 + 60 = 72$

9. $3 \times (4 + 10)$
10. $10 \times (1 + 1)$
11. $6 \times (12 - 7)$
12. $1 \times (6 + 8)$
13. $2 \times (3 + 7)$
14. $3 \times (4 - 2)$
15. $3 \times (10 - 6)$
16. $3 \times 11 + 3 \times 6$
17. $10 \times 6 + 10 \times 8$
18. $5 \times 6 + 5 \times 5$
19. $4 \times 11 + 4 \times 12$

In exercises 20–31, write formulas that perform the indicated functions. (In each case you will need to think of appropriate letters to represent the quantities indicated.)

20. Calculate your pay if you work h hours in one week and are paid $5 per hour.
21. Calculate your pay if you work more than 40 hours and are paid time and one-half for every hour past 40.
22. Calculate how many miles you will travel if you drive 55 miles per hour for h hours.
23. Calculate how many miles you will travel if you drive v miles per hour for h hours.
24. Calculate how long it will take you to travel d miles if you drive at a constant speed of 55 miles per hour.
25. Calculate how long it will take you to travel d miles if you drive at a constant speed of v miles per hour.
26. Calculate the area of a rectangle if you know the lengths of its sides.
27. Convert miles into kilometers.
28. Convert centimeters into inches.
29. Convert pounds into kilograms.
30. Convert degrees Fahrenheit into degrees Celsius.
31. Calculate a baseball player's batting average.
32. Show that $(a + b) \times c = a \times c + b \times c$.
33. If you multiply two even numbers together, is the result even or odd? See if you can prove a rule for that case.
34. If you multiply one even number and one odd number together, is the result even or odd? Explain.
35. If you multiply two odd numbers together, is the result even or odd? Explain.

36. If you add two numbers divisible by a particular number n, will the result be divisible by n? Explain.
37. If you add one number divisible by n and one number not divisible by n, will the result be divisible by n? Explain.
38. If you add two numbers not divisible by n, will the result be divisible by n? Explain.

(Note that Exercises 36 to 38 are generalizations of the rules presented in the chapter for the addition of odd and even numbers. Those rules are special cases of the rules from these exercises, with n having the value 2.)

Calculate the value of each of these expressions, being careful to do the operations in the proper order.

39. $10 + 12 \times 4$
40. $10 + (12 \times 4)$
41. $(10 + 12) \times 4$
42. $10 \times 12 + 4$
43. $10 \times (12 + 4)$
44. $(10 \times 12) + 4$
45. $3 + 4 \times 5 + 6 \times 7 + 8 \times 9 + 10$

2 Equations

When we returned to Capital City, we found that Gerard Macinius Builder, the Royal Construction Engineer, had just completed a huge balance scale. We needed the scale so that we could measure the weight of the friendly giant named Pal who often helped the people of the kingdom when they were in trouble.

"It's time to play on the scale!" Builder called to Pal, who happily jumped on one side of the balance. Then we quickly placed two 100-unit weights on the other side to make the weights on the two sides equal. We figured that Pal must weigh 200 units, since

$$100 + 100 = 200$$

"The balance scale works just like an equation," the king observed. "The equation says that the number on the left is equal to the number on the right, and when the scale is balanced we know that the weight of the object on the left is equal to the weight of the object on the right."

Pal thought it was boring to just sit on one side of the scale, so he jumped over to the other side. To prevent the scale from becoming totally unbalanced, he put the two weights on the other side of the scale:

$$200 = 100 + 100$$

"Stop it!" Recordis shouted to Pal.

"That's all right," Builder responded. "The scale will stay balanced if you switch the weights on the two sides."

"Equations work like that too," the professor said. "It's all right to reverse the two sides of an equation when you want to. We know that

$$100 + 100 = 200$$

means the same thing as

$$200 = 100 + 100$$

"That's obvious!" Recordis said.

Next, Pal jumped off the scale and put two weights on the scale instead. Once again the scale balanced, since

$$100 + 100 = 100 + 100$$

"It's also quite obvious that if the left-hand side is identical to the right-hand side, then the two sides of the equation are equal," Recordis said.

Pal took the weights off the left-hand side and jumped back onto the scale, only this time he was holding his beach ball (which weighed 1 unit). The scale became unbalanced, since

$$200 + 1 \quad \text{does not equal} \quad 100 + 100$$

"Do something! Quickly!" the king shouted.

Builder quickly added a 1-unit weight to the right-hand side. Then the equation again balanced:

$$200 + 1 = 100 + 100 + 1$$

We breathed a sigh of relief. "We were lucky that it balanced," Recordis said.

"Whenever you add the same number to both sides of an equation the equation will still be true," the professor said. "In this case we added 1 to both sides, and the equation remained true."

"Or we could subtract the same number from both sides," the king noted.

$$200 + 1 - 1 = 100 + 100 + 1 - 1$$

"We could multiply both sides of the equation by the same amount," the professor said.

$$2 \times 200 = 2 \times (100 + 100)$$

"Or we could even divide both sides by the same number," the king continued.

$$200 \div 4 = (100 + 100) \div 4$$

"We can summarize these results as a simple rule," Recordis said, writing in his book:

GOLDEN RULE OF EQUATIONS

Whatever you do unto one side of an equation, do the same thing unto the other side of the equation. Then the equation will remain true (assuming that it was true to begin with). In particular, you can:

- add the same number to both sides
- subtract the same number from both sides
- multiply both sides by the same number
- divide both sides by the same number.

However, you cannot divide both sides of an equation by zero, since (as we later found out) you cannot divide anything by zero. Also, although strictly speaking it is legal, it gets you nowhere to multiply both sides of an equation by zero.

If you start with a false equation and do the same thing to both sides of the equation, then of course you will still have a false equation.

"How do we tell the difference between a true equation and a false equation?" Recordis asked.

"That will be easy," the professor said. "Just calculate the value of each side and see if they're the same. For example, $2 + 2 = 5$ is obviously a false equation, since $2 + 2$ is 4, so $2 + 2$ does not equal 5.

We decided to use an equal sign with a slash through it ($\neq$) to stand for *does not equal*. We could then write a false equation like this:

$$2 + 2 \neq 5$$

"I suppose I'll get in trouble for asking this," Recordis said, "but do we have any chance of being able to define the word 'equal'?"

After thinking about this problem, we decided that we would have to leave "equal" (symbolized by =) as an undefined term. However, we were all sure that we intuitively knew what it meant. We wrote out some postulates based on the properties that we had observed while Pal played on the scale. "We'll write each property once in words, just so we're sure we know what it means, and then we'll write it again in symbols to be shorter," Recordis said.

POSTULATES OF EQUALITY

- Any number is equal to itself. In symbols, if a is any number, then

$$a = a$$

  (The technical name for this property is the *reflexive property of equality*.)
- You can reverse the two sides of an equation whenever you feel like it. (Technical name: *symmetric property of equality*.) In symbols, if a and b are any two numbers,

$$a = b \quad \text{means the same thing as} \quad b = a$$

- If two numbers are both equal to a third number, they must be equal to each other. (Technical name: *transitive property*.) In symbols, if a, b, and c are any three numbers, and

$$\text{if } a = c \quad \text{and} \quad b = c, \quad \text{then} \quad a = b$$

- Golden Rule of Equations: Whatever you do to one side of an equation, do exactly the same thing to the other side. In symbols, let a, b, and c be any three numbers. If $a = b$, then

$$a + c = b + c$$

$$a - c = b - c$$

$$a \times c = b \times c$$

$$\frac{a}{c} = \frac{b}{c}$$

(In the last equation, c must not be zero.)

- Substitution Property: If $a = b$, then you can substitute a in the place of b anywhere that b appears in an expression.

"Algebra is still the easiest subject I have ever had to learn!" Recordis said gleefully. "All of these properties are so simple!"

However, that evening we were interrupted by a loud scream from Recordis' room. We went running to see what was the matter. "Some books are missing!" Recordis cried, pointing to a gap in his bookshelf. "I need my record books! I'm totally lost without them!"

"This is serious!" the king said. "We must find out at once who would dare to touch any of Recordis' books! We must send for Ace Sharpeyes, the ace detective!"

The detective quickly was sent for. He wasted no time getting right to business as soon as he arrived. He adjusted his Sherlock Holmes hat as he searched Recordis' room for clues, peering carefully through his magnifying glass. "The first thing we must establish," he pointed out, "is what exactly is missing. How many books are gone?"

"What kind of question is that!" Recordis cried. "I can easily count the books that *are* here—but how am I supposed to count the books that *aren't* here?"

"Calm down," the king said, "You must know how many books there were to begin with."

"I'm supposed to have 62 books," Recordis said.

"There are 54 books left on the shelves," the professor said helpfully.

"Aha!" the detective said. "That is just the clue we need. We can use this information to calculate exactly how many books were taken."

"We can set up an equation," the king said. "We spent this afternoon learning all about equations. We know that:

$$54 + (\text{number of missing books}) = 62$$

We decided to use a question mark (?) as a symbol to represent the number of missing books:

$$54 + ? = 62$$

"Now all we need to do is find the value of the question mark," the professor said.

"Is that a true equation or a false equation?" Recordis asked in puzzlement. "It's not like the equations we were doing this afternoon. The equations we looked at this afternoon contained nothing but numbers, so you could calculate the value of each side of the equation to determine if the equation was true or false."

"We can't say in general," the king finally realized. "That equation might be true sometimes, but it might be false sometimes. For example, suppose we gave ? the value one million. Then we know that the equation is false, since

$$54 + \text{one million} \neq 62$$

"Ah, but the equation *will* be true if ? has the correct value," the detective said. "Our goal is to find the one value of ? that does indeed make the equation into a true equation."

"What we have to do," the professor said shrewdly, "is isolate the question mark so it is by itself on one side of the equation. Then it will be obvious what its true value is."

"How do we do that?" Recordis asked.

"We can use the Golden Rule of Equations," the king said. "Let's subtract 54 from both sides of the equation."

$$54 + ? \mathbf{- 54} = 62 \mathbf{- 54}$$

Recordis agreed that, if the original equation was true, this new equation would also be true. Since $62 - 54 = 8$, we could write the equation:

$$54 + ? - 54 = 8$$

"The two 54's will cancel out," Recordis said helpfully, "since one is added and the other is subtracted." He used shading to clarify what was to be cancelled. That left the equation in this form:

$$? = 8$$

"That's it!" the professor said excitedly. "That must be the solution to the problem! There must have been 8 books taken!"

"Indeed," the detective said. "However, this is only a start. I fear that there will be a long path that we will have to tread until we uncover the identity of the culprit. We will have to investigate many more equations."

"We should find a better symbol than ? to represent the unknown number," Recordis said. "The ? looks too much like a punctuation symbol (probably because it *is* a punctuation symbol)."

"We'll use a letter to represent the unknown number," the king said. "We already know how to use letters to represent numbers."

"Me and my big mouth," Recordis said. "I should have guessed you would say that. I suppose I can get used to using letters to stand for unknown numbers—but mind you, only until we have uncovered their true value and they are no longer unknown."

"We'll let you pick the letter you want to use," the professor said, trying to cheer him up.

"If we're using a letter to represent an unknown number, it needs to be a very dark and mysterious letter," Recordis said. He thought a moment. "We should use x," he decided. "That's the most mysterious-sounding letter I can think of."

We wrote the equation using x to represent the number of missing books.

$$54 + x = 62$$

$$54 + x - 54 = 62 - 54$$

$$x = 8$$

We decided to call this kind of equation a *conditional equation*. If an equation contains expressions involving only numbers, then it is an *arithmetic equation* and it is either true or false. If an equation contains a letter that stands for an unknown number, then it is a conditional equation because it is true only under the condition that the unknown letter has the correct value. We decided to call the value of the unknown that makes the equation true the *solution* of the equation. For example, the equation

$$54 + x = 62$$

has the solution $x = 8$. Finding the solution to conditional equations turns out to be one of the most important problems in all of algebra.

If two equations have the same solution, then they are said to be *equivalent* equations. For example: $x + 7 = 12$ and $x + 3 = 8$ are equivalent equations, because they both have the solution $x = 5$.

"We'll say that when we're finding the correct value of the unknown number we're *solving* the equation," the detective said, "just as I say that I am solving a case when I discover the correct name for the unknown culprit. And, if I were you, I would take this opportunity to gain more practice at solving equations while I continue to look for clues."

"Solving equations isn't hard," Recordis said, "as long as you use the golden rule of equations to isolate x on one side of the equation."

We went to the Main Conference Room to get some more practice with equations. In the Main Conference Room we found Igor waiting for us, ready to draw pictures and symbols. Igor is an amazing machine called a Visiomatic Picture Chalkboard Machine, which is a very unusual mixture of a television screen and a chalkboard. We told Igor we wanted him to draw some equations for us to solve. An equation quickly appeared on the screen:

$$10 + x = 15$$

"That will be easy to solve," Recordis said. "Subtract 10 from both sides."

$$10 + x - 10 = 15 - 10$$

solution: $x = 5$

Igor drew the next equation.

$$x - 32 = 20$$

"Add 32 to both sides," the professor said.

$$x - 32 + 32 = 20 + 32$$

solution: $x = 52$

The next equation was

$$x \times 5 = 35$$

"Divide both sides by 5," the king said.

$$\frac{x \times 5}{5} = \frac{35}{5}$$

solution: $x = 7$

Next equation: $\frac{x}{8} = 6$

Multiply both sides by 8, and the solution is $x = 48$.

The next equation confused us at first, since it contained an x on both sides:

$$3 \times x = 5 + 2 \times x$$

Subtract $2 \times x$ from both sides:

$$(3 \times x) - (2 \times x) = 5 + (2 \times x) - (2 \times x)$$

$$x = 5$$

Next equation:

$$10 \times x + 5 \times x - 35 = x + 7$$

"There's a problem with that expression!" Recordis complained. "The multiplication symbol looks almost the same as the letter x. Sooner or later, we're sure to get them confused. We had better think of a new symbol to stand for multiplication."

"You're suggesting that we replace a time-honored widely-accepted symbol such as the multiplication symbol?" the professor said in shock.

"We don't need to replace it—we just need to think of an alternate symbol to use in cases such as this," the king said. Recordis wanted the symbol to require as little writing as possible. Of all the symbols in the whole world he figured that a little dot (·) required less writing than any of the others. So we decided that the expression $a \cdot x$ would mean the same as the expression $a \times x$.

However, the king had an even more radical idea. "Suppose that, when we multiply two letters, we don't write any symbol between the two letters at all. For example, let's just write ax instead of either $a \cdot x$ or $a \times x$."

"But if there is no symbol between the letters, how do you know that they're supposed to be multiplied?" the professor asked incredulously.

"As much as I like to reduce the amount of writing that we have to do, we can't get away with simply not writing the symbols!" Recordis said. "Suppose I turned in a blank piece of paper this year and said, 'This blank piece of paper represents this year's annual report.'"

But the king was not dissuaded. "We'll have no problem as long as we agree that the absence of a symbol between two letters always represents multiplication, and nothing else." The king listed some examples:

ab means $a \times b$

abc means $a \times b \times c$

$2a$ means $2 \times a$

cx means $c \times x$

$10ahr$ means $10 \times a \times h \times r$

The professor quickly became enthusiastic about this brilliant idea. But Recordis still had objections. "What about *axe*?" he asked. "Does that mean $a \times x \times e$ or does that mean a heavy sharp tool?"

"That obviously depends on whether we're writing words or algebraic expressions!" the professor said.

"What about the expression 23?" Recordis asked. "Does that mean two times three or does it mean twenty-three?"

The king conceded that we could not leave out the multiplication sign when we were multiplying two numbers. "Leaving out the multiplication sign only works when you're multiplying two letters, or when you're multiplying a number times a letter," he admitted.

We were finally all convinced that we should accept the king's plan. It later helped us a lot while doing algebra investigations, and it soon became second nature to us. We called it the *implied multiplication rule*.

IMPLIED MULTIPLICATION RULE

When two letters, or a letter and a number, are written adjacent to each other with no operation symbol between them, then it is implied that they are meant to be multiplied together.

The equation we were trying to solve was:

$$10 \times x + 5 \times x - 35 = x + 7$$

Using implied multiplication, we rewrote the equation:

$$10x + 5x - 35 = x + 7$$

Subtracting x from both sides:

$$10x + 5x - x - 35 = 7$$

Adding 35 to both sides:

$$10x + 5x - x = 7 + 35$$

Simplifying:

$$14x = 42$$

Then we found the final solution: $x = 3$.

For variety, Igor gave us some equations that used y to stand for the unknown variable.

equation	solution
$10y - 20 = 0$	$y = 2$
$6y + 4 = 8y - 10$	$y = 7$
$130y + 6y - 20 = 68y + 184$	$y = 3$
$46y + 38y + 10 = 54y + 10$	$y = 0$

Recordis suggested that we solve the following list of equations:

equation	solution
$3x - 6 = 12$	$x = 6$
$3x - 12 = 12$	$x = 8$
$4x - 4 = 12$	$x = 4$
$4x - 20 = 12$	$x = 8$
$5x - 18 = 12$	$x = 6$

"I had an ulterior motive when I asked you to solve these," Recordis confessed. "I am trying to earn 12 dollars, and I have a choice of five jobs. The wage rates for the jobs are different. However, I also need to take into account the fact that there is a commuting cost for each job. In each of the equations I presented, the first number is the wage rate, and the second number is the commuting cost. The variable x represents the number of hours that I would need to work at each particular job until I earn 12 dollars. I obviously should take the third job, because then I will have to work only four hours to earn 12 dollars."

"We can write a general formula for the solution to your equation," the professor said. "Let's say that p represents the pay rate and c represents the commuting cost. Then we can rewrite your equation like this:

$$px - c = 12$$

And now we can find the solution."

$$px - c + c = 12 + c$$

$$px = 12 + c$$

$$\frac{px}{p} = \frac{12 + c}{p}$$

$$x = \frac{12 + c}{p}$$

"That formula gives the solution!" the professor said.

Recordis stared at this result. "But the solution should be a number!" he protested. "This solution still has two letters in it—p and c."

"I admit that it would do no good to present the solution in a form that depends on letters that represent unknown quantities," the professor said. "But remember that in this case p and c are not unknown, because we know their value in a particular circumstance. An algebraic expression of this form is just as good as a number. In fact, it is even better, because we can use the same expression to calculate the answer for any possible values for p and c. In other words, by just using the formula, we can avoid having to solve the equation each time."

It took Recordis a long time to get used to the difference between letters that represent unknowns (such as x in this example) and letters that represent known quantities (such as p and c). However, he eventually caught on to the idea once he finally began to realize how useful it was to be able to express the solution for an unknown as an expression involving letters representing known quantities.

"I have another problem that I bet we can use equations to find the solution for," Recordis said, trying to take his mind off the previous problem. "I have decided to control precisely the number of calories that I have for lunch. In particular, I have decided that I will have 300 calories—no more and no less. I like to have soups and sandwiches for lunch. Each cup of soup contains 50 calories, and each sandwich contains 100 calories. But I have not as yet been able to figure out how many soups and how many sandwiches I should consume."

We let x stand for the number of sandwiches, and we let y stand for the number of soups. Then the total number of calories that Recordis consumed would be

$$100x + 50y$$

"To make an equation, all we have to do is set that expression equal to 300," the king said.

$$100x + 50y = 300$$

"I know the solution!" the professor said. "Let $x = 3$ and $y = 0$."

"I think the solution is $x = 0$ and $y = 6$," the king said.

"I think the solution is $x = 2$ and $y = 2$," Recordis said.

Before an argument could break out we realized that *all* of these were possible solutions. We also found that $x = 1$, $y = 4$ was a solution.

"How can a problem have more than one right answer?" Recordis asked.

We found that, in general, one equation with two unknowns will usually have more than one solution.

We tried another example,

$$y = 2x$$

Once again we found many possible solutions:

x	y
0	0
1	2
2	4
3	6
4	8
5	10

We took a break from the job of investigating equations. The detective was out looking for clues, but for us the long waiting process had begun.

EQUATIONS

When an equation contains one unknown quantity, then it is often possible to solve for the value of the unknown that makes the equation true. This can be done by using the Golden Rule of Equations (that is, perform exactly the same operation on each side of the equation) until the unknown has been isolated on one side of the equation. (However, some types of equations are too complicated to solve in this manner; see, for example, Chapters 11 and 13.)

When a single equation contains two or more unknown quantities, then it is often possible to find many possible solutions for the equation. In this case a solution consists of a set of values for each of the unknowns in the equation. For example, if the two unknowns are x and y, then a solution to the equation consists of two values: one for x and one for y. (If the number of equations is the same as the number of unknowns that appear, then it is often possible to find a single solution that makes all of the equations true simultaneously; see Chapters 10 and 18.)

Notes to Chapter 2

- Some special equations are true for all possible values of the unknowns they contain. Equations of this kind are called *identities*. Here are some examples of identities.

$$4x = x + x + x + x$$

$$3(a + b) = 3a + 3b$$

$$(a + b)(c + d) = ac + ad + bc + bd$$

Sometimes an equal sign with three bars ($\equiv$) is used in place of the regular equal sign to indicate that an equation is an identity.

- An equation of the form $ax + b = 0$, in which x represents an unknown number and a and b represent known numbers, is said to be a *linear equation*. The reason for this name will become clear later, when we will find that a linear equation can be represented as a line on a graph.
- A statement of the form "x is less than y," written $x < y$, or "a is greater than b," written $a > b$, is called an *inequality*. The arrow in the inequality sign always points to the smaller number. Inequalities containing only numbers will either be true (for example, $10 > 7$) or false (for example, $4 < 3$). Inequalities containing variables (such as $x < 3$) will usually be true for some values of the variable but not for others.

The symbol $\leq$ means "less than or equal to," and the symbol $\geq$ means "greater than or equal to."

A true inequality will still be true if you add or subtract the same quantity from both sides of the inequality. The inequality will still be true if both sides are multiplied by the same positive number. However, we discovered that multiplying both sides of an inequality by a negative number causes the direction of the inequality to become reversed.

Exercises

Solve these equations for x.

1. $3x + 16 = 22$
2. $14 - x = 10$
3. $34 - 10x = 6x + 2$
4. $7x + 5 = 75$
5. $3x - 8 = 16$
6. $6 - 7x + 20 = 5$
7. $9 - x = 1$
8. $x + 2x + 3x + 4x + 5x = 45$
9. $100 - 4x = 60$
10. $10 + 5x = 110$
11. $10 + 5x = 100 - 4x$
12. $12 + ax = 16$
13. $b + ax = 12$
14. $b + ax = c$
15. $ax = bx + c$
16. $ax - c = d$
17. $bx = h$
18. $ax - b = cx + d$
19. If you travel at 45 miles per hour for 5 hours, how far will you travel?
20. If you travel at 30 miles per hour, how long will it take you to travel 3,000 miles?
21. Two people start driving towards each other, starting from two towns 100 miles apart. If one person travels 30 miles per hour and the other person travels 20 miles per hour, how long will it take until they meet?
22. Answer the same question as in the previous problem, only this time assume that the first person travels v_1 miles per hour and the second person travels v_2 miles per hour.
23. Suppose that you are trying to earn $48. You have your choice between working at a hard job that pays $12 per hour, or at an easy job that pays $6 per hour, or you can work some time at both jobs. List some possible ways that you can earn the $48.
24. If CDs cost $12, pizzas cost $6, and you have $48 to spend on these two goods this month, list the possible numbers of pizzas and CDs that you can buy.
25. Farmer Floran has roosters and horses. The animals have a total of 88 feet and 40 wings. How many horses and how many roosters are there?
26. If the city plans to plant a total of 18 trees on two streets, putting twice as many trees on Elm Street as on Maple Street, how many trees will be planted on each street?
27. If Chapter 1 in a book contains five more pages than does Chapter 2, and there are 65 total pages in both chapters, how many pages does each chapter contain?
28. What two consecutive numbers add up to 63?
29. What three consecutive numbers add up to 75?
30. T.J. can type twice as fast as J.R. If they both spend an equal amount of time on a 600-page manuscript, how many pages will they each have typed?
31. A plane is travelling between two cities that are 660 miles apart. The plane can travel at 200 miles per hour relative to the air, but it will go faster relative to the ground if it is helped by the wind. If the plane completes the trip in 3 hours, how fast was the wind blowing?

3 Negative Numbers and Integers

Detective Sharpeyes set out to look for clues in the case of the mysteriously missing books. The next day we received a letter from Mrs. O'Reilly telling about the status of the hotel. "Business has been terrible!" she wrote. "There was a terrible storm yesterday. There were 96 guests the day before, but 96 guests checked out today because of the rain."

"That means there are no guests left," Recordis said, "since 96 − 96 equals nothing."

"We used the symbol 0 in arithmetic to stand for nothing, and we gave it the name 'zero'," the professor said. "That is a number with some peculiar properties that we should investigate now."

"If we list all of the numbers in order, we should put 0 just before 1," Recordis said. (See the number line in Figure 3-1.)

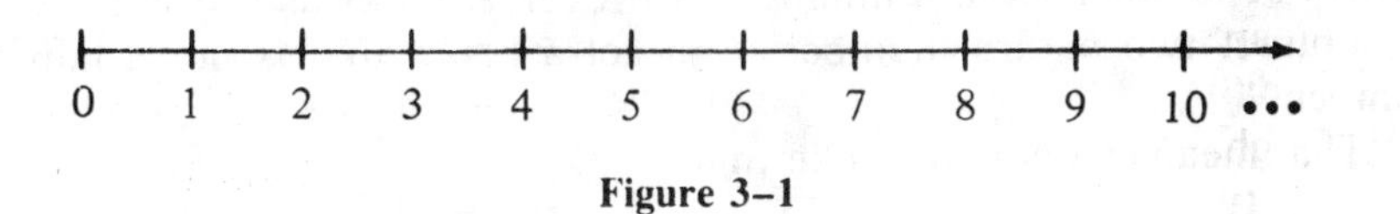

Figure 3–1

We decided to call this set of numbers the *whole numbers*. In other words, the set of whole numbers consists of 0 and all the natural numbers.

"It looks as though zero is powerless if you add it to another number," the professor said. "Look at this: $7 + 0 = 7$; $1 + 0 = 1$; $15 + 0 = 15$. Whenever you add zero to a number, the number remains unchanged."

"But zero is extremely powerful if you multiply it by another number," Recordis noted. "Look at this: $7 \times 0 = 0$; $1 \times 0 = 0$; and $15 \times 0 = 0$. In fact, I bet that if you multiply any number by zero, the result will be zero."

We made a list of the properties of zero.

- Addition-of-zero property:

$$a + 0 = a \qquad \text{for any value of } a$$

- Multiplication-by-zero property:

$$a \times 0 = 0 \qquad \text{for any value of } a$$

We added one more property that seemed important.

- If $ab = 0$, then either $a = 0$ or $b = 0$ or both a and b are equal to zero.

"I'm beginning to like algebra a lot," Recordis said. "All of these properties are so easy."

We continued to read Mrs. O'Reilly's letter. She asked us to calculate her profit for each of several weeks. Calculating the profits was very simple, since we just performed subtraction:

$$\text{profit} = \text{revenue} - \text{expense}$$

For example, one week revenue was 130 dollars and expenses were 111, so the profit was $130 - 111 = 19$. We were proceeding smoothly until suddenly we struck a snag. We came to a week during which revenue was 110 and expenses were 116. Recordis set up the subtraction problem:

$$\text{profit} = 110 - 116$$

"Oh no!" he shouted. "There's no answer to that problem, since you can't subtract a bigger number from a smaller number!"

"How distressing!" the professor said. "We found that whenever you add two natural numbers or multiply two natural numbers, the result will always be another natural number. However, it looks as if it is possible to subtract two natural numbers and get a result that is not a natural number."

"That means an equation like this:

$$x = 110 - 116$$

has no solutions," Recordis said. "We had better make this type of equation illegal."

"Wait a minute," the king said. "We must be able to devise some way to describe a situation in which you lose money, rather than gain it. It's obvious what it means if revenue is 110 and expenses are 116—it means that you *lost* 6 dollars that week. We should be able to think of a way to measure losses."

"But there is absolutely no number that we have thought of up to now that *can* be a solution to the equation $x = 110 - 116$!" Recordis protested.

"Then we must invent a *new* kind of number," the king said.

The professor agreed with the king. "Consider these two situations," she said.

first situation: revenue = 5 expenses = 6

$$\text{profit} = 5 - 6$$

second situation: revenue = 5 expenses = 600

$$\text{profit} = 5 - 600$$

"Recordis says that both of these equations have no solution, which means that we can't tell the difference between the two situations. But, if you had a choice, would you rather be in the first situation or the second situation?"

Recordis had to admit that he would rather be in the first situation, and lose a little, rather than the second situation, and lose a lot.

We tried to find out how numbers that measure losses behave. We tried to find a solution for x in the equation

$$x = 5 - 6$$

We added 6 to both sides:

$$x + 6 = 5$$

We subtracted 5 from both sides:

$$x + 6 - 5 = 0$$
$$x + 1 = 0$$

Finally, we subtracted 1 from both sides:

$$x = 0 - 1$$

"What is $0 - 1$?" Recordis asked.

"We can think of it as the opposite of 1," the professor said. If you gain 1 dollar profit, then your profit is equal to regular 1. If you lose 1 dollar, then we can say that your profit is the opposite of 1, or *negative* 1.

Therefore, we made the definition

$$\text{negative } 1 = 0 - 1$$

"We need a symbol to stand for negative numbers," the king said. "Let's use the same sign we use for subtraction (−), since to subtract means to take away, or lose."

$$\text{negative } 1 = -1$$

$$-1 = 0 - 1$$

"Won't that be confusing?" Recordis asked. "Not that I believe that these so-called negative numbers really exist, but pretending for the moment that I do, how can you tell the difference between the times when the − means subtraction, as in $5 - 4$, and when it means negative numbers, as in -1?"

"We can tell easily," the king said. "If the − sign is between two numbers, then it means subtraction; if it is in front of one number, then it means a negative number."

Recordis still objected to the concept of negative numbers. "What if you want to subtract a negative number, as in 10 − negative 5? Do you write it 10 − − 5?"

"We'll have to use parentheses in a case like that," the king said.

$$10 - (-5)$$

"How many negative numbers are there?" the professor wondered. We made a list of some other negative numbers:

$$\text{negative } 2 = -2 = 0 - 2$$
$$\text{negative } 3 = -3 = 0 - 3$$
$$\text{negative } 4 = -4 = 0 - 4$$

and so on.

"In fact, if we take any natural number, we can find a negative for it," the king said. "That means that there is one negative number for every natural number." We decided to call the natural numbers, and all the numbers greater than zero, the set of positive numbers. Zero is a neutral, fence-sitting number; it is neither a positive number nor a negative number.

We decided that -1 was less than 0 ($-1 < 0$), since a profit of -1 is worse than a profit of 0. For the same reason, $-2 < -1$, $-3 < -2$, and so on. The king showed how we could represent negative numbers on a number line. (See Figure 3-2.) This time the number line went off to infinity in both directions, instead of in just one direction.

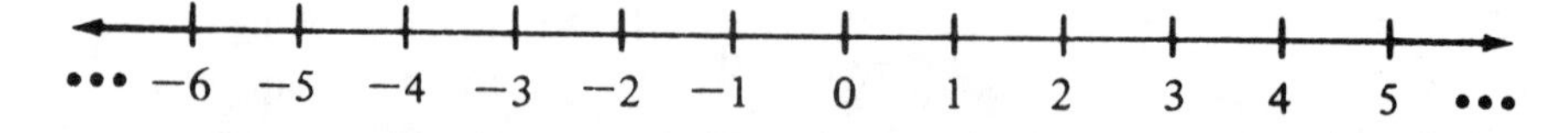

Figure 3–2

"If a is a positive number," the professor said systematically, "then $-a$ is the negative of a and is therefore a negative number.

$$-a = 0 - a$$

"That leads us to an interesting property," she said. "Let's consider the two numbers 5 and -5. We know that $-5 = 0 - 5$. If we add 5 to both sides of that equation, we get:

$$-5 + 5 = 0 \ -5 + 5$$
$$-5 + 5 = 0$$

"See! If you add together 5 and -5, the result is zero." We also found that $-1 + 1 = 0$, $-6 + 6 = 0$, and so on, so we decided to make up a new rule:

MUTUAL ASSURED DESTRUCTION RULE

If you add together any number and its negative, they cancel each other out and the result is zero.

$$a + (-a) = 0 \qquad \text{for any value of } a$$

We also discovered another interesting property: If you take the negative of a negative number, then the result is a positive number:

$$-(-5) = 5$$

since $-(-5) + (-5) = 0$.

Recordis, who had been sulking all along, finally exploded. "You're spending all this time developing properties for things that don't even exist! I can prove to you that there is no such thing as a negative number. Suppose that there are 10 guests in the hotel today, and that 15 guests check out. Then there will be $10 - 15 = -5$ guests in the hotel tomorrow. That is obviously impossible. I can easily show you 5 hotel guests, but I will refuse to believe that there is such a number as -5 until you can show me a room with -5 hotel guests in it."

"Obviously, negative numbers are not the same as regular natural numbers," the king conceded. "You can't use negative numbers in every circumstance where you can use natural numbers. For example, if you count the number of hotel guests, or the number of apples, or the number of people in the kingdom, then the answer must always be a positive number (or perhaps zero). However, there are also some cases where negative numbers are meaningful. For example, if you are talking about the amount of money that you gain, then it is meaningful to say that a negative amount of gain is the same thing as a loss."

"Or, consider the number of yards gained by a football team," the professor said. "That can be negative, because a team can lose yards on a play."

The king mentioned another use for negative numbers. "There is a large ladder that runs up Heartbreak Hill and down into the well below. Now we will be able to put a number on each rung so we can count how far up it is. Obviously, the rung at ground level should be called rung 0, and

Figure 3–3

the rungs above it should be numbered 1, 2, 3, etc. Now we can give negative numbers to the rungs below the ground."

We decided to call the set containing the natural numbers, zero, and the negatives of the natural numbers the set of *integers*.

"I suppose now you're going to tell me that if you climb down 4 rungs you will have gone a total distance of -4 rungs," Recordis said.

"Of course not!" the professor said. "Whenever you measure a distance, such as how far you walk or how many rungs you have climbed, the result must be positive." She showed on the diagram that the distance from rung 0 to rung -4 was equal to 4—just the same as the distance from rung 0 to 4. (See Figure 3-3.) This type of measure seemed important, so we gave it a special name: we called it the *absolute value* of the number.

The rule for calculating absolute value is very simple. If the number is positive or zero, then leave it alone. If the number is negative, then take the negative of it, which turns it into a positive number. We decided to symbolize absolute value with two vertical lines, like this: $|\ \ |$. Here are some examples of absolute values:

$$|0| = 0 \qquad |106| = 106$$

$$|-34| = 34 \qquad |-1| = 1$$

$$|100| = 100 \qquad |-6| = 6$$

Note that the absolute value of any number (except for zero) is positive.

Even Recordis agreed that the absolute value was easy to understand. "I think that it is highly useful—since it turns these peculiar negative numbers into ordinary positive numbers."

ABSOLUTE VALUE OF A NUMBER

The absolute value of a number is the distance along a number line from that number to 0. For example, the absolute value of 0 is 0; the absolute value of 3 is 3; and the absolute value of -5 is 5.

The absolute value of a (written as $|a|$) can be found from this rule:

$$\text{if } a \geq 0, \text{ then } |a| = a$$
$$\text{if } a < 0, \text{ then } |a| = -a$$

(Remember that the negative of a negative number is a positive number!)

"Now we need to make some rules for adding negative numbers," the king said. "Suppose we had a profit of 5 one week and a profit of -3 the next week. Then what will our total profit be for the two weeks?"

"Clearly, it will be 2," the professor said.

$$5 + (-3) = 2$$

"Therefore, adding a negative number is the same as subtracting the absolute value of the number."

In general,

$$a + (-b) = a - b \quad \text{and} \quad -a + b = b - a$$

"Suppose we have $5 + (-8)$?" the king asked. We calculated the answer to be -3.

"When you add a negative number and a positive number, sometimes the result is positive and other times the result is negative. How can you tell which it will be?" Recordis asked.

"Or the result could be zero," the king said. "Suppose we have $10 + (-10)$. That is equal to zero."

"When you add a negative number to a positive number, the nature of the result depends on which number is more powerful," the professor said. "If the positive number has a greater absolute value, then it will overpower the negative number and the result will be positive. If the negative number has the greater absolute value, then it will force its will and the result will be negative. If they have equal absolute values, then the result will be zero."

For example,

$$6 + (-14) = -8$$

$$15 + (-4) = 11$$

$$26 + (-26) = 0$$

"Suppose you add together two negative numbers such as $(-10) + (-5)$," the king said. "It seems that adding together two negative numbers will result in your becoming more and more negative all the time, so in this case the result will be

$$(-10) + (-5) = -15.$$

We also found it helpful to illustrate addition on a number line. Start at the point on the number line corresponding to the first number. Then, if you are adding a positive number, move to the right. If you are adding a negative number, move to the left. (See Figure 3-4.)

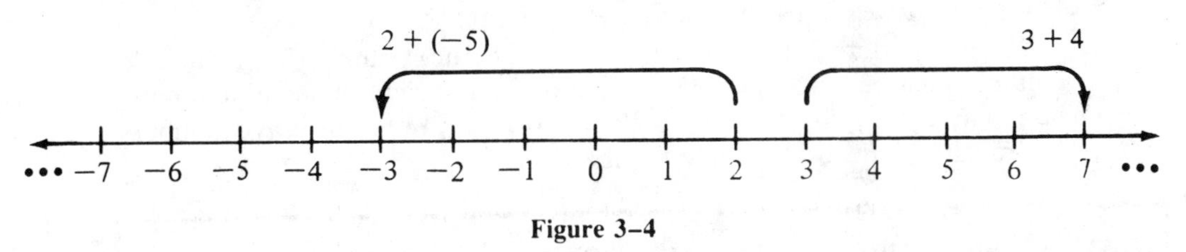

Figure 3–4

We could now find that the distance between any two numbers a and b was $|a - b|$. For example,

distance between 5 and 8: $|8 - 5| = 3$

distance between -5 and 5: $|-5 - 5| = 10$

distance between -3 and -4: $|-3 - (-4)| = 1$

Next we developed rules for multiplying negative numbers. The professor explained that, since multiplication means repeated addition, the expression $3 \times (-5)$ must mean the same thing as $(-5) + (-5) + (-5) = -15$.

Also, by the order-doesn't-make-a-difference property of multiplication, $3 \times (-5)$ must mean the same thing as $(-5) \times 3$, so $(-5) \times 3 = -15$. We were able to establish that, in general, when you multiplied one negative number and one positive number, the result would always be a negative number.

This rule meant that we could state another useful property: Any negative number can be expressed as -1 multiplied by a positive number. For example:

$$-5 = -1 \times 5$$

$$-10 = -1 \times 10$$

$$-24 = -1 \times 24$$

However, we came to a real problem when we tried to find the product of two negative numbers. For example, we tried to find $(-6) \times (-7)$. We rewrote the expression:

$$(-6) \times (-7) = (-1 \times 6) \times (-1 \times 7)$$

Use the commutative property of multiplication to rewrite the expression:

$$= (-1 \times -1) \times (6 \times 7)$$

So the real problem was trying to evaluate $(-1) \times (-1)$. Recordis thought that whenever you multiplied two negative numbers the result would be a new kind of number called a super-duper negative number. (We weren't sure whether he was serious or not.) The professor thought that the effects of the two negative numbers would tend to cancel each other out so that the result would be positive.

"How can we be stuck on so simple a problem?" the king cried in anguish.

"We know that $-1 \times 3 = -3$.
We know that $-1 \times 2 = -2$.
We know that $-1 \times 1 = -1$.
We know that $-1 \times 0 = 0$.
Why can't we figure out what $(-1) \times (-1)$ is?"

"I see a pattern!" the professor said. "The next number in the middle column must be -1, and the next number in the right-hand column must be 1. Therefore, $-1 \times -1 = 1$! When you multiply two negative numbers the result is indeed a positive number!"

ADDITION OF NEGATIVE NUMBERS

If a is positive and b is negative, then

$a + b = a - |b|$.
If $|a| > |b|$, then $a + b$ is positive.
If $|a| < |b|$, then $a + b$ is negative.

If a and b are both negative, then

$$a + b = -(|a| + |b|).$$

MULTIPLICATION OF NEGATIVE NUMBERS

$$|ab| = |a| \times |b|$$

If a is negative and b is positive, then ab is negative.
If b is negative and a is positive, then ab is negative.
If a and b are both negative, then ab is positive.

Exercises

Perform these calculations.

1. $16 - 21$
2. $46 + 3 - 10 - 39$
3. $-6 + 10$
4. $-3 - 18$
5. $-10 - (-6)$
6. $-10 - 6$
7. $5 - 8$
8. $-9 + 5$
9. $32 - 64$
10. $-6 \cdot 5$
11. $(-6) \cdot (-5)$
12. $3(4 - 8)$
13. $-15 \cdot 2$
14. $-13 \cdot (-3)$
15. $(66 - 12) \cdot (-2)$
16. $-3 \cdot (-6 + 6)$

Calculate these absolute values.

17. $|103|$
18. $|-7|$
19. $|0|$
20. $|-16|$
21. $|17|$
22. $|1365.4|$
23. $|8 - 5|$
24. $|5 - 8|$
25. $|11 - 5|$
26. $|165 - 45|$
27. $|13 - 165|$
28. $|-5 - 12|$
29. $|5 - 12|$
30. $|-5 + 12|$
31. $|5||4|$
32. $|-5||4|$
33. $|5||-4|$
34. $|-5||-4|$
35. $|(-5) \times 4|$
36. $|3 \times 10 - 42|$
37. $|11 - 5 - 4 + 3 + 2|$
38. Perform some calculations to verify that $|ab| = |a||b|$.
39. Does $|a + b| = |a| + |b|$? If this rule is not always true, tell under what conditions it will be true.

Solve these equations for x.

40. $16 + 3x = -14$
41. $4(3 + x) + 10 = -11(x - 6) + 1$
42. $10x - 3(x + 1) = 32$
43. $3x + 21 = -2x + 11$
44. $bx = 0$
45. $6x + 4 = 2x$
46. $16x + 48 = -16$
47. $10 - 2x = 90 - x$
48. $12x + 16x - 14x = 14x + 13x + 52$
49. $2x + 936x = 936x - 10$
50. What will the value of $(x + |x|)/2$ be if $x > 0$? if $x < 0$?

4 Fractions and Rational Numbers

We were planning a twelve-mile hike, and after careful measurement we calculated that we could walk 4 miles per hour on the average. The professor explained that we could now calculate how long the hike would take.

"Since

$$(\text{distance travelled}) = (\text{speed}) \times (\text{time})$$

that means that

$$(\text{time}) = (\text{distance travelled}) \div (\text{speed})$$

In our case (distance travelled) is 12 and (speed) is 4, which means it will take us $12 \div 4 = 3$ hours to complete the hike."

"My favorite place is the Scenic Viewpoint Picnic Area, which is 3 miles from the start," Recordis said. "How long will it take us to get there?"

We set up the division problem:

$$(\text{time}) = 3 \div 4$$

"Hold it!" Recordis said. "That's not a natural number! There is no natural number that you can multiply by 4 to get 3." Recordis tried some examples to prove his point: $0 \times 4 = 0$, $1 \times 4 = 4$, $2 \times 4 = 8$, and so on. We were interrupted by the arrival of one of the palace neighbors, who came to ask us for advice about a very thorny problem: She had three guests coming over for lunch, but she only had one orange.

"The solution is very simple," the king said. "Cut the orange in three pieces, and give each guest one of the pieces."

"We invented numbers of a special type to describe that kind of situation!" the professor suddenly remembered. "We called them *fractions*."

Memories came flooding back to us. "In your case you will give each guest *one-third* of an orange, which we symbolized by 1/3. And the time it will take us to reach the viewpoint will be 3/4, or *three-fourths* of an hour," the king said.

"I have vague memories of studying fractions," Recordis said, "but there are a few minor details that I don't quite remember."

"I just happen to have a book by a well-known author that tells all about them," the professor said. She pulled out a book and began reading.

* * *

Complete Guide to Everything
Worth Knowing about Fractions

by Professor A. A. A. Stanislavsky, Ph.D., etc., etc.

Chapter 1: Regular Fractions

A fraction is a number such as 1/2, 1/4, or 2/3. A fraction consists of a top number (called the *numerator*) and a bottom number (called the *denominator*). Suppose you are using a fraction to tell you how much pie you have. Then the denominator (the bottom number) tells you how many equal pieces have been cut out of the pie, and the numerator (the top number) tells how many pieces you have. (See Figure 4-1).

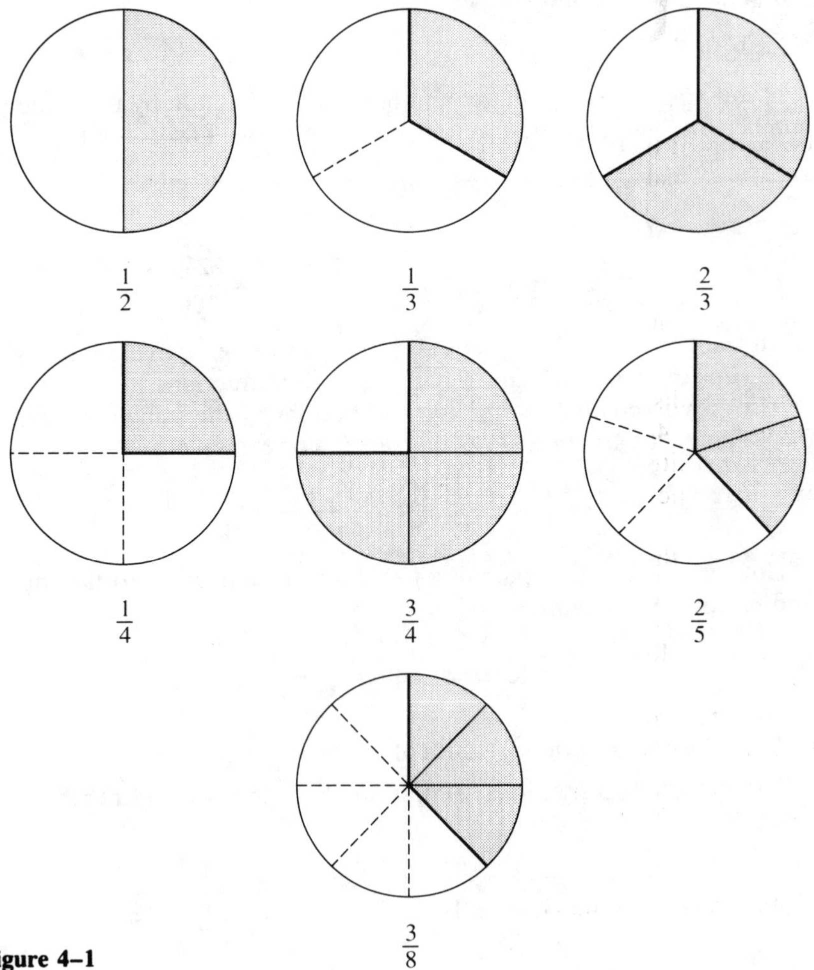

Figure 4–1

Examples of Fractions

If the denominator equals the numerator, then you have the whole pie and the value of the fraction is 1.

Proper Fractions

In a *proper fraction,* the top number is less than the bottom number, and the value of the fraction is between 0 and 1. The number line in Figure 4-2 illustrates some common fractions.

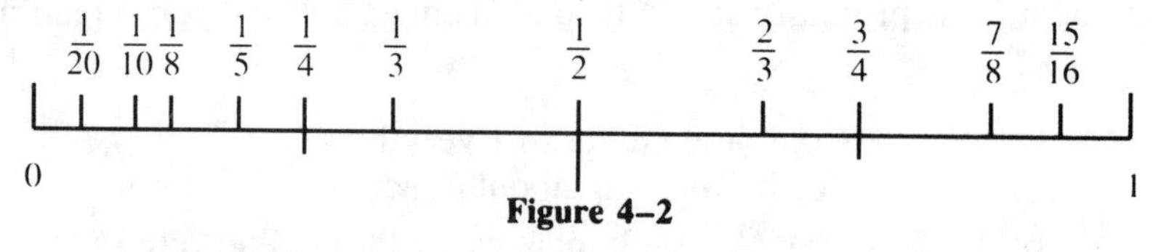

Figure 4–2

If the numerator is bigger than the denominator, you have more than a whole pie and the value of the fraction is greater than 1. For example, 6/5 is greater than 1. In this case the fraction can also be written as the sum of a whole number and a proper fraction. For example, $6/5 = 1 + 1/5$.

If the denominator is 1, then the value of the fraction is equal to the value of the top number. For example, $2/1 = 2$, $3/1 = 3$, and $10/1 = 10$.

Fractions can also be expressed as division problems. For example, $2 \div 3 = 2/3$ and $20 \div 5 = 20/5$.

Equivalent Fractions

If you *multiply* both the top and bottom of a fraction by the same number, the value of the fraction stays the same. For example,

$$\frac{1}{2} = \frac{1 \times 2}{2 \times 2} = \frac{2}{4} \qquad \frac{2}{3} = \frac{2 \times 4}{3 \times 4} = \frac{8}{12}$$

$$\frac{3}{4} = \frac{3 \times 3}{4 \times 3} = \frac{9}{12} \qquad \frac{1}{4} = \frac{1 \times 25}{4 \times 25} = \frac{25}{100}$$

(Two fractions with the same value are said to be *equivalent fractions*. For example, 1/2 and 2/4 are equivalent fractions.)

Also, if you *divide* both the top and bottom by the same number, the value of the fraction stays the same. For example,

$$\frac{6}{9} = \frac{6 \div 3}{9 \div 3} = \frac{2}{3} \qquad \frac{20}{5} = \frac{20 \div 5}{5 \div 5} = \frac{4}{1} = 4$$

However, note that you cannot *add* the same number to the top and bottom. For example,

$$\frac{2}{3} \text{ does } \textit{not} \text{ equal } \frac{2 + 5}{3 + 5}$$

Multiplying and Adding Fractions

To multiply two fractions, simply multiply the two top numbers and the two bottom numbers:

$$\frac{1}{2} \times \frac{1}{2} = \frac{1 \times 1}{2 \times 2} = \frac{1}{4} \qquad \frac{2}{3} \times \frac{4}{5} = \frac{2 \times 4}{3 \times 5} = \frac{8}{15}$$

$$\frac{3}{7} \times \frac{1}{10} = \frac{3}{70} \qquad \frac{7}{16} \times \frac{2}{3} = \frac{14}{48}$$

Adding two fractions is easy if they both have the same bottom number. Then, just add the two top numbers and keep the bottom number the same. For example,

$$\frac{1}{2} + \frac{1}{2} = \frac{1 + 1}{2} = \frac{2}{2} = 1$$

$$\frac{3}{8} + \frac{2}{8} = \frac{3 + 2}{8} = \frac{5}{8}$$

$$\frac{3}{5} + \frac{1}{5} = \frac{3 + 1}{5} = \frac{4}{5}$$

To add two fractions if they don't have the same denominator, you first need to convert them so that they both do have the same denominator. To do this, use the fact that you can multiply both the top and the bottom by the same number and still keep the same value. For example,

$$\frac{1}{2} + \frac{1}{4} = \frac{1 \times 2}{2 \times 2} + \frac{1}{4} = \frac{2}{4} + \frac{1}{4} = \frac{2 + 1}{4} = \frac{3}{4}$$

$$\frac{1}{2} + \frac{1}{3} = \frac{1 \times 3}{2 \times 3} + \frac{1 \times 2}{3 \times 2} = \frac{3}{6} + \frac{2}{6} = \frac{3 + 2}{6} = \frac{5}{6}$$

(See Figure 4-3.)

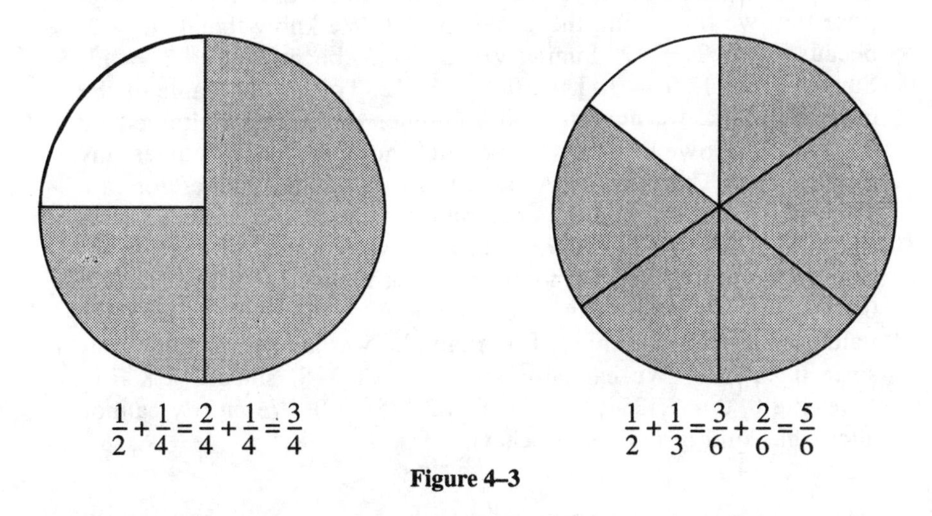

$\frac{1}{2}+\frac{1}{4}=\frac{2}{4}+\frac{1}{4}=\frac{3}{4}$
$\frac{1}{2}+\frac{1}{3}=\frac{3}{6}+\frac{2}{6}=\frac{5}{6}$

Figure 4–3

Reciprocals, Compound Fractions, and Division

The *reciprocal* of a fraction is found simply by turning the fraction upside down (in other words, putting the denominator on top and the numerator on the bottom). For example,

The reciprocal of 2/3 is 3/2.

The reciprocal of 1/2 is 2/1 = 2.

The reciprocal of 5/3 is 3/5.

If you multiply a fraction and its reciprocal together, then the result is 1:

$$\frac{2}{3} \times \frac{3}{2} = \frac{2 \times 3}{3 \times 2} = \frac{6}{6} = 1$$

A *compound fraction* is a fraction that itself has a fraction on both the top and the bottom. Examples are:

$$\frac{1/2}{1/3} \quad \frac{2/3}{3/5} \quad \frac{3/10}{5/14} \quad \frac{1/4}{1/4} \quad \frac{2/6}{5/7}$$

(A compound fraction is also sometimes called a *complex fraction*.)

To simplify a compound fraction (in other words, to convert it into a regular fraction) multiply the top and bottom of the compound fraction by the reciprocal of the fraction in the denominator:

$$\frac{\frac{2}{3}}{\frac{4}{5}} = \frac{\frac{2}{3} \times \frac{5}{4}}{\frac{4}{5} \times \frac{5}{4}} = \frac{\frac{10}{12}}{\frac{20}{20}} = \frac{\frac{10}{12}}{1} = \frac{10}{12}$$

To divide two fractions, multiply the first fraction by the reciprocal of the second fraction:

$$\frac{2}{3} \div \frac{1}{4} = \frac{2}{3} \times \frac{4}{1} = \frac{8}{3}$$

Prohibition: No Zero Denominators

One final word of warning. It is highly illegal to have a fraction whose denominator is zero. 5/0, 2/0, 1/0, and 0/0 are all illegal. Suppose that we try to find the value of 12/0. We know that 12/6 = 2, because $6 \times 2 = 12$. Similarly, 12/4 = 3, because $4 \times 3 = 12$. Suppose that 12/0 = ?. Then $0 \times ? = 12$. To find the value of the question mark, we need to find a number that when multiplied by 0 gives 12. However, there is no such number, since 0 times any number is 0. Therefore, 12/0 does not exist. If the numerator of a fraction is not zero, and the denominator comes very close to zero, then the value of the fraction becomes very large—in fact, it approaches infinity as the denominator approaches zero. The fraction 0/0 is even more peculiar. It can be shown that 0/0 has the same value as *any* other number. For example, we can see that 0/0 = 1, since $0 \times 1 = 0$. We can also see that $0/0 = 8$, since $0 \times 8 = 0$. Likewise, $0/0 = 2{,}135\frac{2}{3}$, because $0 \times 2{,}135\frac{2}{3} = 0$. We clearly cannot allow that kind of indecisive behavior.

* * *

"The fact that we now know algebra will help us to investigate fractions," the professor said. "First, we need to make a place for fractions in our system of numbers—so far we only have integers."

We decided to call the entire collection of all integers and all fractions the set of *rational numbers* (since *ratio* is another word for fraction.) Then the professor suggested this definition for the rational numbers:

RATIONAL NUMBERS

Suppose *a* and *b* are any two integers (except that *b* can't be zero.) Then any number that can be written in the form *a*/*b* is a rational number.

"Note that we've automatically included all the integers in this definition," the professor said, "since if $b = 1$ the value of a/b will be an integer."

"Now that we know about algebra, we can write the general form of the rules for performing operations on fractions," the king said. He wrote:

Let a/b and c/d be fractions, and let x be any rational number. Then:

$$\frac{a}{b} = \frac{ax}{bx}$$

"This rule just says that you can multiply the top and the bottom by the same number." Recordis noticed a neat effect if we worked that rule backwards:

$$\frac{a\,x}{b\,x} = \frac{a}{b}$$

"If there's the same factor in both the numerator and the denominator like that, we can simply cross it out in both places and make the whole fraction simpler." We called this the *cancellation* rule, and we had lots of fun cancelling things. The rest of the rules for fractions are:

RULES FOR FRACTIONS

Multiplication	$\frac{a}{b} \times \frac{c}{d} = \frac{ac}{bd}$
Addition when the denominators are the same	$\frac{a}{b} + \frac{c}{b} = \frac{a + c}{b}$
Addition when the denominators are different	$\frac{a}{b} + \frac{c}{d} = \frac{ad}{bd} + \frac{bc}{bd} = \frac{ad + bc}{bd}$
Subtraction	$\frac{a}{b} - \frac{c}{d} = \frac{ad}{bd} - \frac{bc}{bd} = \frac{ad - bc}{bd}$
Simplification of Compound Fractions	$\frac{a/b}{c/d} = \frac{ad}{bc}$

In all rules for fractions, it is automatically assumed that no denominators are equal to zero.

"I just thought of a complication we didn't have in the old days," the professor said. "What if the fraction is negative?"

"We can put a negative sign in front of a fraction, just as we can put a negative sign in front of an integer," the king said. "$-1/2$ means the negative of $1/2$. We know from the Mutual Assured Destruction Property that $-1/2 + 1/2 = 0$."

We also developed some rules to handle the situation in which either the numerator or the denominator or both are negative. (Assume that a and b are positive integers.)

$$\frac{-a}{-b} = \frac{a}{b} \quad \left(\text{since } \frac{-1}{-1} = 1, \text{ and } \frac{(-1) \times (-a)}{(-1) \times (-b)} = \frac{a}{b}\right)$$

$$\frac{-a}{b} = -\frac{a}{b}$$

$$\frac{a}{-b} = -\frac{a}{b}$$

$$-\frac{-a}{-b} = -\frac{a}{b}$$

(See the exercises for more about these propositions.)

"I remember that when we started performing calculations involving amounts of money, regular fractions were too complicated," Recordis said. "Didn't we develop another kind of fraction?"

"Aha!" the professor said. "We should read chapter 2 of my book."

* * *

Chapter 2: Decimal Fractions

Two fractions with different denominators are hard to compare. For example, at first glance it is difficult to tell whether or not 9/16 is greater than 25/44. Therefore, it often helps to convert fractions into equivalent fractions with the same denominator. The most convenient denominators to use for this purpose are the numbers 10, 100, 1,000, and so on. For example,

$$\frac{1}{2} = \frac{5}{10} \qquad \frac{1}{4} = \frac{25}{100} \qquad \frac{3}{5} = \frac{6}{10}$$

To save writing in these situations, we will not write out the number in the denominator. Instead, we will put a period (.), called a *decimal point*, in front of the numerator. (Often, a zero is placed before the decimal point so that the decimal point will not be overlooked.) For example,

$$\frac{1}{2} = 0.5 \qquad \frac{1}{4} = 0.25 \qquad \frac{3}{5} = 0.6$$

If only one digit is written to the right of the decimal point, then the denominator is 10.

$$0.1 = 1/10 \qquad 0.2 = 2/10 \qquad 0.3 = 3/10$$

$$0.4 = 4/10 \qquad 0.5 = 5/10 \qquad 0.9 = 9/10$$

If there are two digits to the right of the decimal point, then the denominator is 100.

$$0.15 = 15/100 \qquad 0.33 = 33/100 \qquad 0.75 = 75/100$$

And so on:

$$0.123 = 123/1{,}000$$

$$0.1234 = 1234/10{,}000$$

$$0.12345 = 12345/100{,}000$$

If you want to add a whole number to a decimal fraction, then do it like this:

$$2 + 0.5 = 2.5$$

$$10 + 0.54 = 10.54$$

$$16 + 0.378 = 16.378$$

Decimal fractions can be added just the same as whole numbers, provided you make sure that the decimal points stay lined up:

$$\begin{array}{r} 123.4567 \\ +\,234.5678 \\ \hline 358.0245 \end{array}$$

If the two decimal fractions don't have the same number of decimal places, then add some zeros at the end of the shorter one so they both have the same number of decimal places. To add:

$$\begin{array}{r} 788.3672 \\ +\,42.56 \end{array}$$

convert it like this:

$$\begin{array}{r} 788.3672 \\ +\,42.5600 \\ \hline 830.9272 \end{array}$$

To multiply two decimal fractions, multiply them as if they were whole numbers, ignoring the decimal points. The number of decimal places in the result will be the sum of the number of decimal places in each of the two original numbers:

$$\begin{array}{r} 1.2 \\ \times 6 \\ \hline 7.2 \end{array} \quad \begin{array}{r} 1.2 \\ \times .6 \\ \hline .72 \end{array} \quad \begin{array}{r} .12 \\ \times 6 \\ \hline .72 \end{array} \quad \begin{array}{r} .12 \\ \times .6 \\ \hline .072 \end{array} \quad \begin{array}{r} .12 \\ \times .06 \\ \hline .0072 \end{array}$$

$$\begin{array}{r} 12 \\ \times 34 \\ \hline 408 \end{array} \quad \begin{array}{r} 12 \\ \times 3.4 \\ \hline 40.8 \end{array} \quad \begin{array}{r} 1.2 \\ \times 3.4 \\ \hline 4.08 \end{array} \quad \begin{array}{r} 1.2 \\ \times 3.4 \\ \hline .408 \end{array} \quad \begin{array}{r} 1.2 \\ \times .34 \\ \hline .408 \end{array} \quad \begin{array}{r} .12 \\ \times .34 \\ \hline .0408 \end{array}$$

A division problem is easy enough if the numerator, but not the denominator, contains a decimal fraction. To find 7.2/4, set up the division problem like this:

$$4\overline{)7.2}$$

$$\begin{array}{r} 1 \\ 4\overline{)7.2} \end{array}$$

$$\begin{array}{r} 1 \\ 4\overline{)7.2} \\ 4 \\ \hline \end{array}$$

$$\begin{array}{r} 1 \\ 4\overline{)7.2} \\ \underline{4} \\ 3.2 \end{array}$$

$$\begin{array}{r} 1.8 \\ 4\overline{)7.2} \\ \underline{4} \\ 3.2 \end{array}$$

$$\begin{array}{r} 1.8 \\ 4\overline{)7.2} \\ \underline{4} \\ 3.2 \\ \underline{3.2} \\ 0 \end{array}$$

Therefore, 7.2/4 = 1.8. The important thing is to make sure to keep the decimal places lined up.

If the denominator contains a decimal fraction, then it is best to multiply both the numerator and denominator by the appropriate power of ten to remove the decimal fraction from the denominator:

$$\frac{7.2}{.4} = \frac{72}{4} = 18$$

$$\frac{7.2}{.04} = \frac{720}{4} = 180$$

$$\frac{72}{.04} = \frac{7200}{4} = 1800$$

* * *

"Now we have a new way of representing rational numbers," the professor said. "Any rational number can be expressed as a decimal fraction."

"How?" Recordis asked. "If we start with a regular fraction, how do we convert it into a decimal fraction?"

"We have to perform a division problem, like this:

$$4\overline{)3} \qquad 4\overline{)3.00} \qquad \begin{array}{r} .7 \\ 4\overline{)3.00} \end{array} \qquad \begin{array}{r} .7 \\ 4\overline{)3.00} \\ \underline{2\,8} \end{array} \qquad \begin{array}{r} .7 \\ 4\overline{)3.00} \\ \underline{2\,8} \\ 20 \end{array} \qquad \begin{array}{r} .75 \\ 4\overline{)3.00} \\ \underline{2\,8} \\ 20 \\ \underline{20} \end{array}$$

Therefore, $\frac{3}{4} = 0.75$.

(If you have a calculator, then it is even easier to convert fractions into decimal fractions, since the calculator automatically displays its results as decimal fractions. Just type

3 [÷] 4 [=]

and the calculator will display 0.75)

We tried some more examples.

$\frac{3}{5}$:

```
    .6
 5)3.0
   3.0
   ---
     0
```

$\frac{3}{5} = 0.6$

$\frac{5}{8}$:

```
    .625
 8)5.000
   4 8
   ---
     20
     16
     --
      40
      40
      --
       0
```

$\frac{5}{8} = 0.625$

"As I recall, though, there is a problem with this method—the answers don't always work out nice and even," Recordis said. We tried to find the decimal fraction for 1/3:

```
    .3333333
 3)1.0000000
    9
    --
    10
     9
     --
     10
      9
      --
      10
       9
       --
       10
        9
        --
        10
         9
         -
         1
```

"This will go on forever without stopping!" Recordis exclaimed.

"All right, so we can't exactly represent 1/3 as a decimal fraction—but we can come pretty close. We know that 1/3 is almost the same as 0.3333, or 3,333/10,000."

We added an amendment to our rule that any rational number can be represented as a decimal fraction, saying that some rational numbers can be represented only as decimal fractions that endlessly repeat the same pattern of digits. Some examples are:

$$1/6 = 0.166666\ldots$$

$$2/3 = 0.666666\ldots$$

$$1/9 = 0.111111\ldots$$

$$1/11 = 0.09090909\ldots$$

$$15/11 = 1.36363636\ldots$$

$$1/7 = 0.142857\ 142857\ 142857\ldots$$

In cases such as these, you can approximate the true value by considering only a few digits near the beginning. The more digits you decide to take, the more accurate your approximation will be.

Note to Chapter 4

- Notice that dividing by x is the same as multiplying by $1/x$:

$$\frac{a}{x} = a \times \frac{1}{x}$$

Exercises

Perform these calculations:

1. 1/2 + 5/2
2. 5/6 − 2/6
3. 1/2 − 1/3
4. 1/12 + 1/6
5. 1/7 + 1/9
6. 3/7 + 6/11
7. 5/9 × 3/9
8. 1/2 × 4/7
9. 1/3 × 5/3
10. 1/4 × (6/7 − 1/8)
11. 1/2 × (1/2 + 4/5)
12. 1/3 × 4/5 + 1/6 × 7/8
13. 1/2 × 1/3
14. 1/7 × 3/7
15. 1/3 × 5/6

A fraction is in simplest form if the numerator and denominator contain no common factors. For example, 9/12 is not in simplest form, because both 9 and 12 have the common factor 3. The simplest form of this fraction is 3/4. Convert these fractions into simplest form:

16. 2/4
17. 3/12
18. 5/10
19. 6/16
20. 3/39
21. 12/108
22. 64/128
23. 6/9
24. 6/8
25. 9/81
26. 10/64
27. Show that, if a fraction is in simplest form, then either the numerator or the denominator or both must be odd.
28. Show that $a/b = c/d$ means the same as $ad = bc$.
29. Show that $(-a)/b = -(a/b)$.
30. Show that $a/(-b) = -(a/b)$.
31. Show that $-[(-a)/(-b)] = -(a/b)$
32. Express $a + (b/c)$ as a fraction.

Perform these calculations:

33. 0.25 × 0.25
34. 3 × 1.36
35. 4 × 1.14
36. 1.41 × 1.41
37. 3 × 0.333
38. 104.26 + 24.18
39. 14.165 + 2.315

Convert these fractions to decimal fractions by performing division problems:

40. 7/8
41. 5/9
42. 5/8
43. 6/13

Solve these equations for x:

44. $2 + 3x = 0$
45. $14 + (1/2)x = 16$
46. $2x - 12 = 3x + 12$
47. $4x - 8 = 6 + 6x$
48. $3x = 5$
49. $4x = 7$
50. $3x - 11 = 0$
51. $3x - 16 = 14$

Simplify the following expressions. For example, $1 + 1/x = x/x + 1/x = (x + 1)/x$

52. $1 - 1/x$
53. $(1 + 1/x)/(1 - 1/x)$
54. $x/(1 - 1/x)$
55. $(1 + 1/x)/(1/x)$
56. $(x - 1)/[1/(x + 1)]$
57. $(x - 1)/[(x - 5)(x - 1)]$
58. $\dfrac{1}{\dfrac{1}{x}}$
59. $\dfrac{1}{a} + \dfrac{1}{b}$
60. $\dfrac{1}{\dfrac{1}{a} + \dfrac{1}{b}}$
61. $\dfrac{(x - 5)(x + 3)}{(x - 5)(x + 2)}$
62. $\dfrac{(a + 2b)(3a + 6)}{25(a + 2b)}$
63. $\dfrac{x}{5} + \dfrac{x}{4}$
64. $\dfrac{x}{5} + \dfrac{y}{4}$
65. $1/abc + 4$
66. $\dfrac{5}{a + b} + \dfrac{6}{a + b} - \dfrac{2}{a + b}$
67. $\dfrac{1}{a} + \dfrac{1}{b} + \dfrac{1}{c}$

68. The algebraic definition of fractions is: $a/b = x$ if $? \times ? = ?$. Fill in the correct values for the question marks.
69. Are you better off buying detergent in 12-ounce boxes that cost \$1.16 or in 16-ounce boxes that cost \$1.40?
70. If Y is your before-tax income, suppose that the government takes away $\frac{1}{4}Y - 10$ as taxes. What was your pre-tax income if your after-tax income is \$280?

5
Exponents

There still had not been any word from the detective. Recordis tried to take his mind off the missing books by helping with the design for the kingdom's new library. At the moment we were designing one of the walls. The wall was to be made out of tile, but we needed to figure out how many tiles would be needed. Builder pulled out his tape measure. "Each tile is 1 foot wide by 1 foot high," he said. "And this section of the wall is 8 feet across and 8 feet high."

"Then we will just have to draw a pattern of squares and start counting the tiles," Recordis said. (See Figure 5-1.)

8 (across) × 8 (high)

1	2	3	4	5	6	7	8
9	10	11	12	13	14	15	16
17	18	19	20	21	22	23	24
25	26	27	28	29	30	31	32
33	34	35	36	37	38	39	40
41	42	43	44	45	46	47	48
49	50	51	52	53	54	55	56
57	58	59	60	61	62	63	64

Figure 5–1

After completing that calculation, Recordis was dismayed when Builder told him there were many different square walls that needed tiles, with lengths from 1 all the way up to 15. So Recordis dutifully counted the number of tiles for many different sized walls (see Figure 5-2) and then arranged the tiles in different stacks (see Figure 5-3.)

When the professor came by and realized what Recordis was doing, she exclaimed, "We can find the answer much more easily by multiplying! If the wall consists of 8 rows, each of which has 8 tiles, there must be $8 \times 8 = 64$ tiles in the wall."

1	2
3	4

$2 \times 2 = 4$

1	2	3
4	5	6
7	8	9

$3 \times 3 = 9$

1	2	3	4
5	6	7	8
9	10	11	12
13	14	15	16

$4 \times 4 = 16$

1	2	3	4	5
6	7	8	9	10
11	12	13	14	15
16	17	18	19	20
21	22	23	24	25

$5 \times 5 = 25$

1	2	3	4	5	6
7	8	9	10	11	12
13	14	15	16	17	18
19	20	21	22	23	24
25	26	27	28	29	30
31	32	33	34	35	36

$6 \times 6 = 36$

1	2	3	4	5	6	7
8	9	10	11	12	13	14
15	16	17	18	19	20	21
22	23	24	25	26	27	28
29	30	31	32	33	34	35
36	37	38	39	40	41	42
43	44	45	46	47	48	49

$7 \times 7 = 49$

1	2	3	4	5	6	7	8	9
10	11	12	13	14	15	16	17	18
19	20	21	22	23	24	25	26	27
28	29	30	31	32	33	34	35	36
37	38	39	40	41	42	43	44	45
46	47	48	49	50	51	52	53	54
55	56	57	58	59	60	61	62	63
64	65	66	67	68	69	70	71	72
73	74	75	76	77	78	79	80	81

$9 \times 9 = 81$

1	2	3	4	5	6	7	8	9	10
11	12	13	14	15	16	17	18	19	20
21	22	23	24	25	26	27	28	29	30
31	32	33	34	35	36	37	38	39	40
41	42	43	44	45	46	47	48	49	50
51	52	53	54	55	56	57	58	59	60
61	62	63	64	65	66	67	68	69	70
71	72	73	74	75	76	77	78	79	80
81	82	83	84	85	86	87	88	89	90
91	92	93	94	95	96	97	98	99	100

$10 \times 10 = 100$

1	2	3	4	5	6	7	8	9	10	11
12	13	14	15	16	17	18	19	20	21	22
23	24	25	26	27	28	29	30	31	32	33
34	35	36	37	38	39	40	41	42	43	44
45	46	47	48	49	50	51	52	53	54	55
56	57	58	59	60	61	62	63	64	65	66
67	68	69	70	71	72	73	74	75	76	77
78	79	80	81	82	83	84	85	86	87	88
89	90	91	92	93	94	95	96	97	98	99
100	101	102	103	104	105	106	107	108	109	110
111	112	113	114	115	116	117	118	119	120	121

$11 \times 11 = 121$

1	2	3	4	5	6	7	8	9	10	11	12
13	14	15	16	17	18	19	20	21	22	23	24
25	26	27	28	29	30	31	32	33	34	35	36
37	38	39	40	41	42	43	44	45	46	47	48
49	50	51	52	53	54	55	56	57	58	59	60
61	62	63	64	65	66	67	68	69	70	71	72
73	74	75	76	77	78	79	80	81	82	83	84
85	86	87	88	89	90	91	92	93	94	95	96
97	98	99	100	101	102	103	104	105	106	107	108
109	110	111	112	113	114	115	116	117	118	119	120
121	122	123	124	125	126	127	128	129	130	131	132
133	134	135	136	137	138	139	140	141	142	143	144

$12 \times 12 = 144$

Figure 5–2

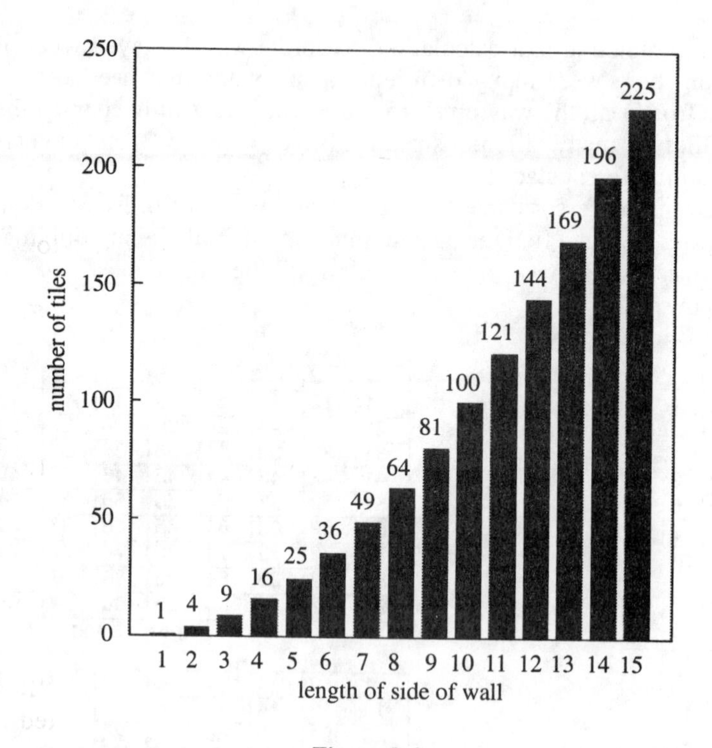

Figure 5–3

"In fact, we can use the same method for all of the square segments in the wall," the king said. "If we let a represent the length of the side of the square (in feet), then the number of tiles that we will need to cover the wall is $a \times a = aa$."

"This calculation is so important that we should give a special name to the number of tiles needed to cover a square with side of length a," Recordis said.

"We'll call it a squared, since we're turning a number into a square," the professor said.

$$a \text{ squared} = a \times a = aa$$

"For example, 2 squared is 4, 9 squared is 81, and so on."

"The next question we have to worry about is how we're going to pay for all these tiles," Recordis said. "It's going to take a lot more than we have in our piggy bank."

"We can open a bank account," the king said. "Then, in addition to the amount that we put in the account, the bank will pay us some interest."

"The bank pays 5 percent interest," Recordis said. "That means that, if we put 100 in the bank at the beginning of the year, and then we don't put anything in or take anything out all year, we will have 105 at the end of the year."

"Or, in general, if we have m at the beginning of the year, then we will have 1.05 m at the end of the year," the professor said.

"But that's not all," the king said. "At the end of two years, we will have

$$\begin{aligned} &1.05 \times (\text{amount at end of one year}) \\ &= 1.05 \times 1.05 \times 100 \\ &= 110.25 \end{aligned}$$

in the account."

The professor went on, "After ten years we will have

$$1.05 \times 1.05 \times 1.05 \times 1.05 \times 1.05 \times 1.05 \times 1.05 \times 1.05 \times 1.05 \times 1.05 \times 100 = 162.889$$

altogether."

"Hold it!" Recordis said. "There must be a shorter way to write that expression!"

"How else could we write it?" the professor said. "We're multiplying 1.05 together 10 times, so we need to write ten 1.05's."

"But if we *add* a number together ten times we don't have to write it ten times," Recordis protested. "If we have

$$1.05 + 1.05 + 1.05 + 1.05 + 1.05 + 1.05 + 1.05 + 1.05 + 1.05 + 1.05 = 10.5$$

we can write it like this."

$$10 \times 1.05 = 10.5$$

"Recordis does have a very good point," the king said. "Remember what multiplication is: Multiplication is the operation of repeated addition. So we should be able to develop an operation that stands for repeated multiplication."

We decided to write a small number above the main line to stand for repeated multiplication. For example, 1.05^{10} means 1.05 multiplied together 10 times. We called the little number the *exponent*, from the Latin words meaning "to place out of (line)." We made a general definition for exponents.

EXPONENTS

If a and n are two numbers, then a^n means n a's multiplied together. For example,

$$11^2 = 11 \times 11 = 121$$

$$5^4 = 5 \times 5 \times 5 \times 5 = 625$$

$$10^3 = 10 \times 10 \times 10 = 1{,}000$$

$$2^{10} = 2 \times 2 \times 2 \times 2 \times 2 \times 2 \times 2 \times 2 \times 2 \times 2 = 1{,}024$$

$$6^1 = 6$$

"We can write a squared with exponents," the professor noted.

$$a \text{ squared} = a^2$$

We also noticed that $a^1 = a$.

Recordis dreamed of stacks of money showing how the bank balance would grow. He calculated the amount using this formula:

$$\text{bank balance in year } n = A(1 + r)^n = 100 \times 1.05^n$$

where A is the initial amount in the account and r is the interest rate. See Figure 5-4. "I'm not unusually greedy," he explained, "it's just that I think it helps to visualize numbers when possible."

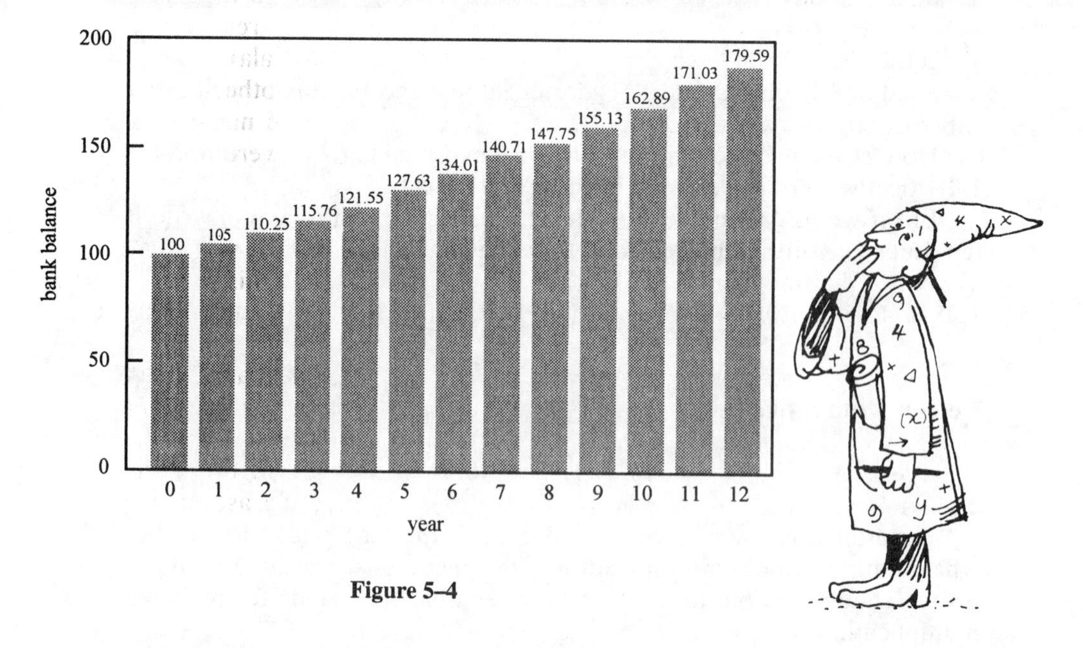

Figure 5–4

"These exponents sure are powerful!" the king said. "Look at how big those numbers are getting." So we decided that another name we could use was *power.* For example, we could say that a^2 was the second power of a, a^3 was the third power, and so on.

"While we're at it, we may as well make a table of different numbers raised to different powers," the professor said.

x	x^2	x^3	x^4	x^5	x^6	x^7
1	1	1	1	1	1	1
2	4	8	16	32	64	128
3	9	27	81	243	729	2,187
4	16	64	256	1,024	4,096	16,384
5	25	125	625	3,125	15,625	78,125
6	36	216	1,296	7,776	46,656	279,936
7	49	343	2,401	16,807	117,649	823,543
8	64	512	4,096	32,768	262,144	2,097,152
9	81	729	6,561	59,049	531,441	4,782,969
10	100	1,000	10,000	100,000	1,000,000	10,000,000

The most striking feature of this table was the interesting property that the powers of 10 had. "Look at the row that lists the powers of 10!" the professor said in awe. "Look at how each number is just a 1 followed by a bunch of zeros."

"I can guess what 10^8 is," Recordis said. "I bet 10^8 is 100,000,000. And I bet that 10^9 is 1,000,000,000, or one billion."

"We can state a general rule," the professor said. "It appears that 10^n is a 1 followed by n zeros."

"This will help a lot with another problem!" the king said. "The Royal Astronomer has lately been telling me that it is very difficult to keep track of some of the horribly large numbers that he needs to keep track of. For instance, he has told me that the distance from the earth to the sun is 149,500,000 kilometers, and that the distance to the nearest star is 40,350,000,000,000 kilometers. The distance to the nearest galaxy is even worse: It is 18,900,000,000,000,000,000 kilometers. Just the other day the astronomer was saying to me, 'If only I could write these numbers in such a way that I could just describe how many zeros there were, without having to write them all out.'"

"How do exponents help?" Recordis asked.

"Let's consider a big number, such as 9,000,000," the king said. "We can write that number as the product of two numbers, like this:

$$9 \times 1{,}000{,}000$$

"I see!" the professor said. "And 1,000,000 can be written as 10^6! Just what we need—the exponent counts the number of zeros for us, which is a lot better than having to write them all out!"

Recordis quickly agreed that exponents were very useful. Using this notation, we wrote out the numbers that were causing the astronomer problems:

$$\text{distance to the sun} = 1.495 \times 10^8$$

$$\text{distance to the nearest star} = 4.035 \times 10^{13}$$

$$\text{distance to the nearest galaxy} = 1.89 \times 10^{19}$$

We decided to call this notation *scientific notation,* since it seemed to be useful for keeping track of the kind of numbers that arose in science. In general, scientific notation is useful whenever you have to write a very large number. (We also found out that it is useful when we need to write down a number that is close to zero, such as 0.0000005.) The normal procedure for writing a number in scientific notation is:

SCIENTIFIC NOTATION

Express the number as the product of a power of 10 and a number between 1 and 10. Examples:

- 8.066×10^{67} (number of ways of arranging a 52-card deck)
- 9.46×10^{15} (number of meters in a light-year)
- 3.16×10^7 (number of seconds in a year)

"I know another use for exponents," Recordis realized. "I am trying to figure out the number of different ways of rolling five dice. For example, one possibility is for the dice to turn up 1, 5, 3, 1, 6. Another possibility is 3, 3, 2, 5, 1. I have started to list all the possible results that can occur."

1 1 1 1 1
1 1 1 1 2
1 1 1 1 3
1 1 1 1 4

and so on.

"There are six possible outcomes if you toss one die," the professor said. "And there must be $6^2 = 36$ outcomes if you toss two dice."

"In general, if you toss n dice, there must be 6^n total possible results," the king said. "In this case, Recordis is tossing 5 dice, so there must be 6^5 results." We calculated 6^5 and came up with the result $6^5 = 7{,}776$.

Recordis whistled. "I hadn't imagined that there were quite that many possibilities!" he said.

Builder came back to us after finishing the calculations for the tiles. "We will need a total of 1,400 tiles," he said. "However, you cannot buy the exact number of tiles you want. They come only in boxes of 500 tiles each. Therefore, we will have to buy three boxes giving us 1,500 tiles, so we will have $1{,}500 - 1{,}400 = 100$ tiles left over. What do you think we should do with them?"

"We could add a tiled patio," the king suggested.

"How large a patio can we make with 100 tiles?" the professor asked.

"We'll have to cut out 100 square cards to represent the tiles, and then lay them out on the floor to see how big a square we can make," Recordis suggested.

"We could calculate the size of the patio directly if we could figure out a way to do exponentiation *backwards*," the king said.

"How could we do exponentiation backwards?" Recordis said. "I have enough trouble doing it forwards."

The king explained. "If we start by knowing the size of the square a, and then calculate a squared, we will know how many tiles we need. If we start by knowing how many tiles we have, and then calculate the square backwards, we should be able to find out the size of the square that uses

that many tiles. For example, if we have a square with sides of length 7, then we will need $7^2 = 49$ tiles. Therefore, if we start with 49 tiles, we know that we will be able to make a square with sides of length 7."

"I can guess how big a square we can make with 100 tiles," Recordis said. "Since a square of size 10 requires $10^2 = 100$ tiles, that obviously means that 100 tiles can be made into a square of size 10."

"That will make a fine-sized patio," the professor said. "We need to think of a name for the operation of taking a backward square."

After a heated discussion, we settled on the term *square root* to stand for backwards square. For example, the square root of 16 is 4, since $4^2 = 16$. We had even more problems deciding on how to symbolize the operation of taking a square root, until finally we decided to adopt a funny-shaped scribble that Pal had accidentally made on the palace wall: $\sqrt{}$. (This symbol is called the *radical* symbol.)

We made a list of square roots:

$$\sqrt{1} = 1 \qquad \sqrt{36} = 6$$

$$\sqrt{4} = 2 \qquad \sqrt{49} = 7$$

$$\sqrt{9} = 3 \qquad \sqrt{64} = 8$$

$$\sqrt{16} = 4 \qquad \sqrt{81} = 9$$

$$\sqrt{25} = 5 \qquad \sqrt{100} = 10$$

"Hold it!" the professor said. "If $a^2 = 9$, then there are two possible values for a: a could be 3 or -3, since $(-3)^2 = (-3) \times (-3) = 9$." We realized that for *any* positive number there were two values that could be squared to give that number, one positive and one negative. For example,

$$1^2 = 1 \qquad (-1)^2 = 1$$

$$2^2 = 4 \qquad (-2)^2 = 4$$

$$5^2 = 25 \qquad (-5)^2 = 25$$

$$6^2 = 36 \qquad (-6)^2 = 36$$

$$10^2 = 100 \qquad (-10)^2 = 100$$

However, we decided that, to avoid confusion, the symbol $\sqrt{}$ would always mean, "Take the positive value of the square root." For example, $\sqrt{49}$ means 7, not -7.

The professor, being very systematic as usual, wanted to look for some more properties of exponents. First, we checked to see whether exponentiation was commutative, and we found that it was not. (In other words, the order *does* make a difference in exponentiation.) Therefore, $a^b \neq b^a$. For example, $3^4 = 81$, which does not equal $4^3 = 64$.

Next, we tried to find a formula for $x^a + x^b$, but we had no luck. So we decided to look for a general formula for $x^a x^b$.

"Let's try a concrete example," the king said. We tried to find an expression for $3^4 \times 3^6$.

"We can write that out the old way," Recordis noted.

$$3^4 = 3 \times 3 \times 3 \times 3$$

$$3^6 = 3 \times 3 \times 3 \times 3 \times 3 \times 3$$

$$3^4 \times 3^6 = 3 \times 3 \times 3 \times 3 \times 3 \times 3 \times 3 \times 3 \times 3 \times 3$$

"Now all we have to do is count the number of 3's," the king said. We counted ten 3's, so that meant

$$3^4 \times 3^6 = 3^{10}$$

We tried some more examples, and found that

$$2^5 \times 2^{10} = 2^{15}$$

$$10^3 \times 10^2 = 10^5$$

"I see a pattern!" Recordis said. "It looks as if $x^a x^b = x^{a+b}$.

Next, we tried to find an expression for x^a/x^b. We tried an example:

$$\frac{5^7}{5^3} = \frac{5 \times 5 \times 5 \times 5 \times 5 \times 5 \times 5}{5 \times 5 \times 5}$$

$$= 5 \times 5 \times 5 \times 5$$

$$= 5^4$$

Some more examples convinced us that in general

$$x^a/x^b = x^{a-b}$$

We also found that $(x^a)^b = x^{ab}$. For example,

$$(3^2)^4 = (3^2) \times (3^2) \times (3^2) \times (3^2)$$

$$= 3 \times 3 \times 3 \times 3 \times 3 \times 3 \times 3 \times 3$$

$$= 3^8$$

The final rule we found was that $(xy)^a = x^a y^a$. For example,

$$(5 \times 3)^2 = (5 \times 3) \times (5 \times 3)$$

$$= (5 \times 5) \times (3 \times 3)$$

$$= 5^2 \times 3^2$$

$$= 225$$

"We have pinned down the behavior of exponents pretty well," Recordis said. "They will have no choice but to obey these four laws."

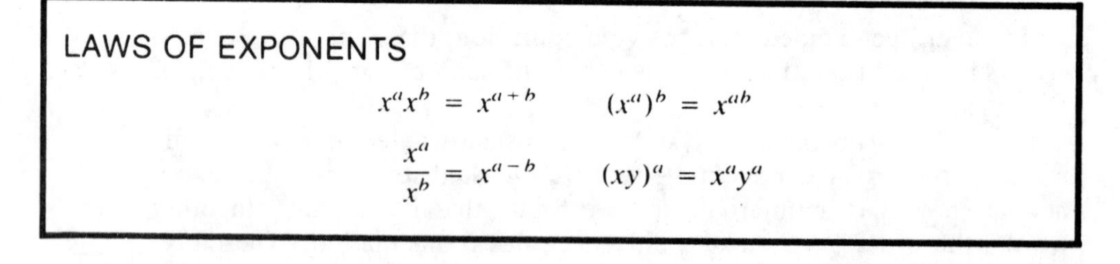

LAWS OF EXPONENTS

$$x^a x^b = x^{a+b} \qquad (x^a)^b = x^{ab}$$

$$\frac{x^a}{x^b} = x^{a-b} \qquad (xy)^a = x^a y^a$$

Before we left for the day, the king was wistfully thinking about another problem. "It sure helps that we have invented scientific notation for very large numbers, but it's too bad we don't have a way to represent very small numbers."

"Very small numbers are no problem," Recordis said. "They're just very negative numbers."

"I didn't mean that," the king said. "I meant numbers with very small absolute values—in other words, numbers very close to zero. For example, the Royal Biologist has told me that she has been looking through her microscope at some amoebas, and she has found that the average width of an amoeba is 0.00002 meters. Once again, we run into the problem of having to keep track of too many zeros."

"We could rewrite the expression like this," Recordis suggested, "although I'm afraid it doesn't help get rid of the zeros."

$$\frac{2}{100{,}000}$$

"Let's write it like this:" the professor said.

$$\frac{2}{10^5}$$

"That's what we need!" the king said. "A way of doing scientific notation for very small numbers!"

"That expression is still rather awkward," the professor went on. "So, I have an ingenious idea. Let's write $1/10^5$ as 10^{-5}."

"A negative exponent?" Recordis asked in bewilderment. "An exponent means repeated multiplication, and it makes no sense to repeatedly multiply something a negative number of times. I still hardly believe in negative numbers anyway, so I'm surely not going to believe in negative exponents."

"A negative exponent means just what I want it to mean," the professor said. "I'm going to make up this definition."

$$x^{-a} = \frac{1}{x^a}$$

"But you can't just make up a definition like that!" Recordis said. "Or at least you never let *me* make up the definitions that I want!"

"Who's going to stop me?" the professor asked.

"But you have to be careful," the king said, "because if we allow you to make up any definition that you want, then you will start making up definitions that are inconsistent with what we've done up to now."

"Aha!" the professor said. "That is the beauty of my plan. If we define negative exponents in the way I have suggested, then we will find that they still obey the laws of exponents. For example, we can write x^{-3} as x^{4-7}. Using the laws of exponents, we can write

$$x^{4-7} = \frac{x^4}{x^7}$$

$$= \frac{xxxx}{xxxxxxx}$$

and after cancelling, we're left with

$$\frac{1}{xxx} = \frac{1}{x^3}$$

so

$$x^{-3} = \frac{1}{x^3}$$

just as I said it should be."

The king was convinced that we should define negative exponents in this way. As usual, Recordis was reluctant to accept the new ideas, but since he could not think of any more objections at the moment he let the matter drop. He started writing in his record book: "We now know how to find x^a, where the exponent a can be an integer (except 0)."

"We can define what x^0 is," the professor interrupted him.

"All right." Recordis wrote: "$x^0 = 0$ for any x."

"No it doesn't!" the professor objected. "$x^0 = 1$."

"What do you mean?" Recordis exploded. "You can't tell me that if you multiply a number together zero times you're going to end up with 1."

"But we must define x^0 to be consistent with the laws of exponents," the professor argued. "Look at this."

$$x^0 = x^{4-4} = \frac{x^4}{x^4} = 1$$

Recordis stared speechless at this result. "Algebra is starting to get hard," he admitted, "but you're going to have to drag me kicking and screaming if you expect me to accept a definition so contrary to intuition as $x^0 = 1$."

"It's actually not as implausible as you might think," the king said. "Suppose you start with a particular number a. If you *add* a total of zero n's to that number, it remains unchanged:

$$a + 0 \times n = a \qquad (\text{meaning } 0 \times n = 0)$$

It seems reasonable that if you *multiply* a total of zero n's by that number, then it should also remain unchanged:

$$a \times n^0 = a \qquad (\text{meaning } n^0 = 1)$$

"Here's another way to look at it. Examine this table."

$$2^4 = 16$$

$$2^3 = 8$$

$$2^2 = 4$$

$$2^1 = 2$$

"Clearly, in order to follow the same pattern, we must have $2^0 = 1$, and $2^{-1} = 1/2$, $2^{-2} = 1/4$, and so on."

Extremely mistrustful, Recordis wrote these two definitions down in his book:

$$x^{-a} = \frac{1}{x^a}$$

$$x^0 = 1$$

(Note that x^n is not defined if $n \leq 0$ and $x = 0$.)

"So now we know what it means when we have an exponent that is any integer," Recordis said. "I'm glad you haven't suggested that we try to find out what happens if we have an exponent that is a fraction!"

"Actually, I was just beginning to think about that," the professor mused.

Note to Chapter 5

- We decided to use the symbol $\pm$ to mean "plus or minus." For example, the expression ± 3 means "positive three or negative three." If $x^2 = 16$, then $x = \pm\sqrt{16} = \pm 4$.

Exercises

1. If you toss a coin n times, how many possible outcomes are there?
2. If you select five cards from a 52-card deck (replacing each card after it is drawn), how many possible ways are there to draw the cards?
3. How many possible four-letter words are there?
4. How many possible three-number lock combinations are there if there are 30 possibilities for each number?
5. Simplify $x/y + y/x$.
6. If a is even, show that a^2 is even.
7. If $x^2 > x$, what can you say about the value of x?
8. If $x^2 < x$, what can you say about the value of x?
9. If $x^2 = x$, then what is the value of x?
10. If $x^2 > | x |$, what can you say about the value of x?
11. If $x^2 < | x |$, what can you say about the value of x?
12. What is $x \cdot | x |$?

Simplify these expressions. Express all of the negative powers as fractions.

13. $3^5/3^2$
14. $2^{10}/2^6$
15. $5/5^2$
16. $(1.16)^2/(1.16)^4$
17. $(3.45)^3/(3.45)^{-2}$
18. x^{-3}/y^{-3}
19. $3r^3/4r^2$
20. $10a^2b^3c/abc$
21. $16a^3b^4c/8a^3b^6c^2$
22. r^2/r
23. $1/x^{-1}$
24. b^{-3}
25. $a^{-2}c^2$
26. $1/(x^{-1} + x^{-2})$
27. If the interest rate is 5 percent, how many years will it be until the amount of money in your bank account doubles?
28. If the interest rate is 10 percent, how many years will it be until your bank balance doubles?
29. If the interest rate is 20 percent, how many years will it be until your bank balance doubles?
30. If x is positive, is x^n positive?
31. If x is negative, is x^n negative?
32. The *kinetic energy* of an object with mass M and velocity v is $\frac{1}{2}Mv^2$. What is the kinetic energy of a bullet with mass 0.001 kilograms and velocity 1,000 meters per second? (In this case energy is measured in a unit called the *joule*.)
33. The volume of a sphere is given by the formula $\frac{4}{3}\pi r^3$, where π is a number about equal to 3.14 and r is the radius of the sphere. The earth is approximately a sphere with radius about 6,400 kilometers. What is the volume of the earth?
34. The speed of light is about 186,000 miles per second. How many miles does light travel in one year? (Use scientific notation.)
35. The force of gravity between two objects is proportional to $1/r^2 = r^{-2}$, where r is the distance between the objects. If the objects move 5 times as far apart, how much weaker will the force of gravity between them become?
36. The amount of energy released per unit area of a star is proportional to T^4, where T is the absolute temperature of the star. If a bright star is twice as hot as a dim star, how much more energy will be released per unit of surface area in the bright star?

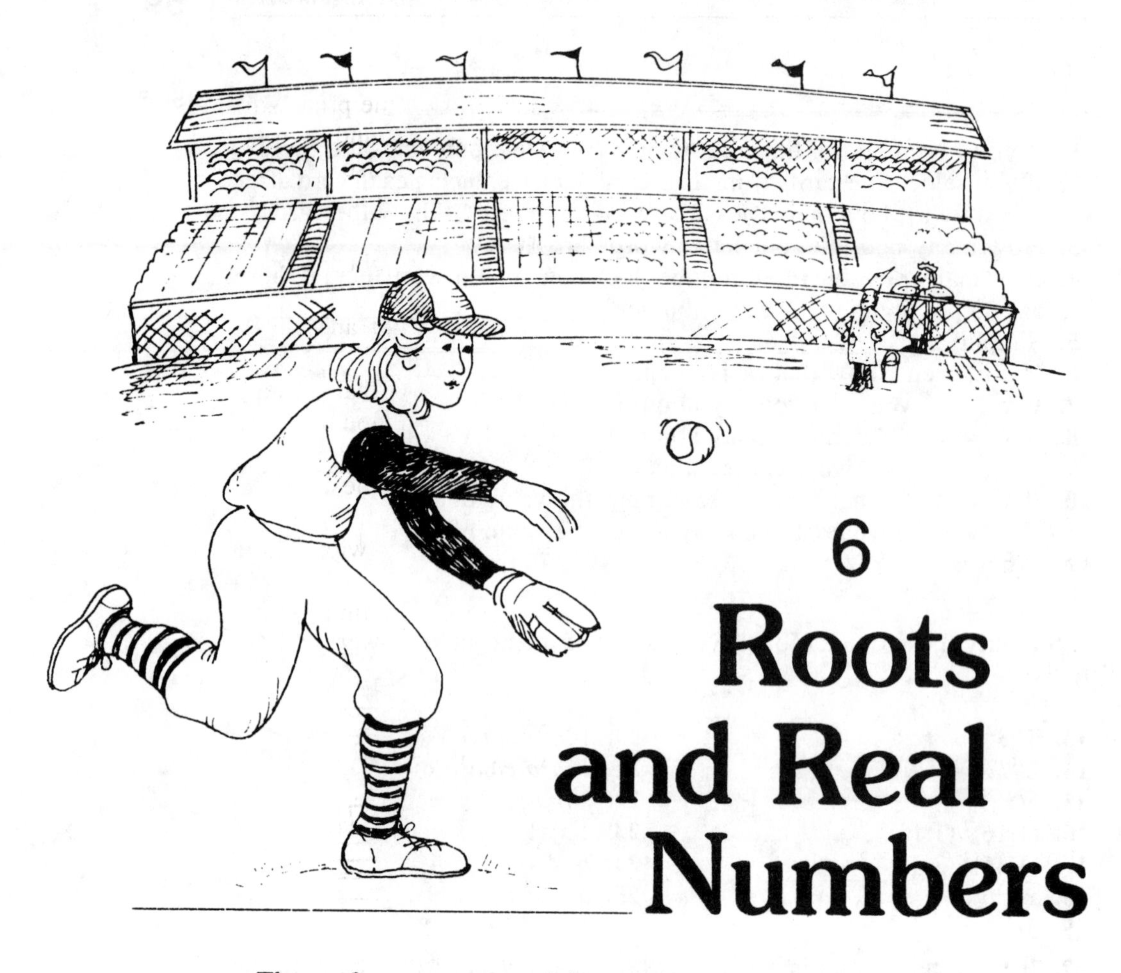

6 Roots and Real Numbers

The professor announced that she was preparing for the opening of the baseball season, when she would be pitcher for the Royal Palace team. "I have to practice my fast ball," she said. "I want the pitch to reach the plate in 0.75 seconds, so that the batter has no time to think. I wonder how fast I need to throw the ball?"

"We need to know the distance from the pitcher's mound to home plate," the king said. Igor drew a diagram (Figure 6-1).

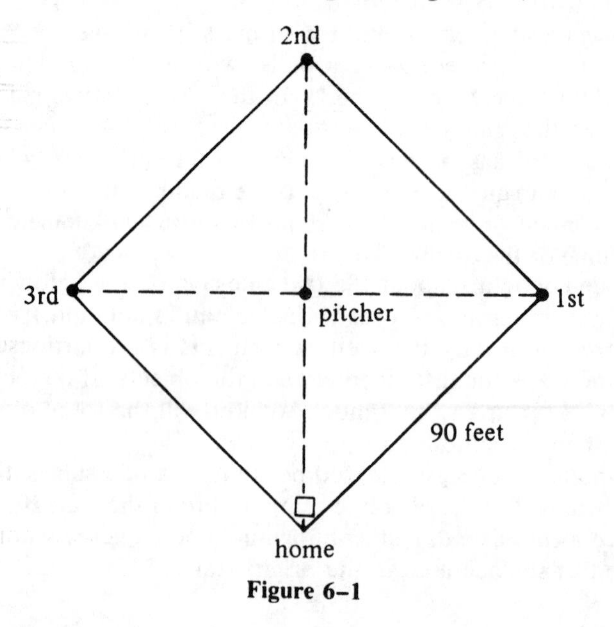

Figure 6–1

"The distance from home plate to first base is 90 feet," the professor said. "I also know that the distance from the pitcher's mound to first base is the same as the distance to home plate, because I also need to practice my sneaky pickoff move. However, I don't know the distance from the mound to the plate."

"We can make a triangle," the king said. "It will be a right triangle, because it contains a right angle."

"I know what right angles are," Recordis said. "A right angle is a square corner. We called them 90-degree angles."

At that moment the detective suddenly walked in. "I have finally found an important clue in the case of the missing books," he announced. "I have found some footprints indicating that the culprit crossed Swamp Field after the incident. The only thing I need to know now is the distance across the field."

"I hope you're not about to suggest that I cross the swamp with a tape measure!" Recordis exclaimed.

"We know that the swamp is a square, with each side one unit long," the professor said. Igor drew a map (Figure 6-2).

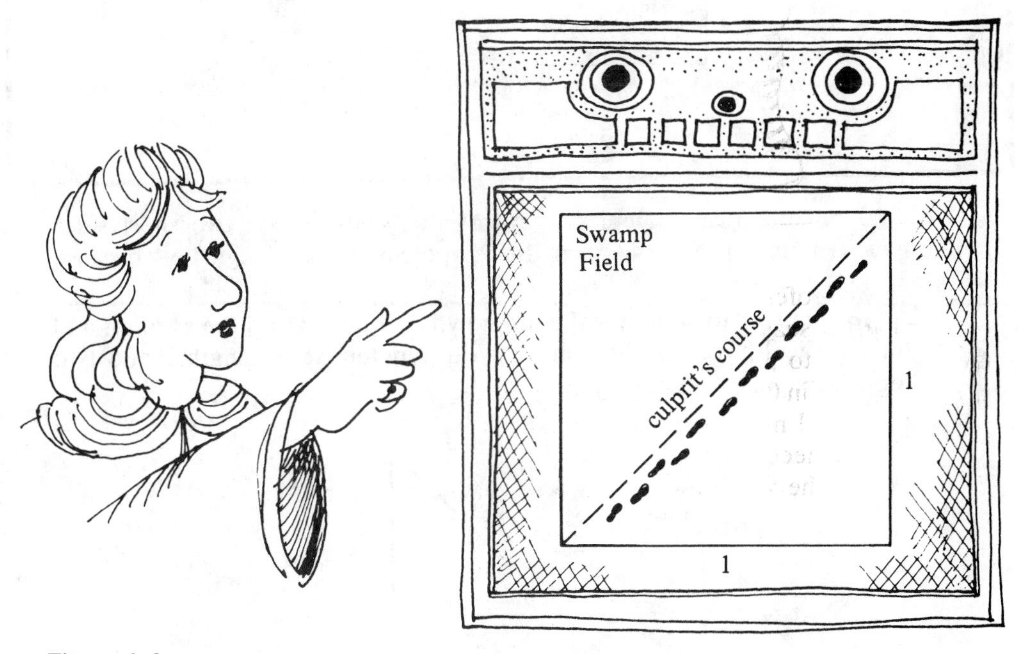

Figure 6–2

"This problem is almost the same as the baseball problem," the king said. "Once again we have a right triangle."

"How do we solve for the length of the unknown side?" Recordis asked.

"In circumstances such as this, it is usually advisable to consult a proposition called the *Pythagorean theorem,*" the detective said.

"The Pythagorean theorem!" the king said in awe.

"The Pythagorean theorem!" the professor said in awe.

"The Pythagorean theorem!" Recordis said in awe. "What's the Pythagorean theorem?"

We consulted an old dust-covered book. "Isn't that the theorem about a hippopotamus that's rectangular?" Recordis asked.

"It's about a hypotenuse that's squared!" the professor said. We found the entry in the book.

THE PYTHAGOREAN THEOREM

Consider a right triangle in which c is the length of the hypotenuse (the side opposite the right angle) and a and b are the lengths of the other two sides (called the *legs*). (See Figure 6-3.) Then

$$a^2 + b^2 = c^2$$

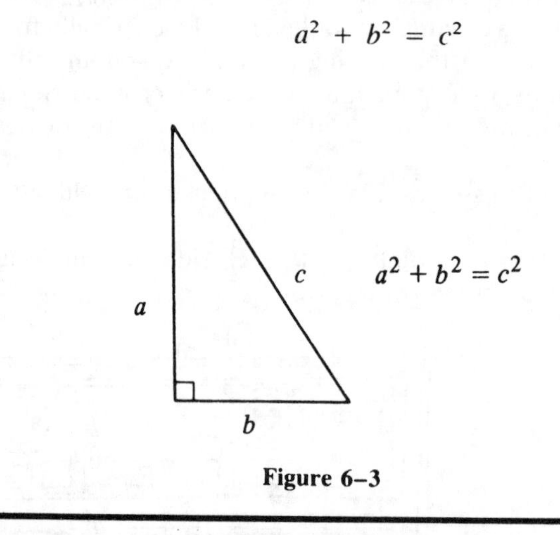

Figure 6–3

"I remember an example," the professor said. "If the two short sides have lengths 3 and 4, then the hypotenuse has length 5, because $3^2 + 4^2 = 25 = 5^2$.

"We'll use d to stand for the unknown distance across the swamp, and then we will have a right triangle with hypotenuse of length d and two legs of length 1." (See Figure 6-4.)

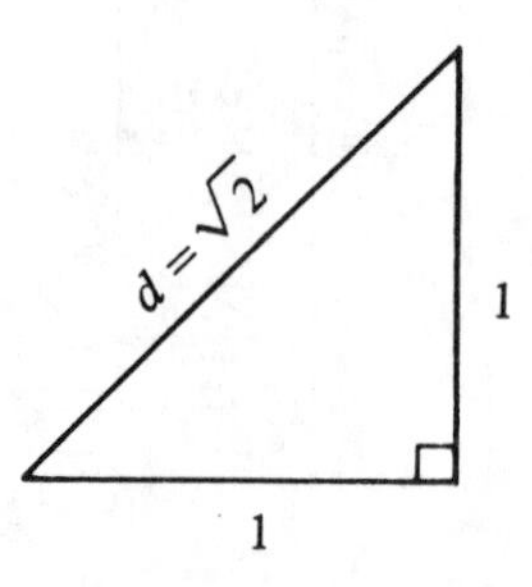

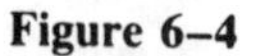

Figure 6–4

Then we applied the Pythagorean theorem:

$$d^2 = 1^2 + 1^2$$

$$d^2 = 1 + 1$$

$$d^2 = 2$$

$$d = \sqrt{2}$$

"So the distance across the swamp is equal to the square root of two," the king said.

"Hold it!" Recordis exclaimed. "There is no such number! We know that $1^2 = 1$ and $2^2 = 4$; 2 is between 1 and 4, so the square root of 2 would have to be between 1 and 2."

"That proves that there is no *natural* number that is the square root of 2," the professor said. "But there must be a rational number—in other words, a fraction—that is equal to the square root of 2. And I have a shrewd guess about what it is." The professor guessed that $\sqrt{2} = 7/5$. We easily calculated that $(7/5)^2 = 49/25 = 1.96$.

"That is tantalizingly close to 2," the king agreed, "since 50/25 is exactly equal to 2. But it is still not quite close enough."

The professor tried some more guesses.

$$(10/7)^2 = 2.04$$

$$(71/50)^2 = 2.016$$

$$(177/125)^2 = 2.005$$

"Give me time and I'll get it," the professor said.

"Sure you will," Recordis taunted. But, as the afternoon dragged on, the professor was unable to come up with a fraction that, when squared, was exactly equal to 2 (although she did come closer with each guess).

"Let's be a bit more general and look for some properties," the king suggested. "Suppose we know that $\sqrt{2} = a/b$, where a and b are two unknown integers that we are trying to solve for, and a/b is the simplest way of expressing that fraction."

We tried to find out whether a was odd or even. Since $a^2/b^2 = 2$, we could write $a^2 = 2b^2$, meaning that a^2 must be even. That meant that a itself must be even. (See Chapter 5, Exercise 6.) We let $c = a/2$, and since a was even we knew that c must be a natural number. Then $a = 2c$ and $a^2 = 4c^2$. Using the substitution property, $2b^2 = 4c^2$. Dividing both sides of this equation by 2 gives $b^2 = 2c^2$. Therefore, b^2 is even, and b must also be even.

The king suddenly realized that something was wrong. "We can't have both a and b be even!" he exclaimed. "We found that, if a fraction is expressed in simplest form, then either the numerator or the denominator or both must be odd." (See Chapter 4, Exercise 27.)

The professor was stunned. "How could we run into a contradiction like that?" she cried. "We must have made a mistake." However, after repeated checking we could not find any mistake in our logic. "This is impossible," the professor went on. "I do not believe my own eyes. Correct logical reasoning cannot lead to a contradiction."

"Then one of our assumptions must be wrong," Recordis said. "As my old teacher used to say, 'if the answer is wrong, then you must have made a mistake somewhere.' And since our reasoning was correct, the only possible place we could have made a mistake was in our initial assumption."

"But that means there are no two integers a and b such that $a/b = \sqrt{2}$!" the king said in distress. "In other words, the assumption that the square root of 2 is even a rational number is wrong!"

"Therefore, we have *proved* that it does not exist," Recordis said.

"But it *must* exist!" the professor insisted. "Look at that triangle! (Figure 6-4.) We could measure the length of that side of the triangle. If the length is not a rational number, then it must be some other kind of number."

"This time you've gone too far!" Recordis exploded. "You've done some weird things since I first met you, but this is the most *irrational* thing you have ever done (at least this month). I let you railroad me into accepting negative numbers, even though I didn't believe in them, but I'm certainly not going to accept a number that is not a rational number."

"I admit that we will have to call this type of number an *irrational* number," the professor agreed. "But how can we find out its value?"

"We can have Builder measure the triangle in the diagram with his ruler," the king said.

Builder was called in and he put his ruler next to the hypotenuse of the triangle Igor had drawn. "The length of this side is 1.41," he said.

"It is a regular rational number after all!" Recordis breathed a sigh of relief.

"But we must check to make sure that $1.41^2 = 2$," the king said. Our calculations showed that $1.41^2 = 1.9881$.

"That's not quite right," the king told Builder. "You will have to measure more accurately."

Builder pulled out a more accurate ruler and came up with the value 1.414, but we checked this value and found that $1.414^2 = 1.999\,396$.

"I see a pattern!" Recordis said. "It will be a regular repeating decimal. The pattern is 1.414141414141. . . ."

However, this idea proved to be incorrect when Builder pulled out a still more accurate ruler coupled to a magnifying glass. His new suggested value was 1.4142, but this still was not exactly right. We found that $1.4142^2 = 1.999\,961\,6$.

We kept begging Builder for more accuracy so he kept coming up with more accurate values:

$$1.41421^2 = 1.999\,989\,9$$

$$1.414213^2 = 1.999\,998\,4$$

$$1.4142135^2 = 1.999\,999\,8$$

"This could go on forever!" Recordis complained.

"Recordis is right," the king said. "We could continue to add more decimal places to our value for the square root of 2, but we could never come up with exactly the correct value. We have already proved that $\sqrt{2}$ is not a rational number, so that means $\sqrt{2}$ cannot be expressed as a decimal number that either stops or continues to repeat the same pattern. So $\sqrt{2}$ can be represented as a decimal fraction, but only one with an infinite number of digits that never repeat any pattern."

"For practical purposes, who cares?" Recordis asked "It doesn't make any difference to me whether the true value is 1.4142 or 1.4143."

"Are there many other irrational numbers?" the king asked.

"There must be a lot of them," the professor said. "I bet $\sqrt{3}$, $\sqrt{5}$, $\sqrt{6}$, $\sqrt{7}$ are all irrational numbers." We calculated some approximate values for these square roots:

$$\sqrt{3} = 1.7321$$

$$\sqrt{5} = 2.2361$$

$$\sqrt{6} = 2.4495$$

$$\sqrt{7} = 2.6458$$

Now we could solve the baseball problem (Figure 6-5).

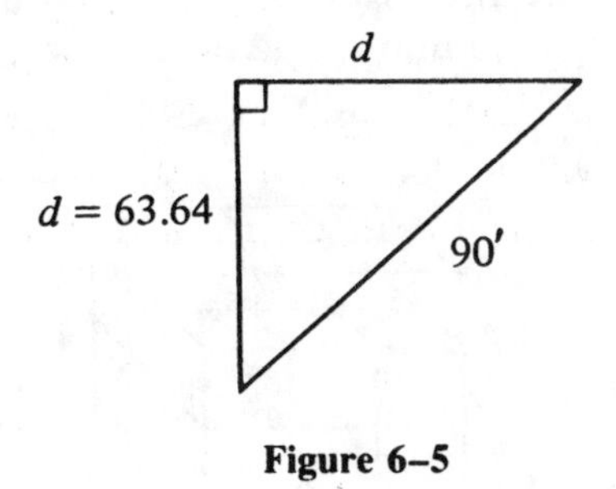

Figure 6–5

$$d^2 + d^2 = 90^2$$
$$2d^2 = 8{,}100$$
$$d^2 = 4{,}050$$
$$d = \sqrt{4{,}050}$$
$$d = 63.64$$

Therefore, in order for the pitch to reach the plate in exactly 0.75 seconds, the speed must be

$$63.64 \text{ feet}/0.75 \text{ seconds} = 84.85 \text{ feet per second}$$
$$= 57.85 \text{ miles per hour}$$

"We should investigate square roots some more to see if we can learn any interesting properties," the professor said. After some experimentation, we found that

$$\sqrt{ab} = \sqrt{a}\sqrt{b} \qquad \text{for any values of } a \text{ and } b$$

For example,

$$\sqrt{4 \times 9} = \sqrt{4} \times \sqrt{9}$$
$$\sqrt{36} = 2 \times 3$$
$$6 = 6$$

Next, we found that

$$\sqrt{\frac{a}{b}} = \frac{\sqrt{a}}{\sqrt{b}}$$

For example,

$$\sqrt{\frac{36}{4}} = \frac{\sqrt{36}}{\sqrt{4}} = \frac{6}{2} = 3$$

However, we found that there was no way to simplify either $\sqrt{a + b}$ or $\sqrt{a} + \sqrt{b}$. Recordis suggested the rule that $\sqrt{a + b} = \sqrt{a} + \sqrt{b}$, but the professor pointed out that this would not work. For example, $\sqrt{9 + 16}$ does *not* equal $\sqrt{9} + \sqrt{16}$. And of course Recordis became upset because the professor seemed to be able to get away with making up outrageous rules, but for some reason he was not allowed to.

"Are there any other kinds of irrational numbers?" the king asked. "So far the only irrational numbers we have found have been square roots."

"Here's an idea," the professor said. "Suppose we take 64 of Pal's letter blocks and form them into a big cube. How tall will the cube be?" (See Figure 6-6.)

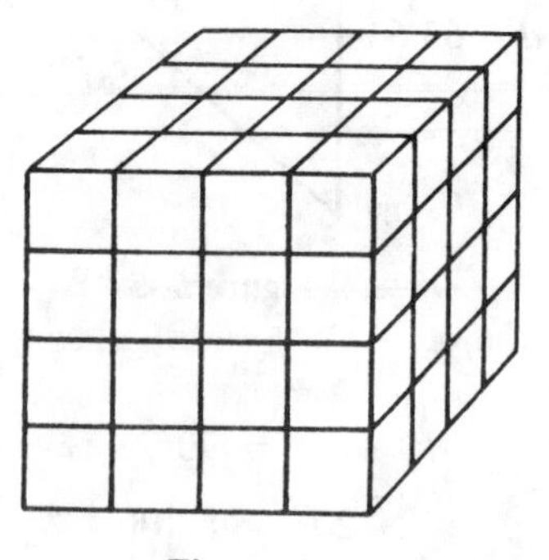

Figure 6–6

"If we build a cube that is a blocks high, a blocks thick, and a blocks wide, then we will need $a \times a \times a = a^3$ blocks," the king said.

"Therefore, if we have a total of 64 blocks, we must find a number a such that $a^3 = 64$." The professor suggested how to solve this equation for a. "If the equation is $a^2 = 64$, then we know that the solution is $a = \sqrt{64} = 8$. We called the answer the square root of 64. So, when the question is $a^3 = 64$, I suggest that we write the answer like this:

$$a = \sqrt[3]{64}$$

and call the answer the *cube root* of 64. We'll use the same radical sign that we used for square roots, only this time we will write a little 3 next to the radical to prevent us from getting cube roots mixed up with square roots."

We guessed that $\sqrt[3]{64} = 4$, and that guess turned out to be right since $4^3 = 4 \times 4 \times 4 = 4 \times 16 = 64$. However, just as we suspected, most values for cube roots turned out to be irrational numbers. We made a list of some approximate values for cube roots:

x	$\sqrt[3]{x}$	x	$\sqrt[3]{x}$
1	1.0000	16	2.5198
2	1.2599	17	2.5713
3	1.4422	18	2.6207
4	1.5874	19	2.6684
5	1.7100	20	2.7144
6	1.8171	21	2.7589
7	1.9129	22	2.8020
8	2.0000	23	2.8439
9	2.0801	24	2.8845
10	2.1544	25	2.9240
11	2.2240	26	2.9625
12	2.2894	27	3.0000
13	2.3513	28	3.0366
14	2.4101	29	3.0723
15	2.4662	30	3.1072

"We could also have fourth roots," the king said.

$$x = \sqrt[4]{a} \text{ if } x^4 = a$$

For example,

$$\sqrt[4]{81} = 3 \text{ since } 3 \times 3 \times 3 \times 3 = 81$$

$$\sqrt[4]{16} = 2 \text{ since } 2 \times 2 \times 2 \times 2 = 16$$

$$\sqrt[4]{625} = 5 \text{ since } 5 \times 5 \times 5 \times 5 = 625$$

"I have another ingenious idea," the professor said modestly. "Exponential notation can be even more versatile than we might imagine. We have not yet defined what we mean by the expression x^a if a is a fraction."

Recordis started to scream. "Don't tell me that you think an expression such as $2^{1/2}$ means anything! There is no way that you can take a number and multiply it together 1/2 times!"

"But there is a natural definition for $2^{1/2}$," the professor said. "We will say that 2 raised to the 1/2 power is equal to the square root of 2:

$$2^{1/2} = \sqrt{2}$$

And, in general,

$$x^{1/2} = \sqrt{x}$$

The professor showed us why this definition was consistent with the laws of exponents:

$$(x^a)^b = x^{ab} \qquad \text{for any values for } x, a, \text{ and } b$$

"Now, let $a = 1/2$, and let $b = 2$. Then we know that

$$(x^{1/2})^2 = x^{1/2 \cdot 2}$$

$$(x^{1/2})^2 = x^1$$

$$(x^{1/2})^2 = x$$

Now take the square root of both sides."

$$x^{1/2} = \sqrt{x}$$

Likewise, we were able to show that $x^{1/3} = \sqrt[3]{x}$, $x^{1/4} = \sqrt[4]{x}$, and, in general, $x^{1/a} = \sqrt[a]{x}$. We also found that $x^{2/3} = (\sqrt[3]{x})^2 = \sqrt[3]{x^2}$, and, in general, $x^{a/b} = \sqrt[b]{x^a} = (\sqrt[b]{x})^a$. (See Exercise 34.)

The professor wanted to use the name *real numbers* for the set containing all of the rational numbers and all of the irrational numbers. Each real number corresponds to exactly one point on a number line. (See Figure 6-7) Real numbers can be represented as decimal fractions that either terminate, endlessly repeat a pattern, or continue endlessly with no pattern. The results of measurements of physical quantities (such as distance, time, mass, or energy) will be real numbers. For a long time we

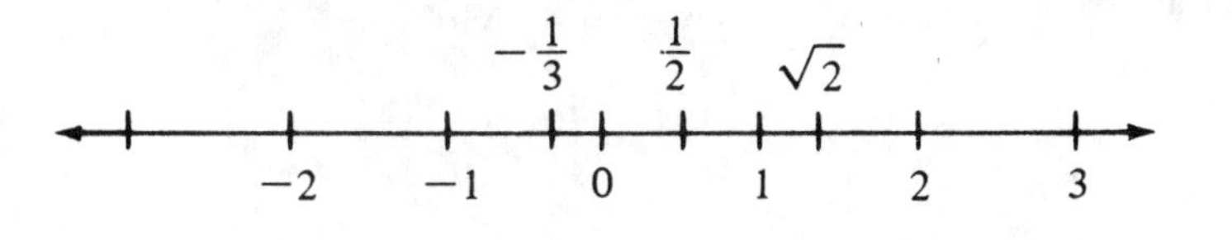

Figure 6–7

thought that the real numbers were the most general possible type of number, but we discovered some very unusual things once we reached Chapter 20.

Unless we are told otherwise, a letter acting as a variable is assumed to represent a real number.

SUMMARY OF ROOTS

The square root of a positive number a (written $\sqrt{a}$ or $a^{1/2}$) is the positive number that when multiplied by itself gives a:

$$\sqrt{a} \times \sqrt{a} = (\sqrt{a})^2 = (a^{1/2})^2 = a$$

The nth root of a (written $\sqrt[n]{a}$ or $a^{1/n}$) is the number that gives a when it is raised to the nth power:

$$(\sqrt[n]{a})^n = (a^{1/n})^n = a$$

Notes to Chapter 6

- If a radical sign appears in the denominator of a fraction, you can rewrite the fraction so that there are no radicals in the denominator. For example,

$$\frac{1}{\sqrt{2}} = \frac{1}{\sqrt{2}} \times \frac{\sqrt{2}}{\sqrt{2}} = \frac{\sqrt{2}}{2}$$

Matters are slightly trickier if there is a sum involving a radical in the denominator. For example, consider the fraction

$$\frac{3}{6 + \sqrt{5}}$$

To simplify, multiply the top and bottom by $(6 - \sqrt{5})$:

$$\frac{3}{(6 + \sqrt{5})} \frac{(6 - \sqrt{5})}{(6 - \sqrt{5})} = \frac{3(6 - \sqrt{5})}{6^2 - 5} = \frac{18 - 3\sqrt{5}}{31}$$

In general,

$$\frac{1}{\sqrt{a} + \sqrt{b}} = \frac{\sqrt{a} - \sqrt{b}}{a - b} \quad \text{and} \quad \frac{1}{\sqrt{a} - \sqrt{b}} = \frac{\sqrt{a} + \sqrt{b}}{a - b}$$

Exercises

Simplify these radicals if possible. For example, $\sqrt{18} = \sqrt{9 \cdot 2} = \sqrt{9}\sqrt{2} = 3\sqrt{2}$.

1. $\sqrt{12}$
2. $\sqrt{100}$
3. $\sqrt{9/16}$
4. $\sqrt{8}$
5. $\sqrt{72}$
6. $\sqrt{32}$
7. $\sqrt{44}$
8. $\sqrt{a^4x^2}$
9. $\sqrt{4x^6}$
10. $\sqrt{12y^2}$
11. $\sqrt{4a^2}$
12. Verify the results that are presented in the Note at the end of the chapter.

Remove the radicals from the denominators in these expressions:

13. $1/\sqrt{5}$
14. $1/(\sqrt{5} + 4)$
15. $10/\sqrt{5}$
16. $5/(\sqrt{6} + 9)$
17. $(1 + \sqrt{2})/(1 - \sqrt{2})$
18. $(1 - \sqrt{5})/(1 - \sqrt{2})$
19. $1/(\sqrt{2} + \sqrt{8})$
20. $5/(\sqrt{6} + \sqrt{54})$

21. Suppose you think that x_1 is a good guess for $\sqrt{a}$. A closer guess (x_2) can be calculated from the formula $x_2 = \frac{1}{2}(x_1 + a/x_1)$. You can keep repeating the same formula to come up with still closer guesses. Use this formula to calculate a value for $\sqrt{3}$. Start with the initial guess $x_1 = 1.5$ and perform three repetitions of the process.

Simplify the expressions:

22. $\sqrt[10]{1024}$
23. $\sqrt[4]{81}$
24. $\sqrt[5]{100,000}$
25. $\sqrt[9]{1}$
26. $\sqrt[4]{1/256}$
27. $\sqrt[4]{4}$
28. $\sqrt[4]{9}$
29. $\sqrt[10]{32}$
30. $\sqrt[3]{729}$
31. $\sqrt[6]{a^3b^6c^{12}}$
32. $\sqrt[4]{a^{12}/b^{24}}$

33. According to the theory of relativity, if an object is moving with velocity v it will become shorter. If the length of the object is L when it is not moving then its length while moving will be $L\sqrt{1 - v^2/c^2}$. In this formula c is the speed of light (about 186,000 miles per second). How much shorter will a 1-meter rod become if it is moving at 60 miles per hour? At 600 miles per second? At 1/10 the speed of light? At 1/2 the speed of light? At 9/10 the speed of light?

34. Show that the definition $x^{a/b} = (\sqrt[b]{x})^a$ is consistent with the laws of exponents.

7 Algebraic Expressions

Recordis asked Builder to help him design a new pigpen. "I have 40 feet of fencing," Recordis explained. "I would like the pen built in the shape of a rectangle, but I want the rectangle to have as much area as possible. I don't know whether to make the pen very long and narrow, or very short and wide, or square-shaped." (See Figure 7-1.)

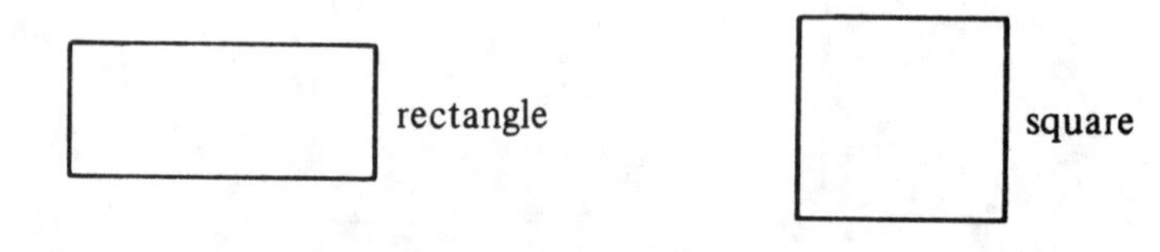

Figure 7–1

"We can easily calculate the area if you make it square-shaped," the professor said. "Since you have 40 feet of fencing, each side of the square would be $40/4 = 10$ feet long. In that case the area of the pen would be $10 \times 10 = 100$ square feet." (Figure 7-2.)

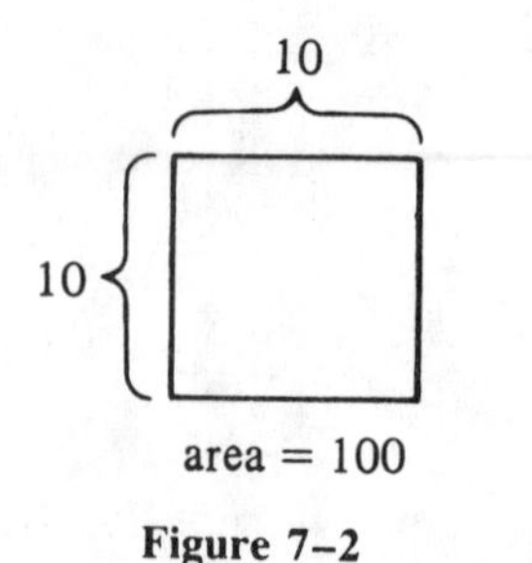

Figure 7–2

"I could also make it $10\frac{1}{4}$ by $9\frac{3}{4}$ or $10\frac{1}{2}$ by $9\frac{1}{2}$ or 11 by 9 or 5 by 15. There are so many possibilities," Recordis said. "It will take a long time to calculate the area of each of the possibilities to find the one with the largest area."

"I'm sure we don't have to do that much work!" the king said. "That is the point of developing algebra—to help us avoid a lot of calculations by making it possible to look at things in general terms. Let's look at your problem this way. We can call the width of the pen $10 - x$, where x is unknown, and then the length of the pen will be $10 + x$." (See Figure 7-3.)

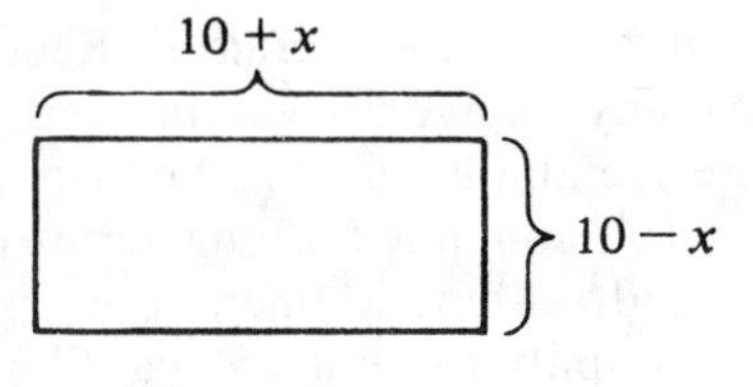

Figure 7–3

"Since

$$(10 + x) + (10 - x) + (10 + x) + (10 - x) = 40$$

it is clear that we are still using exactly 40 feet of fencing," the professor pointed out.

We wrote the formula for the area of the pen:

$$(10 - x)(10 + x)$$

"What do we do with that weird expression?" Recordis asked. "It contains both letters and numbers."

"I bet we can simplify it using the distributive property," the professor said.

$$(10 - x)(10 + x) = 10(10 + x) - x(10 + x)$$

"Now we can multiply out each of the parts on the right hand side, using the distributive property in reverse."

$$(10 - x)(10 + x) = 10 \times 10 + 10x - 10x - x^2$$

We simplified that last expression to

$$(10 - x)(10 + x) = 100 - x^2$$

"If x equals zero then you are building a square pen, and the area will be 100, just as we predicted it would be," the professor said.

"Also, it is pretty clear that the bigger x is, the smaller the area will be," the king said, "which means that the largest possible area occurs when x equals 0—in other words, when the pen is built in the shape of a square."

Recordis was still skeptical. "I'll concede that $(10 - x)(10 + x)$ is the area of the pen, but I'm still not sure that $(10 - x)(10 + x)$ is the same as $100 - x^2$."

"All we have done is turn a complicated algebraic expression into a simpler one," the professor said.

At that moment there was a loud thunderclap and suddenly there stood in front of us once again our vile enemy—the gremlin!

"Fools!" he cackled. "It will take you forever to develop algebra if you proceed one problem at a time like this. At this rate you will develop new rules only after you have found a use for them, and you will *never* develop enough practice to become comfortable with algebra! So long as algebraic manipulations seem strange and uncomfortable to you, you will have no hope. If someone presented you with a long list of algebraic expressions to simplify all at once, you would never be able to stand it!"

"I dare someone to present us with a long list of problems!" Recordis said, despite his fear.

"Since you asked for it—!" The gremlin tossed a large scroll of problems onto the table.

"You didn't have to take me literally!" Recordis protested.

"When you're ready to give up, call me and I will come and take over the kingdom!" the gremlin laughed as he disappeared.

The gremlin's scroll contained a long series of markings consisting of numbers, letters, and arithmetic operations signs. At the top were the ominous words: "Simplify these if you can!" The first entry was

$$10 + 5$$

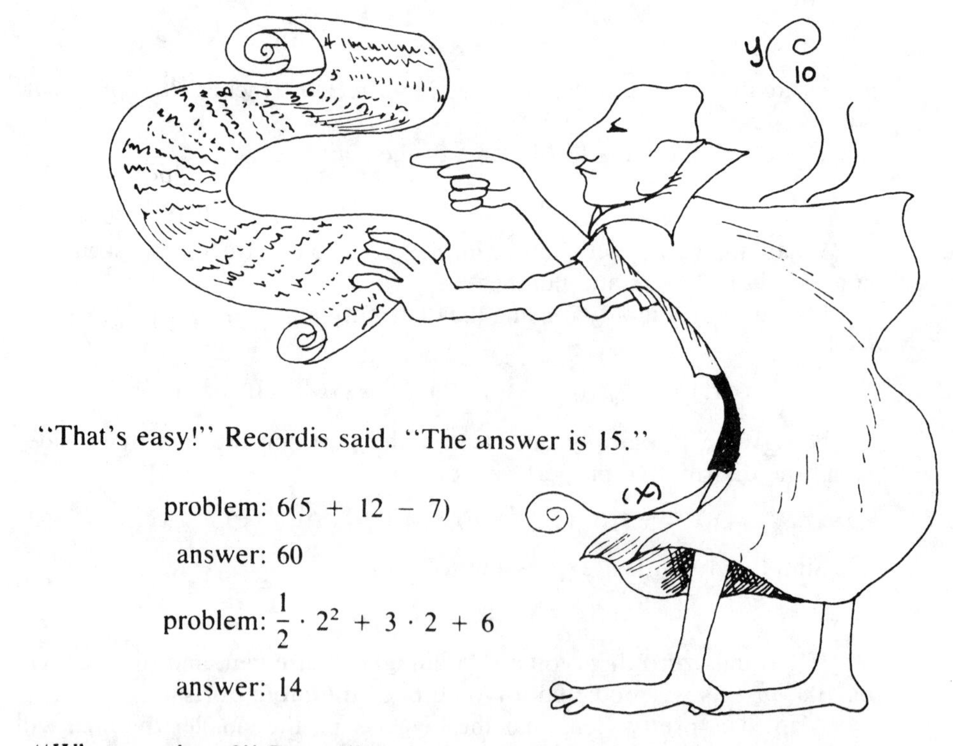

"That's easy!" Recordis said. "The answer is 15."

problem: $6(5 + 12 - 7)$

answer: 60

problem: $\frac{1}{2} \cdot 2^2 + 3 \cdot 2 + 6$

answer: 14

"What are these?" Recordis asked.

"The gremlin called them *algebraic expressions*," the professor said, "We can define algebraic expressions this way: An algebraic expression is a collection of numbers and letters representing numbers that are connected by arithmetic operation signs, such as $+$, $-$, /, or $\cdot$."

"The first few examples don't even contain any letters," Recordis said.

We decided that an algebraic expression containing only numbers and operation signs (in other words, no letters) would be called an *arithmetic expression*. The way to simplify an arithmetic expression is to calculate the value of the expression and then express the result as a number. Here are some more examples of arithmetic expressions that were on the gremlin's list:

$$\frac{1}{1/2 + 1/3} = \frac{6}{5}$$

$$\frac{(2 + 5)(3 - 4)}{(8 + 10)(3 - 4)} = \frac{7}{18}$$

$$\frac{2^5 + 1}{2^5 - 1} = \frac{33}{31}$$

"We hardly even need algebra to simplify arithmetic expressions," Recordis said. However, the next item on the list contained letters, and then suddenly everything became different.

problem: $10x + 5x$

"When an algebraic expression contains a letter, we can no longer calculate an explicit numerical value for the expression," the professor said. "However, we still should be able to simplify it."

The king suggested that we try using the distributive property:

$$10x + 5x = (10 + 5)x$$

We added the 10 and the 5 together to get 15, and the final result was

$$10x + 5x = 15x$$

"It's clear that those two expressions have the same value," Recordis said. "If you start with ten x's and then you get five more x's, you will end up with fifteen x's altogether."

$$(x+x+x+x+x+x+x+x+x+x)+(x+x+x+x+x)$$
$$=(x+x+x+x+x+x+x+x+x+x + x+x+x+x+x)$$

We simplified some more of the expressions on the gremlin's list:

gremlin's expression	simplified form
$3x + 24x$	$27x$
$6a + 3a$	$9a$
$6a + \frac{1}{2}a$	$6\frac{1}{2}a$
$12c - 3c$	$9c$

The next problem was:

Calculate $9a$, given that $a = 4$

"The answer clearly is 36," Recordis said.

"This proves that we can come up with a numerical answer if we're given the value of the letter in the algebraic expression," the professor said. We called this process the act of *evaluating* the algebraic expression for a particular value of the variable. We performed some more evaluations:

$3x + 5$, given $x = 8$

result: 29

$x^5 - x$, given $x = 2$

result: 30

$$\frac{1}{1/x + 1/x}, \quad \text{given } x = 3$$

result: 3/2

We had no problem with any of these, but then we had a problem when we came to:

problem: $10a + 6b$

"I don't know any way to simplify that expression," Recordis moaned. "If we had $10a + 6a$ then we could easily simplify the result to $16a$. And if we had numerical values for a and b, then we could evaluate the expression."

"That expression is as simple as it can get," the king said. We all reluctantly agreed, and went back to work on the other expressions.

gremlin's expression	simplified form
$10x + 2$	$10x + 2$
$3x + 25x + b$	$28x + b$
$4a + 7b + 6a$	$10a + 7b$
$3b + 2a + 1$	$3b + 2a + 1$
$5x - 2b + 7x$	$12x - 2b$
$-15a - 6a - 4b - 11b$	$-21a - 15b$
$12ab + 3a + 4b$	$12ab + 3a + 4b$

The next expression was

$$6ab + 4a - 2b + 10ab - 2a + b$$

We decided that the expression could be simplified so that it contained three parts: one part involving an a, one part involving a b, and one part involving an ab. We drew lines connecting the parts that we wanted to combine:

$$6ab + 4a - 2b + 10ab - 2a + b$$

and the result was

$$16ab + 2a - b$$

"I am beginning to see some patterns," the professor said in an organized sort of mood. "The gremlin said that we should start being systematic. I have discovered one obvious way to classify algebraic expressions: We can classify them by the number of little clumps that are added together. For example, both $15x$ and $6ab$ are one-part expressions; $10a + 7b$ and $5x + 7$ are two-part expressions; and $16ab + 2a - b$ and $16x + 2y - 1$ are three-part expressions."

We decided that we would use the word *term* to stand for the parts of an algebraic expression that were separated by plus or minus signs. For example, $\frac{4}{3}pr^3$ contains one term; $2a^2 + 3b^3$ contains two terms ($2a^2$ is one term and $3b^3$ is the other term); and the expression $ax^2 - bx + c$ contains three terms.

"Now we can classify expressions by the number of terms they contain," the professor said. "I suggest that we call a one-term expression a *monomial*, a two-term expression a *binomial*, and a three-term expression a *trinomial*. We know that in Greek 'mono' means 'one', as in 'monolith' or 'monorail', that 'bi' means 'two,' as in 'bicycle,' and that 'tri' means 'three,' as in 'tricycle.'" (The professor always liked to impress people with the fact that she had studied the ancient Greek language.) We agreed, and also decided that any expression containing two or more terms would be called a *multinomial*.

"Now we can appreciate the inherent beauty of my plan," the professor said. "Any term consists of a group of things multiplied together, such as $6ab$ or $5x$ or $10x^2$.

"We had a name for each part of an expression like those," the king said. "When several things were multiplied together, we said that each part was a *factor*. For example, in the expression 4×3, 4 and 3 are both factors; and in the expression $16ab$, 16, a, and b are all factors."

"Right," the professor agreed. "A term will consist of a group of factors. The important thing is that a term cannot consist of two things added together or subtracted."

"Why not?" Recordis demanded.

"Because if it did, then it would really be more than one term," the king said. "For example, the expression $10x + 4$ contains two terms, but the expression $10abcdxyz$ contains only one term."

"Now I suggest the following," the professor said. "We can divide any term into two parts, the number part and the letter part. For example, in the term $6x$, 6 is the number part and x is the letter part. In the term $13ab$, 13 is the number part and ab is the letter part. Now let's consider two terms that have the same letter part. We know that we can combine those into one term:

$$6x + 7x = 13x$$

In fact, whenever two terms have the same letter part we can combine them into one term."

Examples:

$$12x + 4x = 16x$$

$$13a^2b^2 + 15a^2b^2 = 28a^2b^2$$

"Two terms with the same letter part must like each other," Recordis said. "They seem to belong together."

"All right, we'll say that two terms are *like terms* if they have the same letter part," the king said. "So we may as well say that two terms with different letter parts are *unlike terms*. For example, let's consider the expression

$$6a + 7b$$

In this case the two terms $6a$ and $7b$ have different letter parts, so they are unlike terms and therefore cannot be combined to form one term."

"Here is another example of an expression with two unlike terms," the professor said

$$3a + 4$$

"In this case the letter part of the first term is a, and the second term doesn't even have a letter part."

"How about this expression?" Recordis asked.

$$2x^2 + 3x$$

"Both of the terms have an x in them."

"But the letter parts are not *exactly* the same," the professor said. "Therefore, they are unlike terms. The letter part of the first term is x^2, but the letter part for the second term is x."

ALGEBRAIC EXPRESSION RULE: LIKE TERMS

Two terms are said to be like terms if their letter parts are the same. In order to simplify an algebraic expression, combine all of the like terms into one term.

We added a new word to our vocabulary. Since the professor often presented papers at distinguished societies she liked to impress people by using long words. We invented the four-syllable word *coefficient* to mean "number part." For example, in the expression $6a$, 6 is the coefficient of a; in the expression $2x^2$, 2 is the coefficient of x^2; and in the expression $23a^2b^2$, 23 is the coefficient of a^2b^2.

"What about the expression a?" Recordis asked. "It doesn't have a coefficient at all."

"Actually, its coefficient is 1," the king said, "since $1 \times a = a$."

"It looks as if the number 1 is an *invisible coefficient,*" Recordis said. "When it's there, you usually don't see it."

Next on the gremlin's list we came to several different examples where we had to add together two different multinomials. Now that we were used to the method of combining like terms, we found that the calculations proceeded swiftly:

$$(6x + 4a) + (-x - a) = 5x + 3a$$

$$(12x^2 + 3x + 4) + (x^2 - x - 2) = 13x^2 + 2x + 2$$

$$(10a + 3c) + (4a + 2b) = 14a + 2b + 3c$$

$$(4a^2b + 2ab^2) + (a + 2ab + b) = 4a^2b + 2ab^2 + 2ab + a + b$$

Next, the gremlin gave us some examples in which we had to multiply two multinomials. We found that multiplying monomials was easy, so long as we remembered the laws of exponents:

$$(3x) \times (2a) = 6ax$$

$$(x) \times (7x) = 7x^2$$

$$(3a^2b) \times (4ab) = 12a^3b^2$$

"Unfortunately, matters will become more complicated if we have to multiply together two expressions that contain more than one term," Recordis warned.

"We already found one such example," the king said. "We found that $(10 - x)(10 + x) = 100 - x^2$. All we have to do is remember the distributive property."

DISTRIBUTIVE PROPERTY

$$a(b + c) = ab + ac$$

"It will be easy to multiply a monomial by a binomial," the professor said, "since we can use the distributive property directly and obtain the answer in one fell swoop."

Examples:

$$10 \times (2a + 4) = 20a + 40$$

$$x \times (3x + 5) = 3x^2 + 5x$$

$$2 \times (ax^2 + bx + c) = 2ax^2 + 2bx + 2c$$

Next we came to a harder problem: multiplying two binomials together. The next problem was $(x - 3)(x + 2)$. "We have to use the distributive law twice," the king said. First, we used the distributive law to tell us that

$$(x - 3)(x + 2) = (x - 3)x + (x - 3)2$$

Then we used the distributive law on each of the two terms on the right:

$$(x - 3)x + (x - 3)2 = x^2 - 3x + 2x - 6$$

Then, since $-3x$ and $2x$ are like terms, we were able to combine them in order to get the final answer:

$$(x - 3)(x + 2) = x^2 - x - 6$$

We worked some more of the gremlin's problems that involved multiplying binomials:

gremlin's expression	simplified form
$(a + b)(a + 2)$	$a^2 + 2a + ab + 2b$
$(a + 4)(b - 2)$	$ab - 2a + 4b - 8$
$(2 - y)(2 + y)$	$4 - y^2$

The professor, again trying to be systematic, decided that we should look for a general expression for the product of two binomials. We decided to call the first binomial $(a + b)$ and the second binomial $(c + d)$. Then we calculated the product:

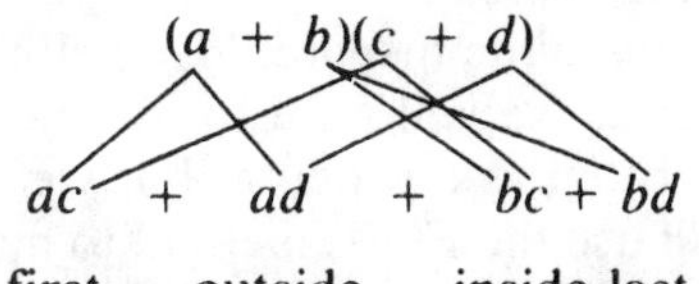

first outside inside last

"When you multiply two binomials, you will end up with four terms," the professor said. "One term in the answer consists of the two *first* terms (a and c) multiplied together; another term consists of the two *outside* terms (a and d) multiplied together; another term consists of the two *inside* terms (b and c) multiplied together; and the final one consists of the *last* terms (b and d) multiplied together."

"That's a lot to remember," Recordis complained. He closed his eyes and started repeating to himself: "first times first, outside times outside, inside times inside, last times last." He opened his eyes again. "I fear that we will be foiled just by the amount of memorization," he said sadly.

"We can use the word 'foil' to remember the procedure!" the professor suddenly realized. "We need to remember:

First times first
Outside times outside
Inside times inside
Last times last

"Look at the first letter of each phrase! If you put them together they spell F-O-I-L—or FOIL."

We decided to call this the FOIL method for multiplying binomials.

FOIL METHOD FOR MULTIPLYING BINOMIALS

$$(a + b)(c + d) = ac + ad + bc + bd$$

The result has four terms: first times first, outside times outside, inside times inside, and last times last.

"I hate to be too picky," Recordis said, "but I have one more question: Is this really an improvement? We started with the expression:

$$(a + b)(c + d)$$

and we turned it into the expression

$$ac + ad + bc + bd$$

If you want my opinion, the second expression is more complicated than the first expression, so how can we say we have simplified the expression?"

We thought about Recordis' point. "It appears that we have our choice between two different ways of writing this type of expression," the king said. "In the first case, we have written the expression as the *product* of two *factors*—the two factors are $(a + b)$ and $(c + d)$. In the second case we have written the expression as the *sum* of four *terms*—ac, ad, bc, and bd. I wonder which form is generally easier to deal with?"

Since there was no obvious answer, there was a heated argument between Recordis and the professor.

The king had to mediate the dispute. "Sometimes we will want to use one form and other times we will want to use the other form. If we start with an expression written as a product, but we want to have it written as a sum, then we will use the FOIL method to multiply out the product. If we start with an expression written as a sum, but we want it written as a product, then we will have to figure out some way to factor the expression."

"That might be hard!" Recordis suddenly exclaimed. "I think it will be a lot more trouble to turn a sum into a product then it is to turn a product into a sum. I can see how to turn the expression $(a + b)(c + d)$ into $ac + ad + bc + bd$. However, if you gave me the expression

$ac + ad + bc + bd$, then I'm sure I couldn't tell you that it really was the same thing as $(a + b)(c + d)$ unless I already happened to know the answer, which luckily in this case I do."

Recordis was right—it did turn out to be more difficult to factor an expression. However, we later developed some methods for doing this. (See Chapters 11 and 13.)

The gremlin had given us some more examples of multiplying together binomials. (Notice that whenever you multiply two binomials you initially end up with four terms, but it often happens that some of these terms are like terms that can be combined, allowing you to express the final answer in fewer than four terms.)

gremlin's problem	F O I L	final answer
$(a - 4)(a - 3)$	$a^2 - 3a - 4a + 12$	$a^2 - 7a + 12$
$(a - b)(2 + c)$	$2a + ac - 2b - bc$	$2a + ac - 2b - bc$
$(a + 5)(a + b)$	$a^2 + ab + 5a + 5b$	$a^2 + ab + 5a + 5b$
$(x - 2)(x - 2)$	$x^2 - 2x - 2x + 4$	$x^2 - 4x + 4$

"In that last case we can write the original expression as $(x - 2)^2$, since $(x - 2)$ is being multiplied by itself."

We decided that it would be useful to develop a general rule for the result when we were taking the square of a binomial—in other words, when we were raising a binomial to the second power. We tried a general example:

$$(a + b)^2 = (a + b)(a + b)$$
$$= a^2 + ab + ab + b^2$$
$$= a^2 + 2ab + b^2$$

We wrote this result as a general rule, along with a couple other interesting results that we discovered:

GENERAL RULES WORTH REMEMBERING

$$(a + b)^2 = a^2 + 2ab + b^2$$
$$(a - b)^2 = a^2 - 2ab + b^2$$
$$(a - b)(a + b) = a^2 - b^2$$

By now we had so much confidence in our ability to simplify algebraic expressions that we cruised through the rest of the gremlin's list without any trouble. The rest of the results are included in the exercises at the end of the chapter. There were two more results that seemed especially interesting. One was the cube (that is, the third power) of a binomial:

$$(a + b)^3 = (a + b)(a + b)^2$$
$$= (a + b)(a^2 + 2ab + b^2)$$
$$= a(a^2 + 2ab + b^2) + b(a^2 + 2ab + b^2)$$
$$= a^3 + 2a^2b + ab^2 + a^2b + 2ab^2 + b^3$$

(Notice how we were able to use the commutative property of multiplication to write the factors of each term in alphabetical order.) The final result was

$$(a+b)^3 = a^3 + 3a^2b + 3ab^2 + b^3$$

(We later investigated how to calculate $(a + b)^4$, $(a + b)^5$, and so on. See Chapter 15.)

Another interesting problem asked us to calculate the square of a trinomial:

$$\begin{aligned}(a + b + c)^2 &= (a + b + c)(a + b + c)\\ &= a(a + b + c) + b(a + b + c) + c(a + b + c)\\ &= a^2 + ab + ac + ab + b^2 + bc + ac + bc + c^2\\ &= a^2 + b^2 + c^2 + 2ab + 2ac + 2bc\end{aligned}$$

"I can see now that algebra isn't *really* very hard," Recordis said. "It's just that it is so boring if you have to do too much of it at once."

Note to Chapter 7

- The term coefficient has a more general meaning than the one used in the chapter. In an expression consisting of several factors multiplied together, the coefficient of any one factor is the product of all of the other factors in the expression. For example, in the expression $15x$, 15 is the coefficient of x and x is the coefficient of 15. In the expression $3a^2b^3$, $3b^3$ is the coefficient of a^2, $3a^2$ is the coefficient of b^3, and 3 is the coefficient of a^2b^3. The number part of an expression can be called the *numerical coefficient*.

Exercises

1. Show that $(a - b)(a + b) = a^2 - b^2$.

Multiply:

2. $(2y + x)(2y - x)$
3. $(2y + 2x)(2y - x)$
4. $(a + 4)(a + 4)$
5. $(abc - 2)(abc - 1)$
6. $(V - b)(P + a/V^2)$
7. $(wL/p)(T - L)$
8. $(s^2 - c^2)(s^2 + c^2)$
9. $(x - kx^2)^2$
10. $(x - 1/2)^2$
11. $(x - b/2a)^2$
12. $(x - y - 1)(x + y)$
13. $(x - y - 1)(x - y)$
14. $(x + y - 1)(x - y)$
15. $(x + y + 1)^2$
16. $(x + y + 1)(x + y - 1)$
17. $(x + y - 1)(x - y - 1)$
18. $(x^3 + x^2)(x + 1)$
19. $(x^3 + x^2)(x^2 + x + 1)$
20. $(x^3 + x^2 + x)(x^2 + x + 1)$
21. $(x^2 + x + 1)^2$
22. $[(2x + 1) + a][3x + 1]$
23. $[(2x + 1) + (2a + 1)][(3x - 1) + (3a - 1)]$

Multiply out each of the following:

24. $(1 + x)^2$
25. $(1 + x)^3$
26. $(1 + x)^4$
27. $(1 + x)^5$

28. Now, suppose that x is quite small. Can you suggest an approximation for $(1 + x)^n$?

Write each of the following expressions in the form $Ax^2 + Bxy + Cy^2 + Dx + Ey + F = 0$:

29. $(x - h)^2 + (y - k)^2 = r^2$
30. $(x - y)(x + y) = 1$
31. $x^2/16 + y^2/9 = 1$
32. $(x - 3)^2/16 + (y + 3)^2/9 = 1$
33. $y = \frac{1}{4}x^2$
34. $(y - 5) = \frac{1}{4}(x + 10)^2$
35. Calculate $155^2 - 154^2$ (without using a calculator).

Multiply each of these:

36. $(a - b)(a^2 + ab + b^2)$
37. $(a + b)(a^2 - ab + b^2)$

8 Functions

Finally the detective had reached a crucial point in the case of the missing books. "In a very short time I will be able to narrow down the current location of the culprit. However, I have some problems that need to be solved. They involve tedious calculations, so I need your help."

"Recordis is great with calculations," the professor said. "The more tedious, the better."

First, the detective told us that the culprit had escaped by skating at a constant speed across Frozen Lake. The detective told us that at a particular time t, the location of the culprit would be given by the expression

$$6 + 10t$$

"That calculation will be no problem," the professor said. "It just involves multiplication and addition." Recordis started the calculation process. He organized the results in a table:

time	distance
0	6
0.1	7
0.2	8
0.3	9

"Here is the next question," the detective said. "I am sure that the culprit escaped the palace by jumping from one of the windows in the north wall. However, there are many different windows, all at different heights. I need to know how much time elapsed from the time the culprit jumped from the window until he or she or it hit the ground. My investigations reveal that if somebody jumps out of a window at a height h

meters above the ground then the time (measured in seconds) until that person hits the ground is given by the formula

$$t = \sqrt{\frac{h}{4.9}}$$

"For example, if you jump out of a window that is 10 meters high, then it will take $\sqrt{10/4.9} = \sqrt{2.04} = 1.43$ seconds until you hit the ground."

"Calculations with square roots are hard!" Recordis moaned. "I'm beginning to dislike square roots as much as I dislike red tape."

h	t
0	0
5	1.0102
10	1.4286
15	1.7496
20	2.0203

Recordis continued with both sets of calculations as quickly as possible. However, he was becoming more and more tired.

The detective was nervously pacing the floor. "We must have these results quickly," he finally said, "or else the culprit may escape."

The king realized that it was time to take decisive action. "We must get some help for Recordis," he said. He summoned Builder and explained the problem to him. "Do you think you can build some sort of machine that can do these calculations?"

Builder went back to his workroom and returned a while later with two strange-looking machines. (See Figure 8–1.) "These are called function machines," he announced. Each function machine had a large funnel at the top labelled IN and a large upside-down funnel at the bottom labeled OUT. One function machine was labelled SKATER, and it had these instructions:

(1) Put the number representing the time (t) in the IN slot.

(2) The number $6 + 10t$ will fall out of the OUT slot.

The other function machine was labelled FALLTIME, with these instructions:

(1) Put the number representing the height (h) in the IN slot.

(2) The number $\sqrt{h/4.9}$ will fall out of the OUT slot.

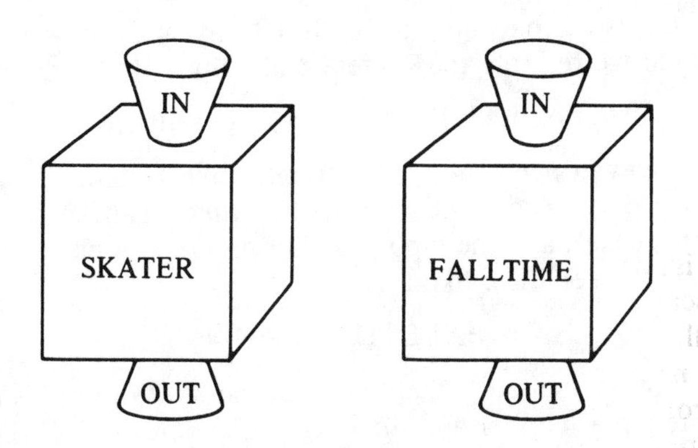

Figure 8–1

Pal wanted to try the machines first. He wrote the number 2 on a slip of paper and threw it into the SKATER function machine. The machine hummed for a second and then out of the bottom of the machine dropped a piece of paper with the number 26 (since $26 = 6 + 10 \times 2$).

"It works!" the king said triumphantly. (Of course, in our country we do not have function machines that take numbers written on slips of paper as input. However, you can program a computer to perform the same task. The computer will require you to type the input number on a keyboard, and the output number will be displayed on a television screen.)

Pal tried some more numbers in the SKATER machine, and we made a table of the results:

time	position
1	16
2	26
3	36
4	46

The king said, "Evidently these machines operate by some fixed rule, which we can call a *function,* that determines a certain output number for whatever input number we choose. In the case of the SKATER function, for example, the rule is (output) = 6 + 10 (input)."

The professor wanted to make up a new term to use for the input number. Recordis disagreed with him, and they had a vehement argument. Finally the king intervened. "If you're going to argue so much we're just going to have to call the input number to a function the *argument.*"

Next, we tried some examples of the FALLTIME function:

input number, or argument (height of window)	output number (time it takes to fall)
1	0.4518
2	0.6387
3	0.7825
4	0.9035

We decided to develop a new notation to represent functions. We wanted to find a shorter way to write the sentence, "If the input number is 10 in the FALLTIME function, then the output number is 1.43." The professor suggested that we write the sentence like this:

FALLTIME FUNCTION: input number 10 output number 1.43

Recordis suggested that the words "function" and "number" were unnecessary, so we rewrote the sentence like this:

FALLTIME: input 10 output 1.43

"We don't even need the words 'input' and 'output'," the king suggested. "We only need three things: the name of the function (FALLTIME in this case), and the input number and the output number. So we tried writing the sentence like this:

FALLTIME 10 1.43

"That's not very good," the professor said. "We need some way to separate the input number and the output number, or else we will be too likely to confuse them."

"We could write the input number inside something, since it is being put inside the function machine," the king suggested. We wrote the input number in a box:

FALLTIME [10] 1.43

Since boxes took too long to draw, we decided that we could write the input number inside a pair of parentheses:

FALLTIME (10) 1.43

That expression still looked incomplete, so we added an = sign. The final form that we agreed on was:

FALLTIME (10) = 1.43

"If we're reading that sentence out loud, then we will say, 'FALLTIME of 10 equals 1.43,'" Recordis said. "And remember that the expression is short for, 'If you are using the FALLTIME function and the input number is 10, then the output number will be 1.43.'"

We decided to call this notation *function notation*, and we wrote some more examples of function notation:

FALLTIME (8) = 1.2778 FALLTIME (17) = 1.8626

SKATER (0.5) = 11 SKATER (10) = 106

"Or, in general, if we use the letter h to stand for the input number, then we can say that

$$\text{FALLTIME}(h) = \sqrt{\frac{h}{4.9}}$$

This equation is read, "FALLTIME of h is the square root of h over 4.9." Since the falltime depends on the height h, we will also say that

FALLTIME *is a function of* h

"I think function machines will be useful for many other types of problems," Recordis said. "So far we have used only two function machines—the ones for the FALLTIME function and the SKATER function."

We made a list of some quantities that could be expressed as functions of other quantities:

- fuel used as a function of the length of the trip
- heat loss as a function of temperature
- quantity demanded as a function of price
- brightness of a star as a function of its distance.

"But how much do function machines cost?" the king wondered. "It would be very expensive if Builder needed to build a new function machine for each calculation we do."

Builder thought a moment. "I have the solution," he said. "I will build a *general* function machine that you can set to calculate whatever specific function you need." Builder quickly went to work. Soon he brought back

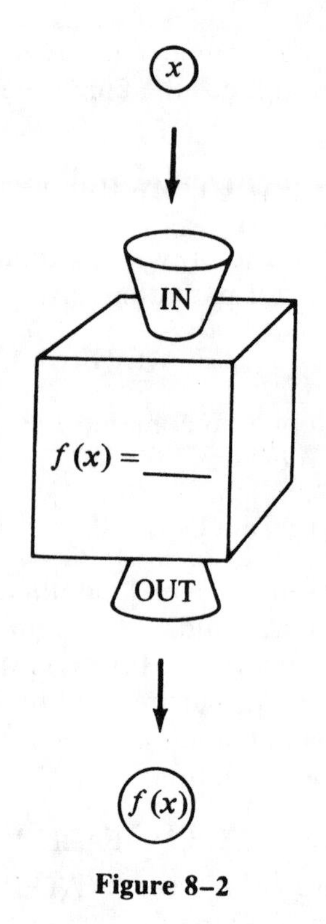

Figure 8–2

a new function machine (Figure 8–2) that looked like the others except that the front of the machine looked like this:

$$f(x) = \_\_\_\_$$

"I call this machine f, for function," Builder said. "All you have to do is write on the front of the machine the instructions that describe the function you want it to calculate."

"It would help to have a function machine that calculates the square of numbers," Recordis said.

Builder wrote on the blank part of the screen:

$$f(x) = x^2$$

"That means we now want the f function to give us squares. We can read that sentence, 'f of x is x squared.'"

Recordis had an objection. "Normally, when we put two letters together with no operation sign between them, that means to multiply them. How do we know that the expression $f(x)$ means 'f of x' and not 'f times x'?" Unfortunately, we found no way to prevent that confusion totally, but we decided that function notation was so useful that we would use it despite the slight danger of that type of mistaken identity. Normally, the expression $f(x)$ will mean "f of x." Pal threw some numbers into the f machine and we wrote down the results:

$f(0) = 0$ (read "f of zero equals zero.")
$f(1) = 1$
$f(2) = 4$
$f(3) = 9$
$f(10) = 100$
$f(12) = 144$

Next, Recordis wanted to try the function x^3. The professor wrote the new instructions on the screen of the function machine.

$$f(x) = x^3$$

"Wait!" Recordis protested. "We already used $f(x)$ to stand for the x-squared function. Won't it be confusing if we now let $f(x)$ stand for the x-cubed function?"

"We don't have enough letters to give each function we use a different letter," the professor said. "We decided that we can use the same letter to mean different things in different problems, so there is no reason why we can't use the letter f to stand for different functions in different problems. The only important thing is that we must remember not to use f to stand for two different functions at the same time."

We made a list of some results for the x-cubed function:

$$f(x) = x^3$$
$$f(1) = 1$$
$$f(2) = 8$$
$$f(3) = 27$$
$$f(4) = 64$$
$$f(10) = 1{,}000$$

Next, we worked on several functions representing the positions of different runners in the annual race. In each case we could represent the distance that the racer had run from the starting line as a function of time. It turned out that we could determine the position function if we knew the speed of the racer. For example, if the speed of the racer was 2 yards per second, then the position function was $f(t) = 2t$. We calculated some examples:

speed = 2	speed = 5
$f(t) = 2t$	$f(t) = 5t$
$f(3) = 6$	$f(3) = 15$
$f(4) = 8$	$f(4) = 20$
$f(5) = 10$	$f(5) = 25$
$f(6) = 12$	$f(6) = 30$

speed = 4.5
$f(t) = 4.5t$
$f(3) = 13.5$
$f(4) = 18$
$f(5) = 22.5$
$f(6) = 27$

In general, if the speed of the racer is v, the position function is $f(t) = vt$.

"However, sometimes we give slow racers a head start," Recordis reminded us. We found that if a racer of speed 2 was given a head start of 10, the position function would be $f(t) = 10 + 2t$. We calculated some examples:

speed = 2, head start = 10	speed = 1.5, head start = 15
$f(t) = 10 + 2t$	$f(t) = 15 + 1.5t$
$f(1) = 12$	$f(1) = 16.5$
$f(2) = 14$	$f(2) = 18$
$f(3) = 16$	$f(3) = 19.5$
$f(4) = 18$	$f(4) = 21$
$f(5) = 20$	

In general, we found that if a racer with speed v is given a head start of h, then the position function is

$$f(t) = h + vt$$

We set the function machine to $f(x) = 10 + 2x$ and put some more numbers in it. However, at one point Pal threw a number into the machine before Recordis had a chance to write it down. The number that came out was 45.

"Now you've done it!" Recordis scolded. "We know that $f(x) = 45$ for some number x, but we don't know what x is. How are we going to find out?"

Builder went to his workroom and quickly came back with a new function machine. This one was labelled "$g(x) =$ ______." "If you need to use two different functions in the same problem, then you can use this g machine. Like the f function, you can set it however you want by writing down the instructions on the front screen. In your case, we need to work the function $f(x) = 10 + 2x$ *backwards*."

"I know how we can work that function backwards!" the professor said. "Let's use y to represent the output number from the function: $y = 10 + 2x$. Then we can say that $y = f(x)$ (read as y equals f of x). Now we can solve for x in terms of y: $2x = y - 10$; $x = (y - 10)/2 = \frac{1}{2}y - 5$. Now, if we define the function g as follows: $g(y) = \frac{1}{2}y - 5$; then we have $x = g(y)$. That means that the g function does the opposite of the f function. If we put the value of x into the f function, then out comes the value of y; but if we put the value of y into the g function, then out comes the value of x! In our case, we know $y = 45$, so in order to find the value of x we need to calculate $g(45) = \frac{1}{2} \times 45 - 5 = 22\frac{1}{2} - 5 = 17\frac{1}{2}$." To double-check this result, we put the number $17\frac{1}{2}$ into the f function and we were indeed able to show that $f(17\frac{1}{2}) = 45$.

We decided that if two functions did the opposite of each other, then we would call them *inverse functions*. Builder showed what happened if you connected a function and its inverse together. (Figure 8–3.) In symbols: $g(f(x)) = x$.

Pal became very frustrated when the number that he got out of the combined $g(f(x))$ function was always exactly the same as the number he put in, but there wasn't anything he could do about it.

We made a list of some more functions and their inverses.

function	inverse
$y = f(x) = 3x$	$x = g(y) = y/3$
$y = f(x) = x^2$	$x = g(y) = \sqrt{y}$
$y = f(x) = 1/x$	$x = g(y) = 1/y$
$y = f(x) = \frac{5}{9}(x - 32)$	$x = g(y) = \frac{9}{5}y + 32$
$d = f(t) = 6 + 10t$	$t = g(d) = (d - 6)/10$
$t = f(h) = \sqrt{h/4.9}$	$h = g(t) = 4.9t^2$

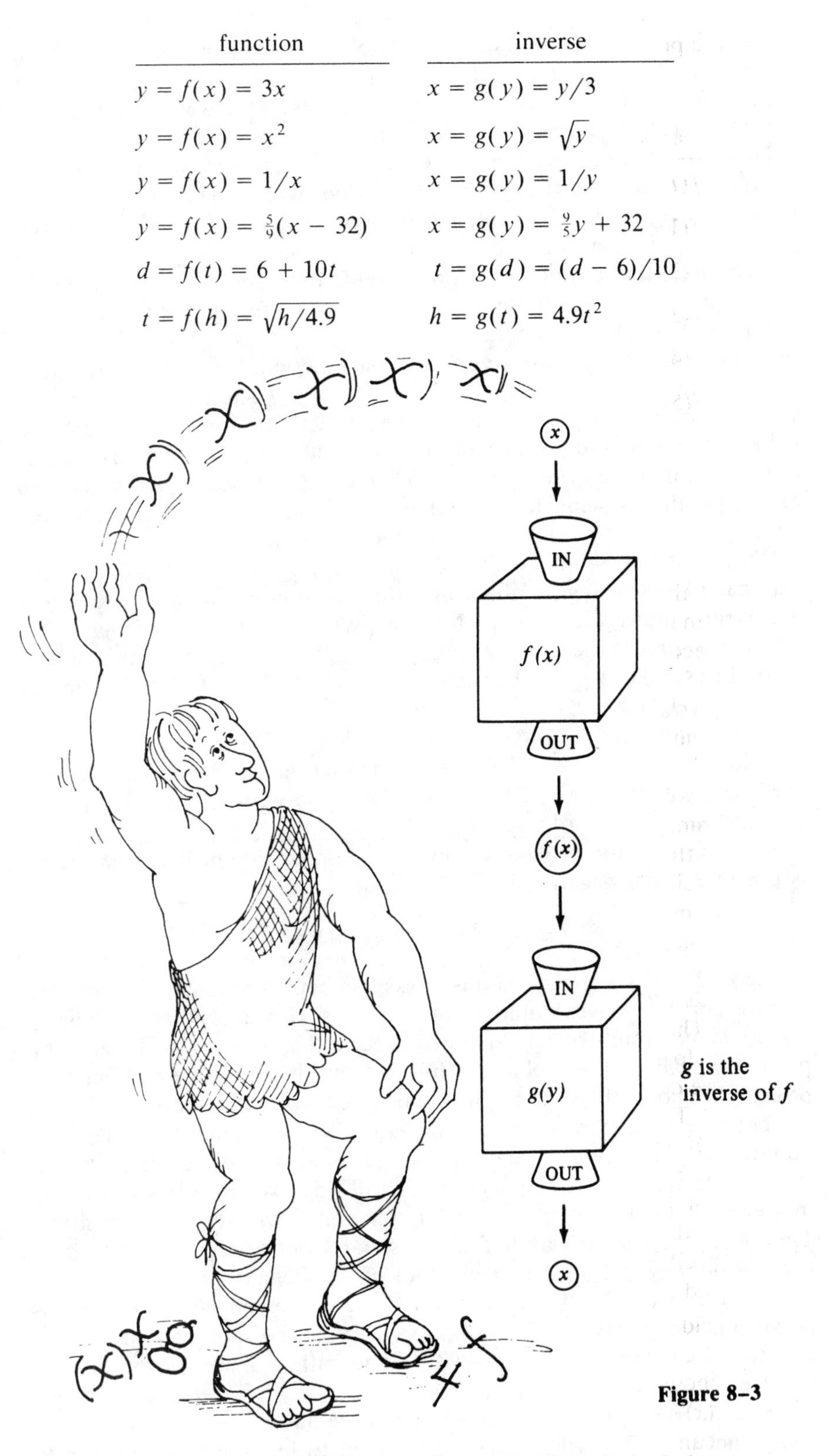

Figure 8–3

Next, Builder set the function to $f(x) = 2$. Then Pal tried some examples:

$$f(0) = 2 \qquad f(3) = 2$$

$$f(-10) = 2 \qquad f(8) = 2$$

Pal quickly became tired of 2's, but no matter what number he put into the function, the number that came out was always a 2. We called this type of function a *constant* function, since its value was constantly the same, regardless of what the input number was.

However, the professor noticed a distressing feature of this function: It had no inverse function. Since the value of the function is always 2, then knowing the value of the output number does not give you a shred of information about the input number.

(We found that a function has an inverse function only if the values of the output number for the function are always different for two different input numbers. For example, if we find that $f(6) = 12$ and $f(7) = 12$, then the function f cannot have an inverse function. If the inverse function was g, then would $g(12) = 6$ or would $g(12) = 7$? There is no way to tell, so g cannot be a function, because for a function there must be just one value of the output for every value of the input. The object g is an example of something more general called a *relation*. A relation is like a function except that it is possible for one value of the input number to be associated with more than one value of the output number.)

We investigated the square root function $g(x) = \sqrt{x}$ some more. Pal was happily throwing numbers into the function machine and Recordis was writing down the results. However, when Pal threw the number -1 into the machine, there was a violent wrenching sound and the machine started to smoke. No number came out of the OUT slot, but the number -1 was coughed back out the IN slot.

"What happened?" Recordis cried in alarm. "Why can't the machine calculate $g(-1) = \sqrt{-1}$?" While Builder set out to fix the machine, we tried to calculate the $\sqrt{-1}$ by hand. "The square root of negative 1 must be negative 1," Recordis said confidently.

"No it isn't!" the professor said. "Remember that negative 1 squared is positive 1, not negative 1."

$$(-1)^2 = (-1) \times (-1) = 1$$

Recordis was surprised that his guess had been wrong, but he was sure that the correct answer would now be obvious. However, try as we might, we could not think of any number x such that $x^2 = -1$. Finally the professor realized the problem. "If you multiply a positive number by a positive number, then the result is positive, correct?" she said. "So, if x is positive, then x^2 must be positive. However, if you multiply a negative number by a negative number, the result is also positive, correct? So, that means that if x is negative, x^2 is still positive. In other words, no matter what kind of number x is, the number x^2 will always be positive. Therefore, there is no way to find the square root of negative 1—in fact, there is no way to find the square root of any negative number!"

We were struck by the significance of what she had said. "So there is no such number as $\sqrt{-1}$!" the king said. (We were later to return to the question of looking at $\sqrt{-1}$—with some rather amazing results. See Chapter 20.)

"So that means that it is not legal to put the number -1 into the square root function," Recordis said. We decided to use the term *domain* to stand for all of the legal values for the input number (argument) of a function. Recordis felt that we should also have a term for all of the possible values of the output number, so we used the term *range* to stand

for that. We made a list of the domains and ranges of some different functions:

function	domain	range
$f(x) = 3x$	real numbers	real numbers
$f(x) = 1/x$	real numbers (except 0)	real numbers (except 0)
$f(x) = x^2$	real numbers	$f(x) \geq 0$
$f(x) = \sqrt{x}$	$x \geq 0$	$f(x) \geq 0$
$f(x) = \|x\|$	real numbers	$f(x) \geq 0$
$f(x) = \sqrt{(1 - x^2)}$	$-1 \leq x \leq 1$	$0 \leq x \leq 1$

If f and g are inverse functions, then the domain of f is the same as the range of g, and vice versa.

The detective set off. "I'll have results very soon," he promised.

FUNCTION NOTATION

The notation $f(x)$ (which is read as "f of x") means, "f is a function of a variable x." In a particular circumstance you need to specify which function you mean by listing a formula; for example, $f(x) = 4x^2$. Then the result (or output) of the function depends on the specific number (or expression) that is used as the argument (or input) to the function. For this example, $f(3) = 4 \times 3^2 = 36$, $f(12) = 4 \times 12^2 = 576$, and $f(ab) = 4(ab)^2$.

The set of all possible values for the input to the function is called the domain; the set of all possible values for the output is called the range.

Exercises

Let $f(x) = x^2$. Then evaluate:

1. $f(0)$
2. $f(1)$
3. $f(2)$
4. $f(a)$
5. $f(q)$
6. $f(a + b)$
7. $f(x + h)$
8. $f(x + 2)$
9. $f(-1)$
10. $f(x + h) - f(x)$

Let $f(x) = x^2 + 2x + 2$. Then evaluate:

11. $f(0)$
12. $f(1)$
13. $f(-1)$
14. $f(3)$
15. $f(a)$
16. $f(x + h)$
17. $f(x + h) - f(x)$

Let $f(x) = \sqrt{x}$. Then evaluate:

18. $f(4)$
19. $f(7)$
20. $f(3a^2b^2)$
21. $f(ab^2c^3)$

Let $f(x) = \sqrt{1 + x}$. Then evaluate:

22. $f(15)$
23. $f(10)$
24. $f(5)$
25. $f(1)$
26. $f(0.5)$
27. $f(0.1)$
28. $f(0.01)$
29. $f(0.001)$

Let $f(x) = 1 + \frac{1}{2}x$. Then evaluate:

30. $f(15)$
31. $f(10)$
32. $f(5)$
33. $f(1)$
34. $f(0.5)$
35. $f(0.1)$
36. $f(0.01)$
37. $f(0.001)$

38. Do you notice any relationship between the two functions $f(x) = \sqrt{1 + x}$ and $f(x) = 1 + \frac{1}{2}x$ as the value of x becomes small?

39. Strictly speaking, it is not correct to say that the function $g(x) = \sqrt{x}$ is the inverse of the function $f(x) = x^2$ unless you add one restriction. What is that restriction?

A function is said to be an *odd* function if $f(-x) = -f(x)$ for all values of x. A function is said to be an *even* function if $f(-x) = f(x)$ for all values of x. Are the following functions odd, even, or neither?

40. $f(x) = 20x$
41. $f(x) = x^2$
42. $f(x) = x^3$
43. $f(x) = x^2 + x$
44. $f(x) = |x|$
45. $f(x) = 6x^5 + 4x^3 + 6x$

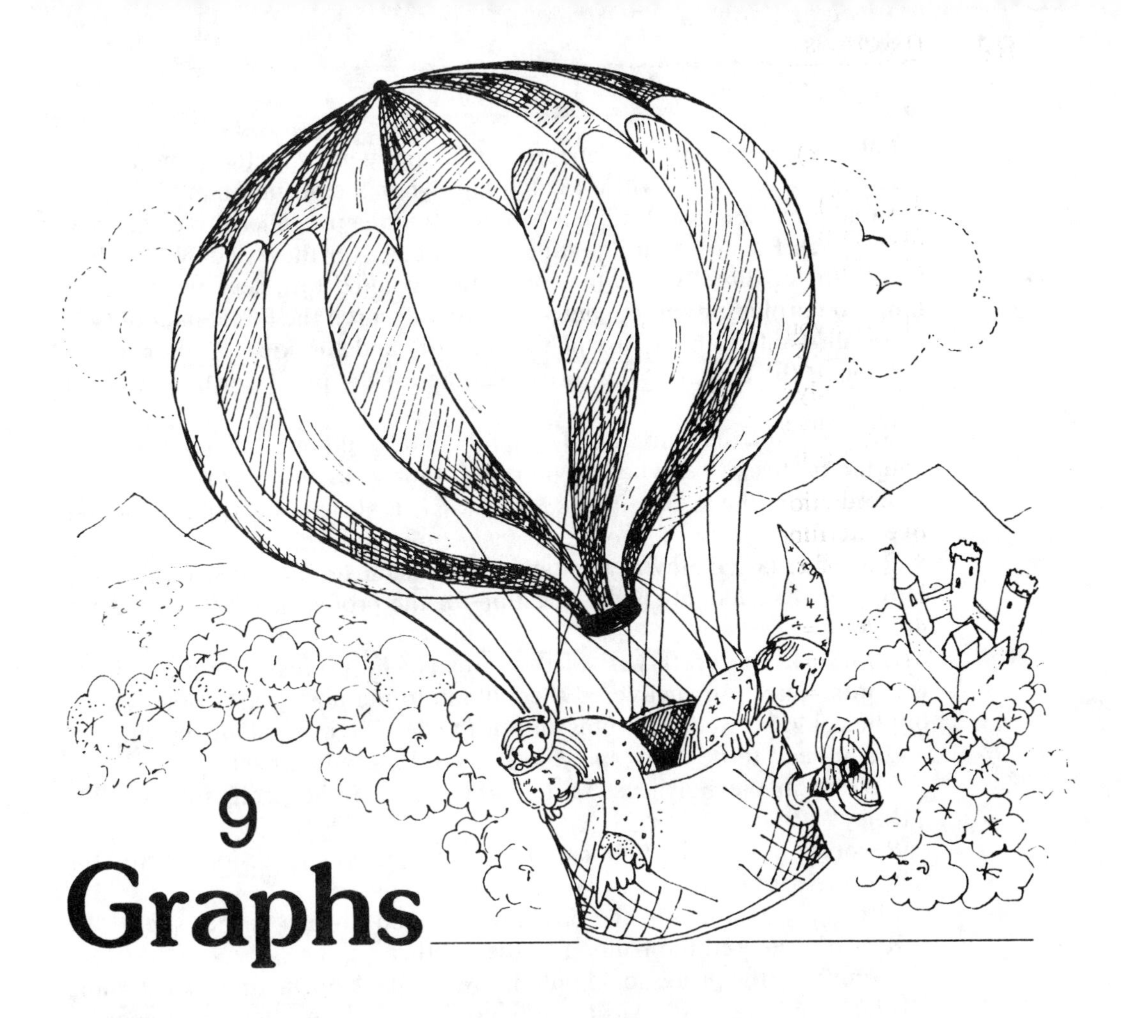

9
Graphs

"I have narrowed down the current location of the culprit," the detective announced the next day. "I know that the culprit is somewhere in the middle of Gloomy Wood. All we have to do is find out exactly where." He showed us a map of the wood. (Figure 9–1.)

Figure 9–1

"We could be facing a dangerous criminal who will stop at nothing," Recordis said. "How will we catch him when we do find him?"

Builder had an idea. He quickly built a floating propeller-driven balloon with which Recordis and the king could float over the wood. When they were situated directly over the culprit they could drop a net down to catch him. The professor and the detective would search the forest on foot. We were all excited about this plan. Recordis and the king positioned the balloon at the corner of the forest waiting for the professor to give them directions.

Recordis and the king waited patiently in the balloon while the detective and the professor searched all morning. Finally the professor shouted to Recordis over the walkie-talkie, "We've found the culprit! Get the balloon over here!"

Recordis started moving the balloon across the forest as fast as possible. However, he suddenly had a question for the professor: "Where exactly are you?"

"We're in the north part of the forest," the professor said urgently. Recordis quickly maneuvered the balloon to the north part of the forest, but he could not see the professor nor the detective, let alone the culprit.

"You need to be more specific!" Recordis shouted back.

"We're in the northeast part of the forest!" the professor radioed. "Hurry!"

Recordis moved the balloon to the east, but the directions were still too vague for him to know where the others were

"I can see you now!" the professor radioed "Move farther south!"

Recordis steered the balloon to the south.

"Too far!" the professor shouted. "Move back north—now west—no, too far—go back east—west—now north—," but by the time Recordis was finally clear about the location the detective discovered that the culprit had fled.

"We would have had him if the balloon had been here in time," the detective said.

Recordis and the professor started to argue about whose fault it had been. The king intervened. "The problem is that we need a better system to identify locations in the forest," he said. "General terms such as 'northwest' are too vague."

Recordis suggested that we could give every tree in the forest a name, but then he quickly realized that this plan would be far too much work. The king finally had an idea. "Instead of telling the operator of the balloon to go east or go north, we will tell him exactly how far east and how far north he needs to go."

We tried an example of the king's idea. The professor went into the thickest part of the forest, and Recordis returned the balloon to the corner. Then the professor radioed, "To reach me, travel 45 yards east and 30 yards north." (See Figure 9–2.) Recordis followed these instructions. Sure enough, the balloon was directly over the professor.

Recordis returned the balloon to the corner while the professor went to a new point. They repeated this process several times, and each time the king was able to find the professor perfectly. First, the professor stood at the point 10 yards east and 5 yards north; then she went to the point 100 yards east and 35 yards north; then to the point 0 yards east and 15 yards north. (See Figure 9–3.)

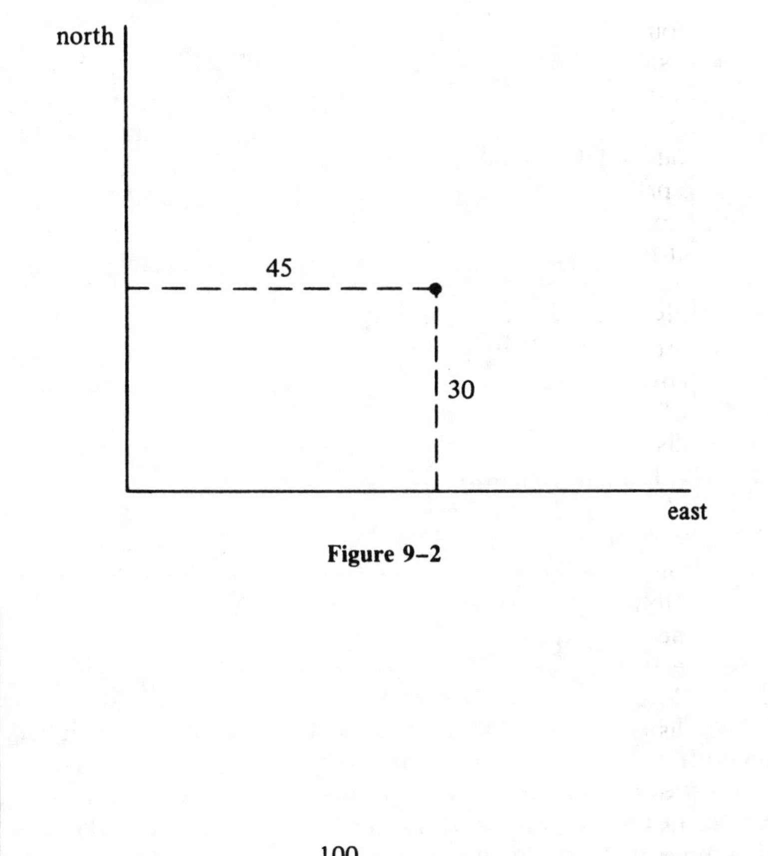

Figure 9–2

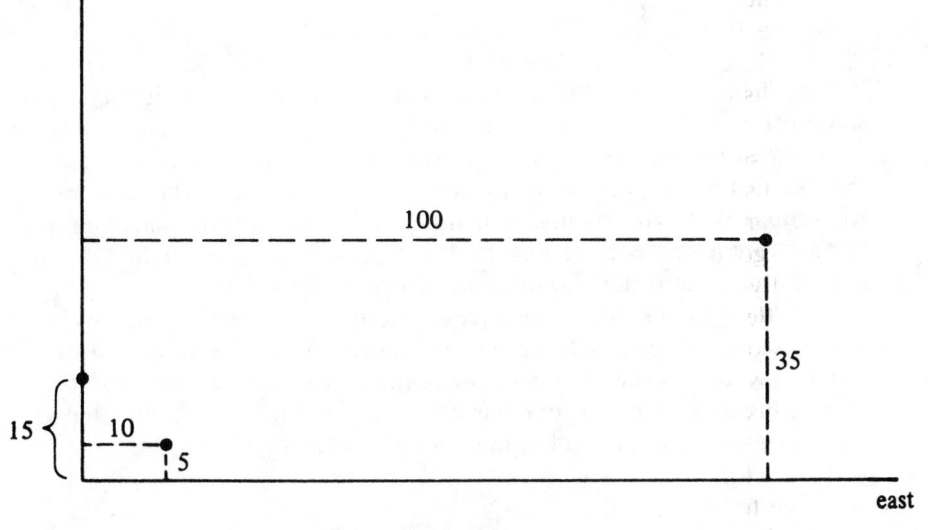

Figure 9–3

"We can now locate any point in the entire forest!" the king said jubilantly. "In order to identify points in the forest, we need two numbers: the *east* distance and the *north* distance." We drew a diagram of the forest and marked some more points. (See Figure 9–4.) We called the corner of the forest the *origin,* since all flights originated there. The origin can be identified as the point (0 east, 0 north.)

As would be expected, Recordis complained that there was too much writing involved. "Why do we have to write the words 'east' and 'north' all the time?" he asked. "Why can't we just write down the numbers?"

"If we just wrote down the numbers, how would we know which was north and which was east?" the professor protested. "Suppose I told you to take the balloon to the point (5, 30). How would you know whether that means 5 east and 30 north or 30 east and 5 north? There is a big difference between those two points!"

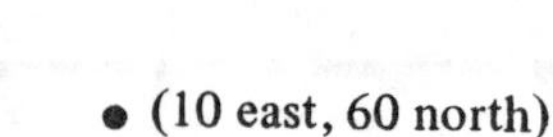

Figure 9–4

"We'll have no trouble as long as we write the two numbers in the same order every time," the king said. "For example, let's agree that we'll always write the east number first and the north number second." We decided to adopt the king's plan. There was no particular reason why we should write the numbers in this order, rather than the other way around, but it was very important that we agree always to write the numbers in the same order. So the king issued a royal decree.

Be it hereby decreed: When a pair of numbers is used to identify points in the forest, the first number always indicates the distance *east* of the origin, and the second number always indicates the distance *north* of the origin. For example, the expression (x, y) identifies the point that is x units east of the origin and y units north." (See Figure 9–5.)

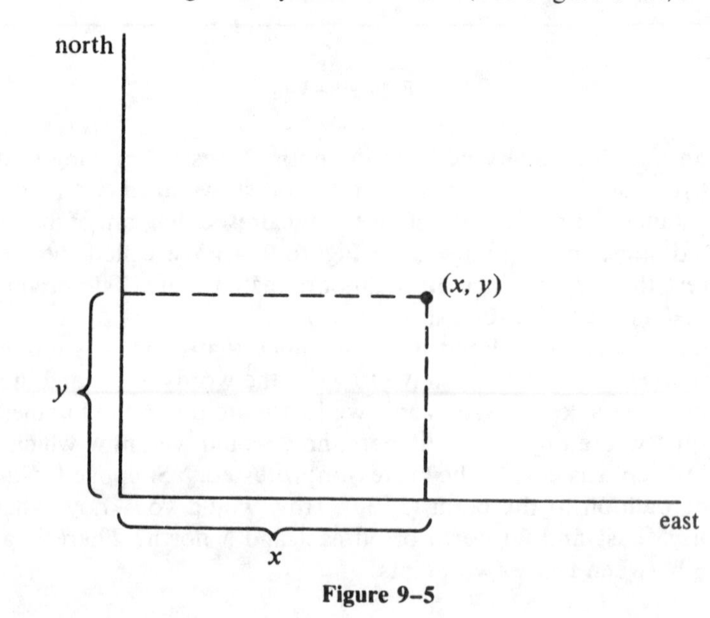

Figure 9–5

We decided to call a pair of numbers in which the order of the numbers matters an *ordered pair*. For example, the ordered pair (2, 3) means something different from the ordered pair (3, 2). We also decided that we could call the numbers in an ordered pair the *coordinates* of the point. For example, the point labelled * in Figure 9–6 can be identified by the two coordinates (8, 5). The east coordinate is 8 and the north coordinate is 5.

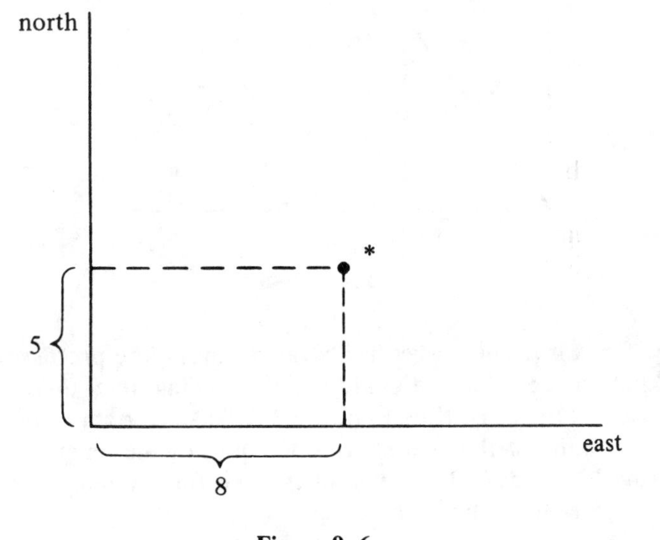

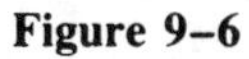
Figure 9–6

We tried some more examples of locating points in the forest. The professor told Recordis to go to the point (3, 4). However, Recordis had an idea. "Wouldn't it be easier to just travel in a straight line from the origin to the point (3, 4)?" On the map of the forest he drew a straight line illustrating his proposed course. (See Figure 9–7.)

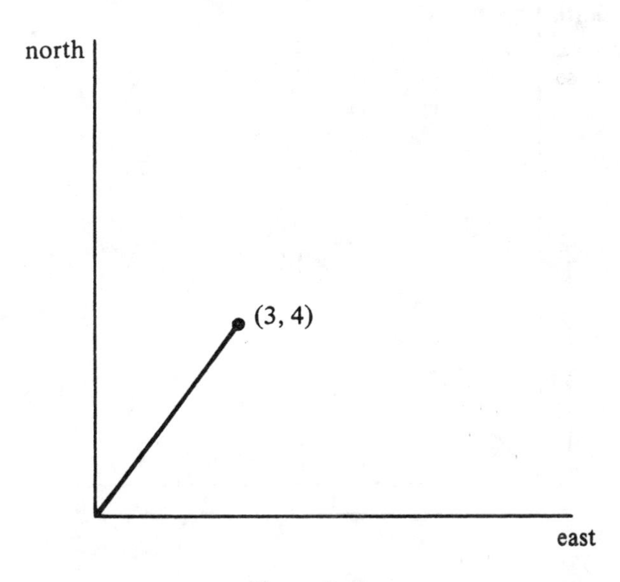

Figure 9–7

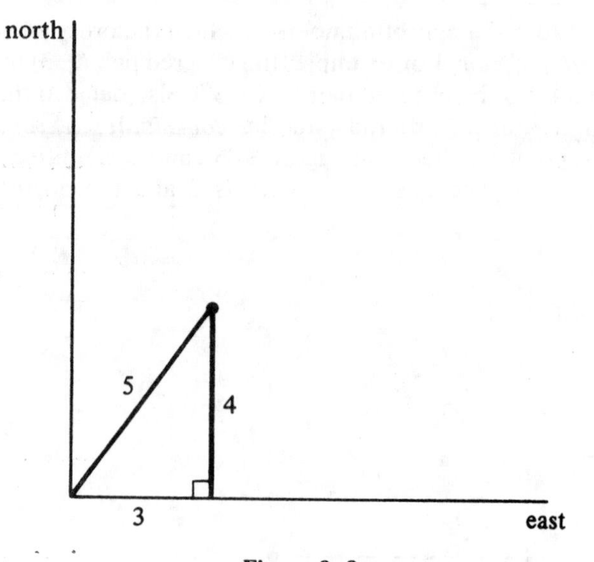

Figure 9–8

"How far is it if you travel in a straight line?" the professor asked.

The king realized that we could use the Pythagorean theorem. "Let's draw a triangle like this. (See Figure 9–8.) This is a right triangle, and we know the lengths of the two short sides but not the length of the hypotenuse (the long side). But if we let d stand for the length of the hypotenuse, we can find d from the equation

$$d^2 = 3^2 + 4^2 = 9 + 16 = 25$$

$$d = 5$$

We wrote a general formula for the distance from the origin to the point (x, y) (See Figure 9–9.):

$$d = \sqrt{x^2 + y^2}$$

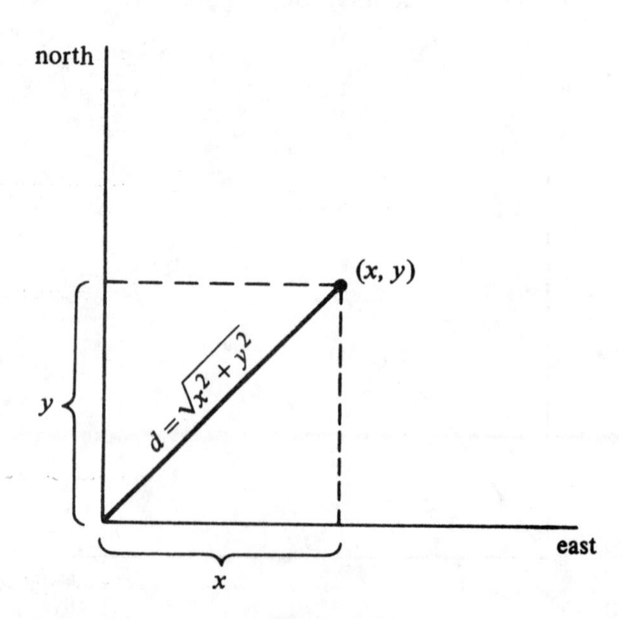

Figure 9–9

The professor also realized that we could find the distance between any two points even if the origin wasn't one of the points. For example, suppose we want to find the distance between the points (6, 10) and (11, 22). Then we can set up a right triangle which will have legs of length 5 and 12. (See Figure 9–10.) The distance between the two points will be

$$d = \sqrt{5^2 + 12^2} = \sqrt{25 + 144} = \sqrt{169} = 13$$

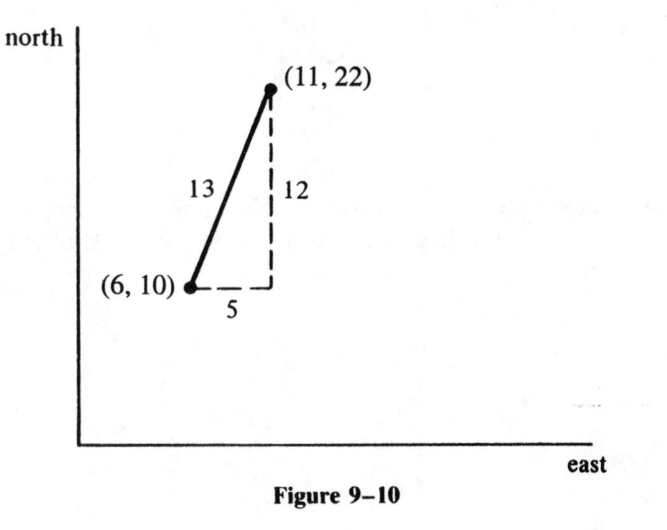

Figure 9–10

In general, the distance between the two points (x_1, y_1) and (x_2, y_2) is

$$d = \sqrt{(x_1 - x_2)^2 + (y_1 - y_2)^2}$$

Next, the professor moved to the point (6, 12) and Recordis started to move the balloon in a straight line from the origin to that point. To help him keep the balloon on a straight course he used a ruler to draw the

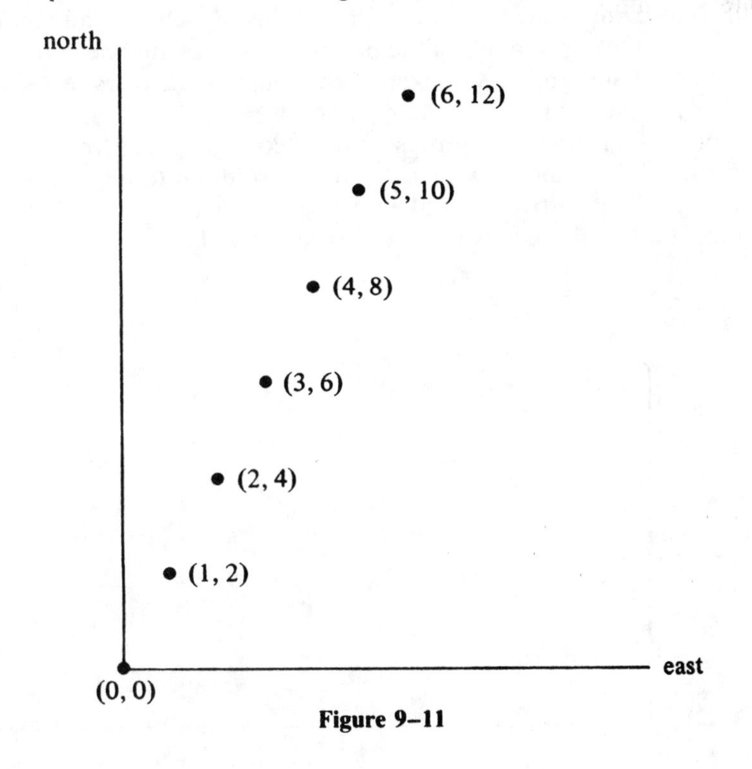

Figure 9–11

course on the map of the forest. Then he wrote down several points that he knew he must go over in order to maintain his straight line course. (See Figure 9–11.)

Recordis' straight line course from (0, 0) to (6, 12):

(0, 0)
(1, 2)
(2, 4)
(3, 6)
(4, 8)
(5, 10)
(6, 12)

Recordis maneuvered the balloon smoothly along the course. However, a sudden gust of wind blew his notebook shut. He quickly shuffled through the book to find the right page again:

x	y
0	0
1	2
2	4
3	6
4	8
5	10

The balloon had just about reached the professor's position when the king suddenly noticed something. "That's not the right page!" he said. Recordis looked down at the page. The heading at the top said:

"Values of x and y that are solutions to the equation $y = 2x$." (We had performed these calculations in Chapter 2).

"But the pairs of numbers are exactly the same!" Recordis said.

"This is amazing!" the king said. "It looks as though the pairs of numbers that form the coordinates of the points on the straight line connecting the points (0, 0) and (6, 12) are exactly the same as the pairs of numbers (x, y) that are the solutions of the equation $y = 2x$."

As soon as we landed, the professor decided to help us investigate this phenomenon. "We can draw pictures of the solution to any equation," she said. "Let's draw two lines like this:" (See Figure 9–12.) "We'll call the horizontal line the x axis and the vertical line the y axis."

Figure 9–12

"Why not the other way around?" Recordis asked.

"It doesn't matter which way we do it, so long as we do it the same way every time," the professor responded. So the king issued another royal decree.

"*Be it decreed*: The horizontal axis will always be called the x axis, and the vertical axis will always be called the y axis."

"Now let's put numbers on the axes, like this" the professor said. ("Axes" is the plural of axis.) See Figure 9–13.

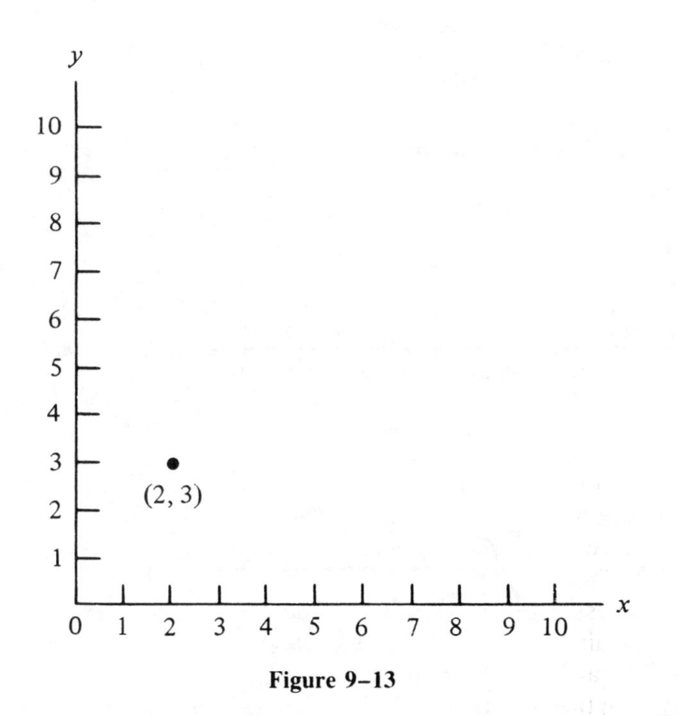

Figure 9–13

"Now we can draw pictures to represent ordered pairs of numbers," the professor said. We can represent the ordered pair (2, 3) on the graph by putting a dot at the point 2 units to the right and 3 units up. So this plan will help us both ways. The fact that we can draw pictures of equations will help us understand the equations, and the fact that we can use equations to describe pictures will help us understand the pictures. Now suppose we start with a list of points, such as all the points that are solutions to the equation $y = 2x$. We can put a dot at the point representing each of these ordered pairs. (See Figure 9–14.) And we know that these points are not the only solutions to the equation $y = 2x$. We know that $(\frac{1}{2}, 1)$ is a solution; $(\frac{1}{4}, \frac{1}{2})$ is a solution; and, in fact, if we consider all real numbers, there is an infinite number of solutions. Any point lying along the line will be a solution." (See Figure 9–15.) We decided to call the set of points representing the solutions of an equation the *graph* of that equation.

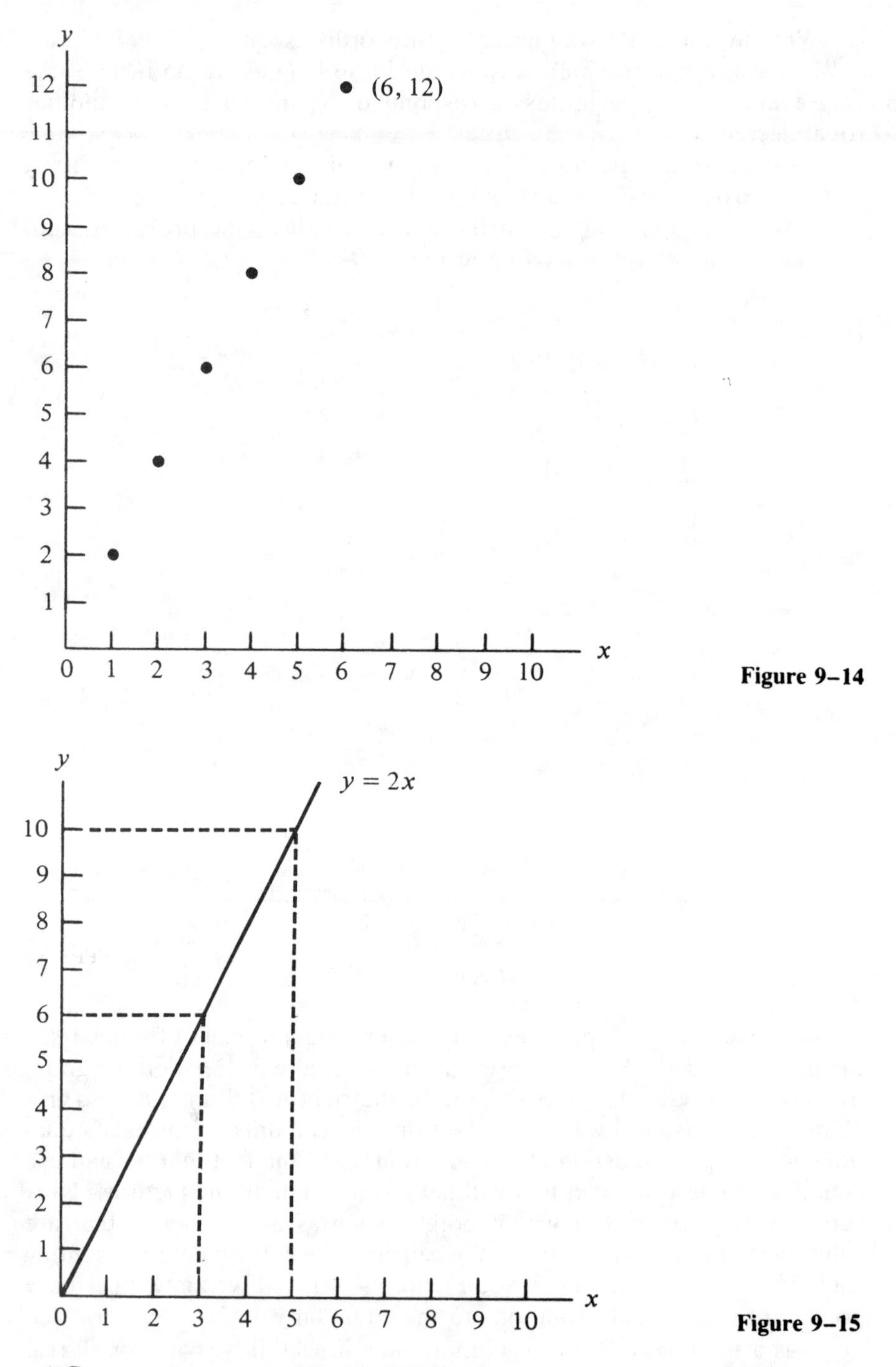

Figure 9–14

Figure 9–15

"Can we represent solutions that are negative numbers?" the king asked. "For example, the ordered pair $(-1, -2)$ is also a solution of the equation $y = 2x$." The professor logically figured that if positive values of x meant "move to the right," then negative values of x should mean "move to the left." Also, since positive values of y meant "move up," negative values of y should mean "move down." She showed us how we could represent negative numbers. (See Figure 9–16.)

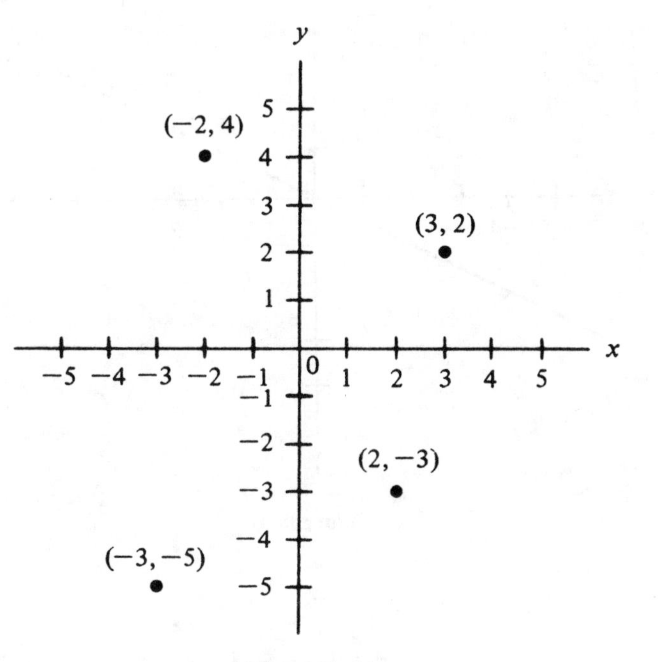

Figure 9–16

We drew the graphs of some more equations. (See Figures 9–17, 9–18, and 9–19.)

"We can even draw graphs of functions!" the king said. "Let's suppose that x represents time, and that $y = f(x)$ represents the position of a racer.

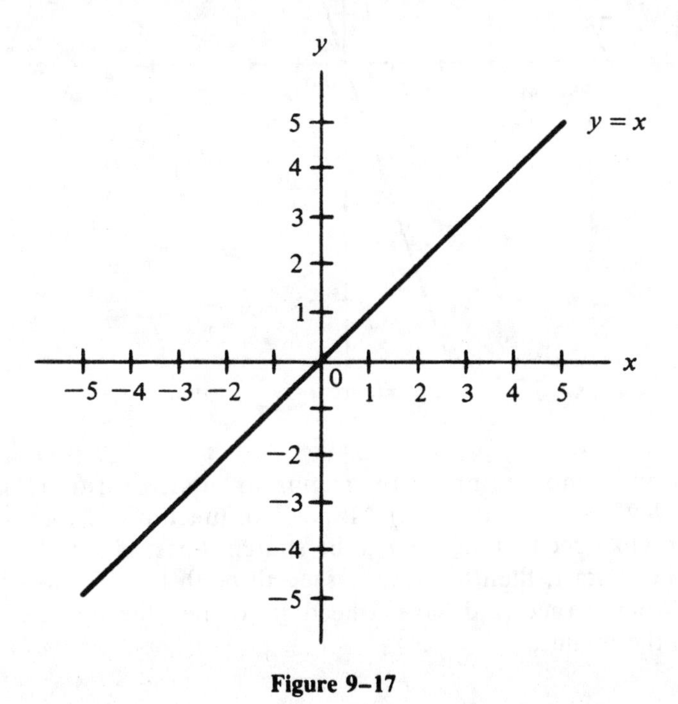

Figure 9–17

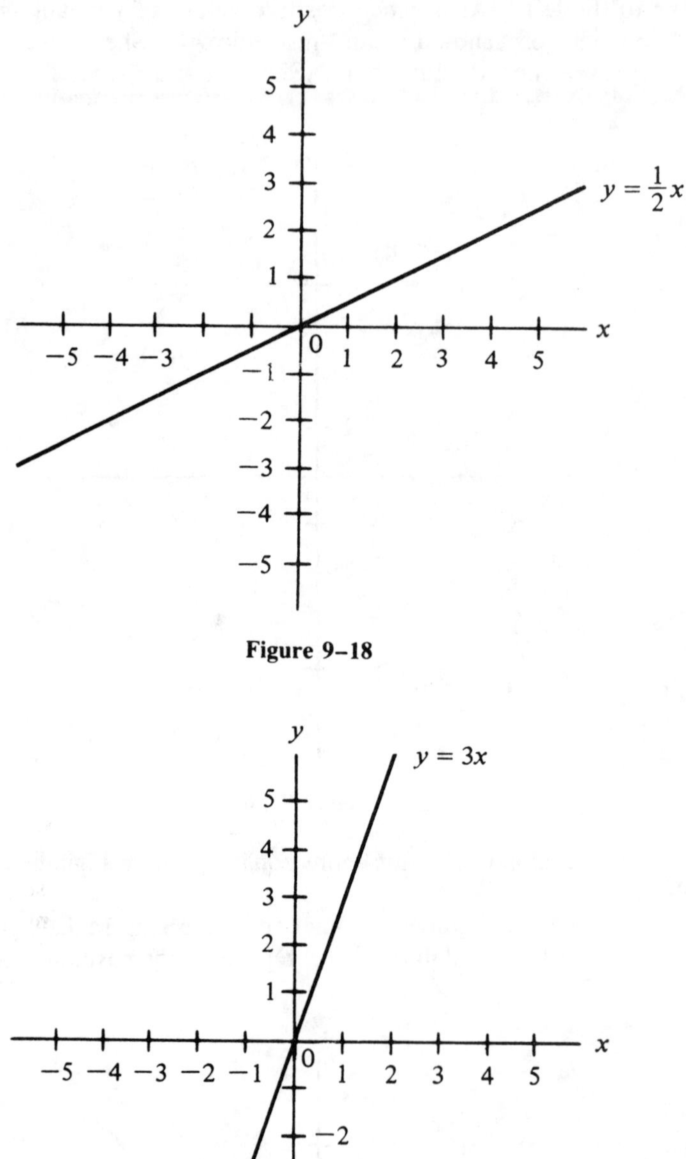

Figure 9–18

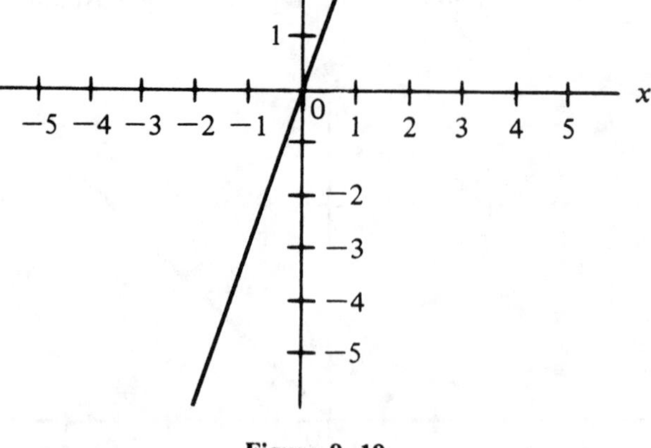

Figure 9–19

We drew some examples of graphs of position functions. (See Figures 9–20, 9–21, and 9–22.) All position functions of racers moving with constant speeds turned out to be straight lines. If the racer did not have a head start, then the line passed through the origin—the point (0, 0). When a racer did have a head start, then the line did not pass through the origin.

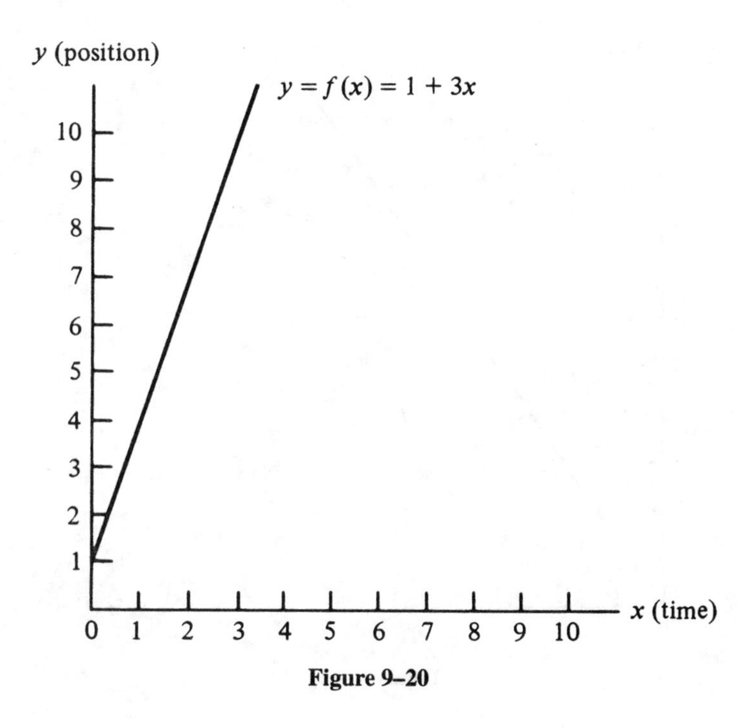

Figure 9–20

"Let's call a function of the form $f(x) = ax + b$ a *linear function*," the professor said, "since its graph is a line." (a and b represent known constant numbers.)

By this time Recordis was getting tired of plotting points. "How many points do we need to calculate in order to determine which line is the correct line?" he asked. We tried to graph the line represented by the equation $y = 2 + 3x$. We calculated that (1, 5) was one point on the line.

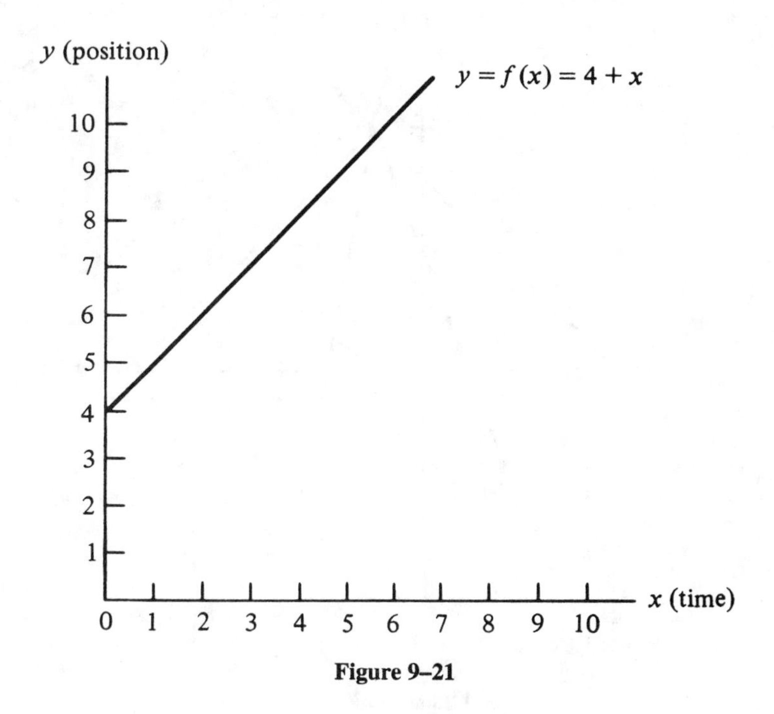

Figure 9–21

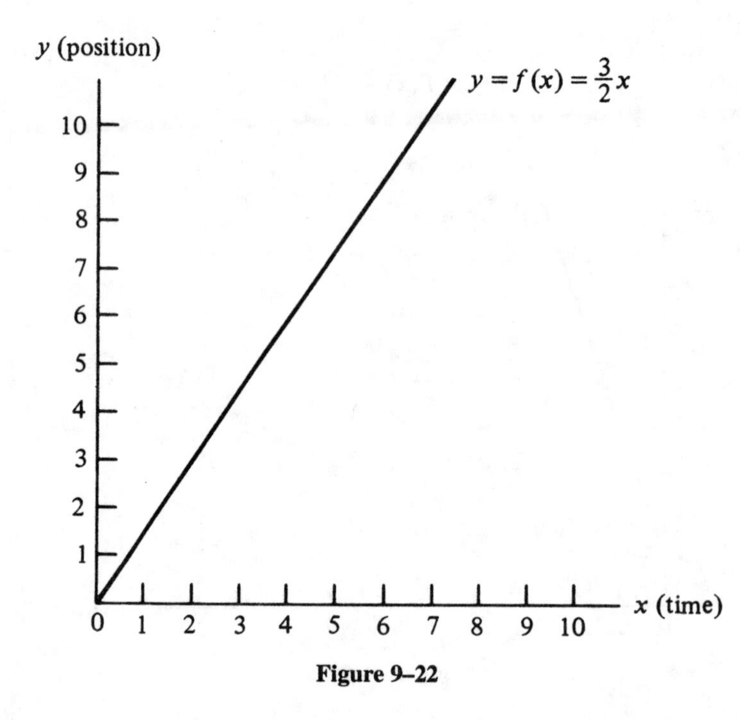

Figure 9–22

However, we quickly found out that this one point alone did not tell us which was the correct line, since there were many different lines that could be drawn through this point. (See Figure 9–23.) We realized that we needed to calculate another point, so we determined that the point (2, 8) was also a solution to the equation. Recordis set his ruler on the graph and drew a line that connected those two points. (See Figure 9–24.)

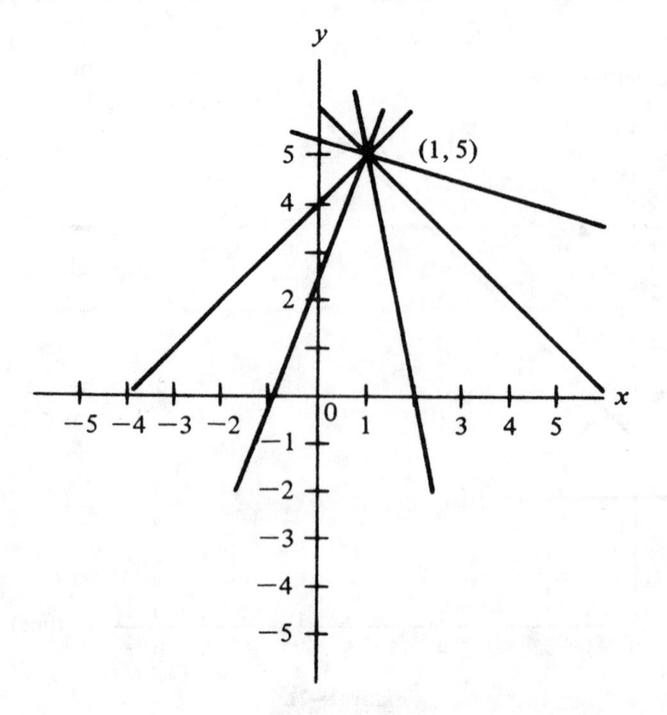

Figure 9–23

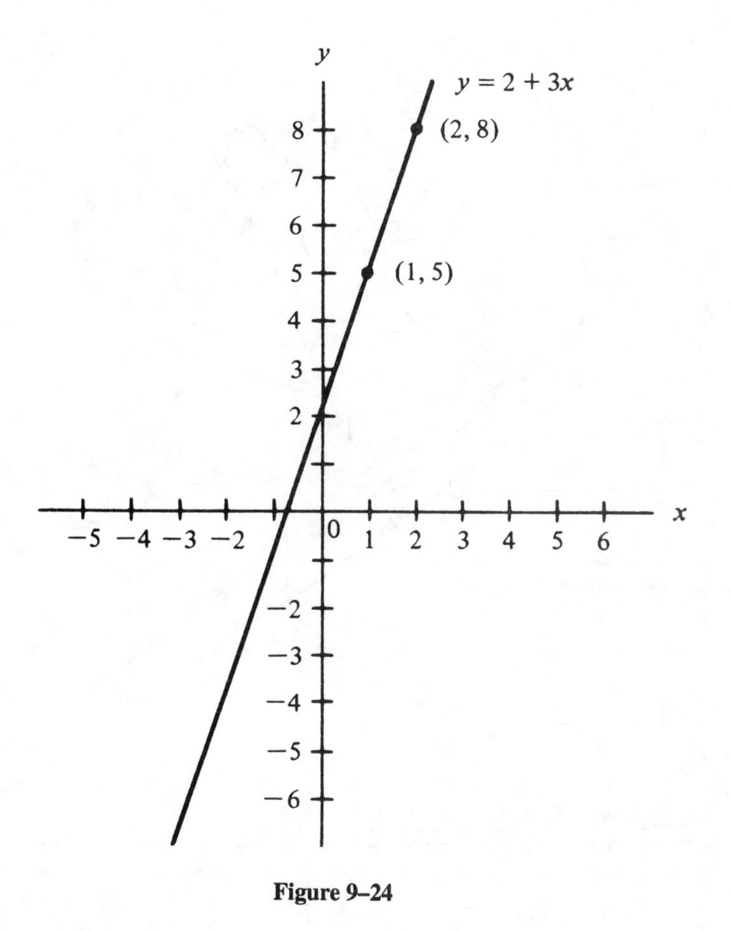

Figure 9–24

"Are we sure that this is the correct line?" the professor asked. No matter how hard he tried, Recordis was unable to find another line that passed through both of those points. We decided that we could make a general rule:

> There is one and only one straight line that passes through two points.

We were now able to start with an equation and draw the graph of that equation. However, we still did not know how to do the opposite: to start with a graph that is a line and then figure out the equation that represents it. We were stuck when we tried to figure out the equation of the line that passes through the points (3, 4) and (7, 10).

While we were working on this problem Builder arrived to demonstrate the latest toy he had built for Pal: an adjustable sliding and climbing ramp. "The ramp can be turned into a slide if you adjust it to go down," Builder said, "or it can be turned into a climbing ramp or a flat walking ramp." (See Figure 9–25.)

"We must have a way to measure steepness," the professor said. Recordis suggested one obvious fact: A flat ramp should have a steepness

of 0, while an upward-sloping ramp should have a positive steepness and a downward-sloping ramp should have a negative steepness. (See Figure 9–26.)

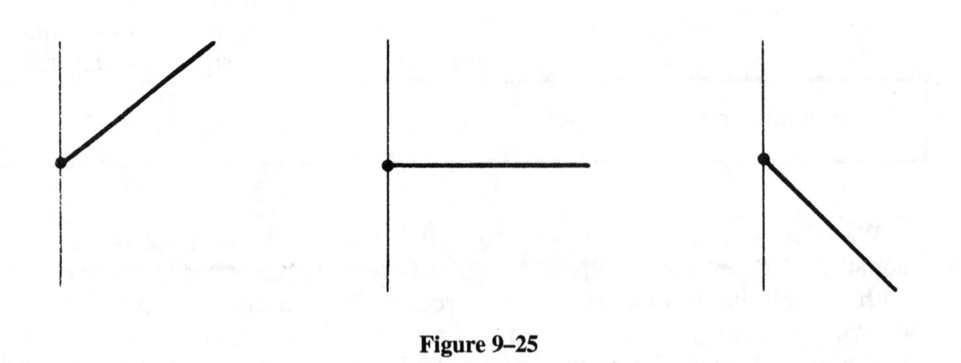

Figure 9–25

After some experimentation we found a suitable measure of the steepness of a line. We called it the *slope*.

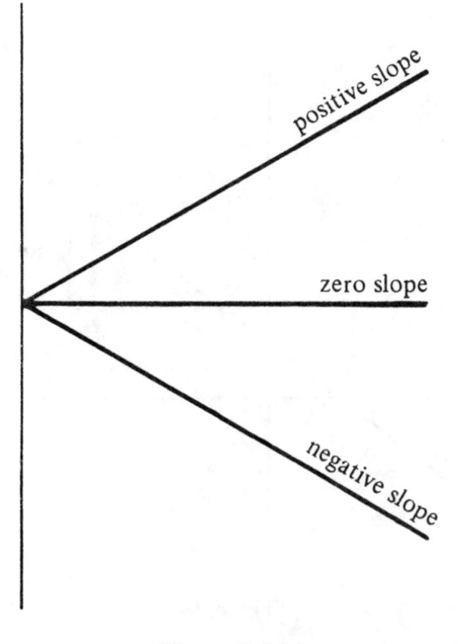

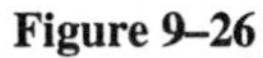
Figure 9–26

TO CALCULATE THE SLOPE OF A LINE:

- Pick any two points on the line—call them (x_1, y_1) and (x_2, y_2).
- Calculate the distance the line goes up between those two points.
- Divide by the distance it goes sideways:

$$\text{Slope} = \frac{(\text{distance up})}{(\text{distance sideways})}$$

The above rule works for a line with a positive slope. If the line has a negative slope, then remember to take the negative of the ratio between the two distances. The slope will automatically have the correct sign if you use this formula:

$$\text{slope} = \frac{y_2 - y_1}{x_2 - x_1}$$

See Figure 9–27.

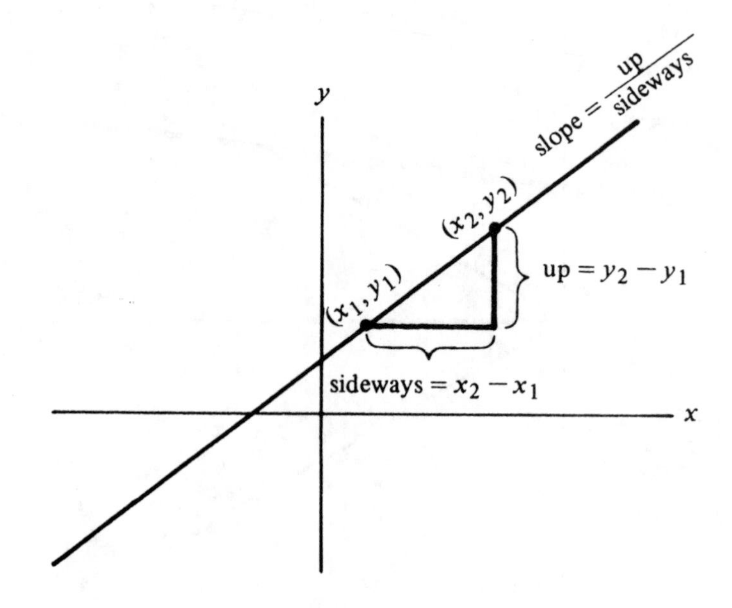

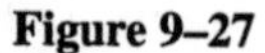
Figure 9–27

We calculated the slopes of several of the lines that we had investigated:

line: $y = 2x$
points: (0, 0) and (2, 4)

$$\text{slope: } \frac{4-0}{2-0} = 2$$

---

line: $y = 3x + 5$
points: (1, 8) and (3, 14)

$$\text{slope: } \frac{14-8}{3-1} = \frac{6}{2} = 3$$

---

line: $y = 10x + 15$
points: (2, 35) and (10, 115)

$$\text{slope: } \frac{115-35}{10-2} = \frac{80}{8} = 10$$

Suddenly we noticed a pattern. "Look!" Recordis said. "The slope of the line $y = 2x$ is 2; the slope of the line $y = 3x + 5$ is 3; and the slope of the line $y = 10x + 15$ is 10,"

"I bet we can say that the line $y = mx$ (where m is any real number) will have a slope of m," the professor said.

Igor drew several lines all with a slope of 2. (See Figure 9–28.)

We decided that we could identify a line by specifying two numbers: the slope of the line, and the y coordinate of the point where the line crosses the y axis (called the y-intercept). Therefore, the general form of the equation of a line is

$$y = mx + b$$

$$m = \text{slope}, \ b = y\text{-intercept}$$

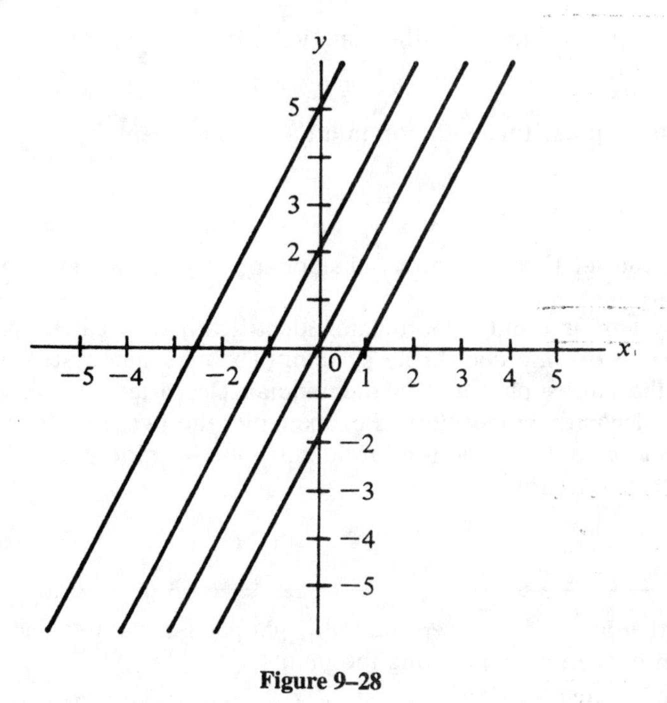

Figure 9–28

Suddenly the detective said, "I have a new clue! Hurry!" Before we had time to board the balloon the detective led us back into the woods.

Notes to Chapter 9

- A flat surface extending off to infinity is called a *plane*. The x axis and the y axis divide a plane into four regions called *quadrants*. The quadrant where both x and y are positive is called quadrant I. The other three quadrants are labelled as shown in Figure 9–29.

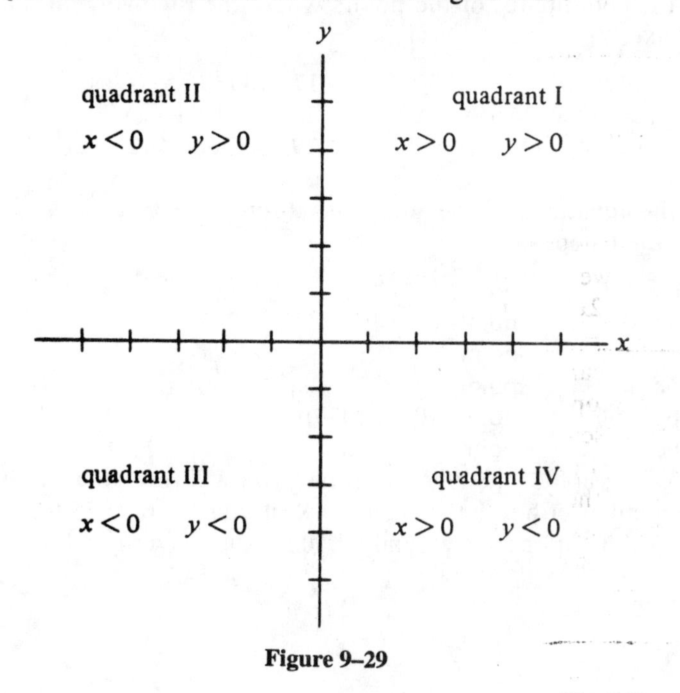

Figure 9–29

- Another general form for the equation of a line is

$$ax + by = c$$

We can express this equation in a more convenient form:

$$y = -\frac{a}{b}x + \frac{c}{b}$$

so we can see that this line has a slope of $-a/b$ and a y-intercept of c/b.

- The system of x and y coordinates identified here is called the *rectangular* or *cartesian* coordinate system. (The cartesian system is named after the philosopher and mathematician Descartes.) There are other types of coordinate systems. For example, the two coordinates longitude and latitude can be used to identify the location of a point on the surface of the earth.

Exercises

Find the distance between the following pairs of points. Then find the equation of the line connecting the points.

1. (0, 0) and (6, 8)
2. (2, 3) and (6, 10)
3. (−1, 0) and (1, 1)
4. (−4, 6) and (2, 10)
5. (9, 11) and (8, 17)
6. (16, 4) and (0, −3)
7. (2, 4) and (2, 12)
8. (3, 10) and (10, 10)

Draw the graphs of these functions:

9. $y = |x|$
10. $y = |x + 2|$
11. $y = |x| + 2$
12. $y = |x| - 2$
13. y is the greatest integer that is less than or equal to x.

Find the coordinates of the points where the following lines cross the x and y axes:

14. $5y - 10x = 0$
15. $x + y = 1$
16. $2x + y = 10$
17. $100x - y = 1{,}000$
18. $\frac{1}{4}x + \frac{1}{3}y = 1$
19. $0.37x - 4.12y = 6.34$

Find the equations of lines that pass through the given points and have the indicated slopes:

20. slope: 1/2 point: (6, 4)
21. slope: 0 point: (5, 10)
22. slope: m point: (1, 2)
23. slope: m point (a, 5)
24. slope: 100 point (1025, 1030)
25. slope: m point (−5, −6)
26. Make a graph of the amount of pay you will receive as a function of the number of hours that you work, if you are paid $5 for each hour up to 40 hours per week and $7.50 for each overtime hour.

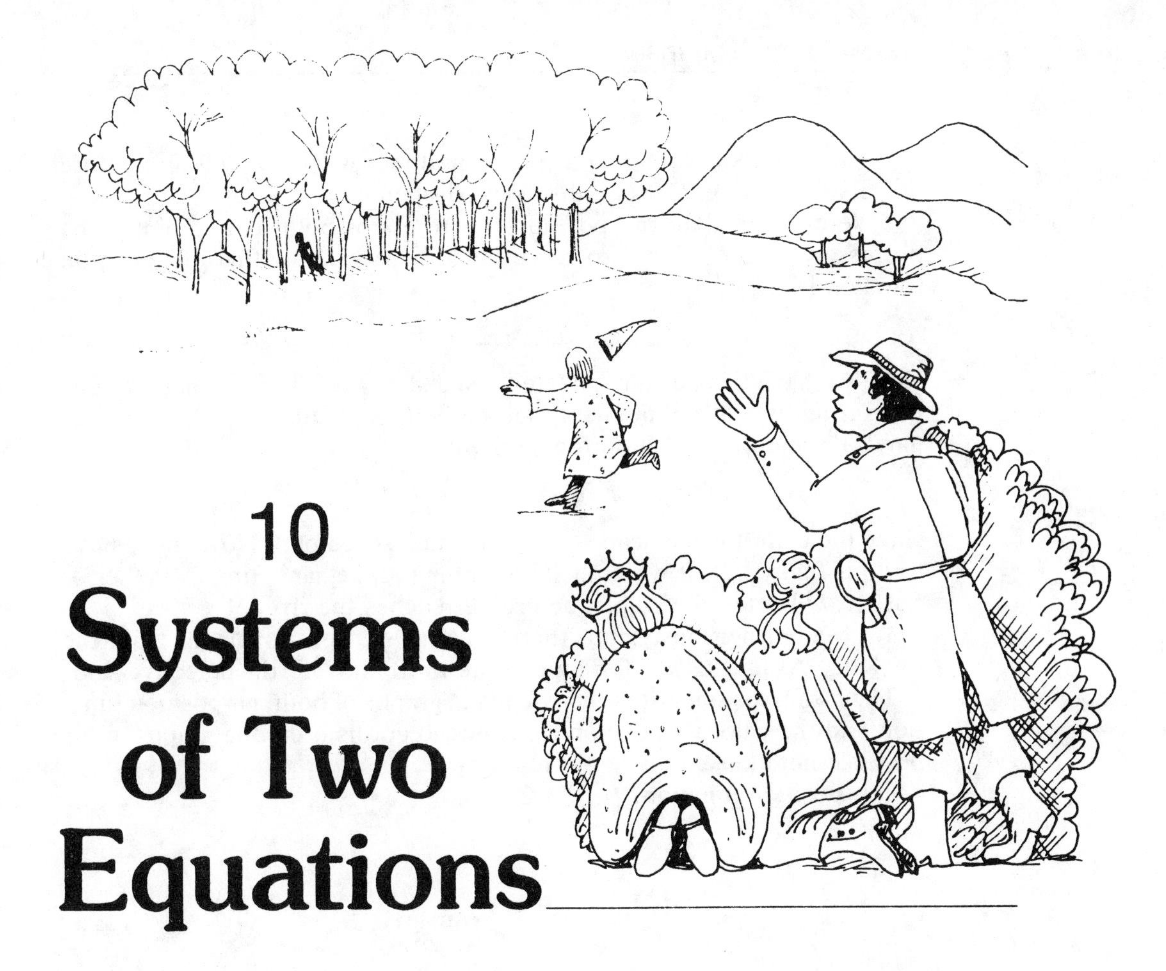

10 Systems of Two Equations

We quickly followed the detective into the woods. "We will catch him red-handed," the detective promised. Suddenly he stopped and told us to all hide behind a bush 2 yards from the road. "There!" he said, pointing to a thicket 8 yards away. Recordis started to run to the thicket.

"Wait!" the detective said. "The culprit can run 3 yards per second. We can run 4 yards per second, but remember that the culprit has a bigger head start then we do. If the culprit beats us to the river, then he can escape." (See Figure 10–1.)

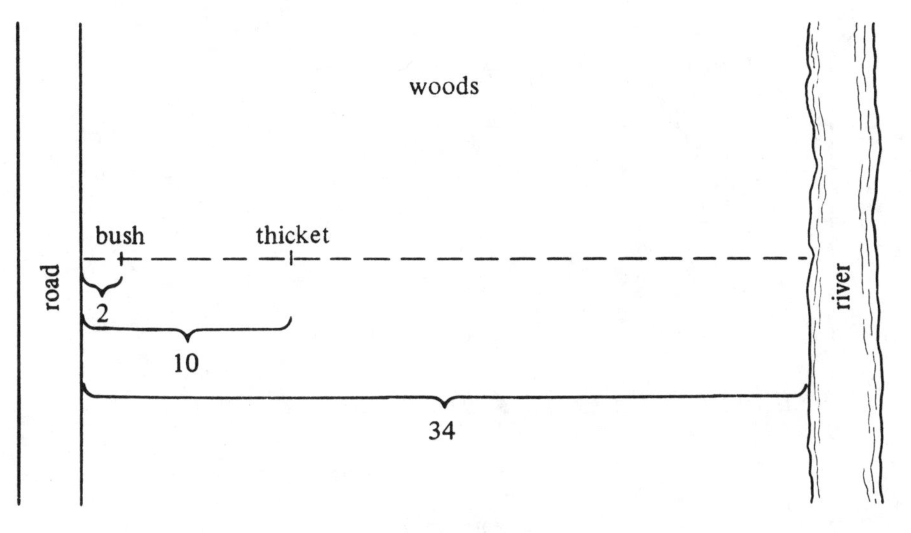

Figure 10–1

"If we both start running at the same time, will we catch him before he gets to the river?" Recordis asked desperately.

"We can calculate the position functions for both us and him," the professor said. "Using x to represent time, we know that our position function is

$$f(x) = 2 + 4x$$

since we have a head start of 2 and a speed of 4. (Assuming that we start measuring time at the moment that we both start running.) On the other hand, the culprit's position function will be

$$g(x) = 10 + 3x$$

since the culprit has a head start of 10 and a speed of 3. Now all we have to do is figure out, if we both start running at the same time, whether or not we will catch the culprit before he reaches the river."

We waited nervously, but there was no sign of movement from the thicket. "As long as we don't move, he won't move," the detective said.

The professor suggested that we make graphs of both our position function and the culprit's position function. Recordis drew the graphs in his portable notebook:

Our position function, $f(x) = 2 + 4x$

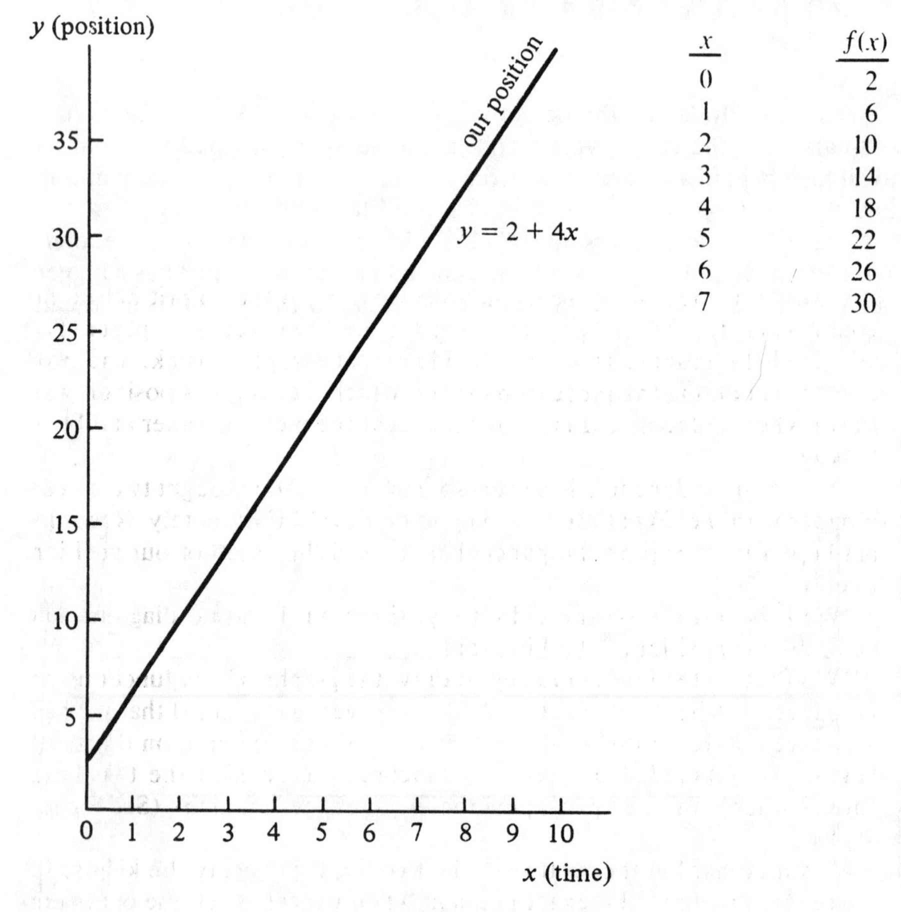

x	$f(x)$
0	2
1	6
2	10
3	14
4	18
5	22
6	26
7	30

Figure 10–2

Culprit's position function, $g(x) = 10 + 3x$

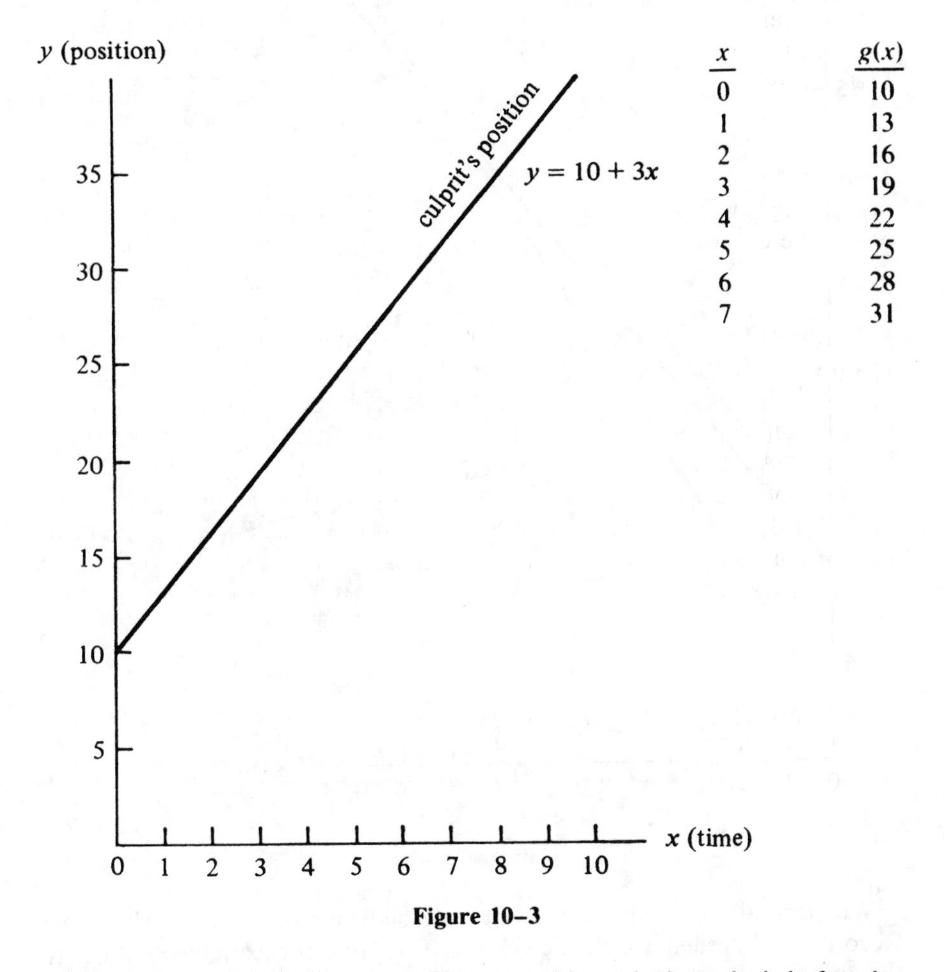

x	$g(x)$
0	10
1	13
2	16
3	19
4	22
5	25
6	28
7	31

Figure 10–3

Next we needed to figure out if we would catch the culprit before he would reach the river. We stared hard at the two pieces of paper that contained the graphs. However, suddenly catastrophe struck. The professor was holding the piece of paper on which the culprit's position was drawn when suddenly a gust of wind caught the piece of paper and blew it away.

"Now you've done it!" Recordis screamed. "I only brought two pieces of paper with me! What are we going to do now?" Fortunately, Recordis had kept a tight grip on the paper that showed the graph of our position function.

"We'll have to draw the culprit's position on the same diagram that we drew our position," the king said.

"Won't it be terribly confusing to draw the graphs of *two* functions on one piece of paper?" Recordis asked. However, we realized that we had no choice. So Recordis drew the culprit's position function on the same diagram that included our position function. He labelled the two lines "our position" and "culprit's position" to keep them straight. (See Figure 10–4.)

We stared hard at the picture of the two lines. Suddenly the king said: "I see the solution! The exact moment when we will catch the opponent is shown on the graph as the point where the two lines cross!"

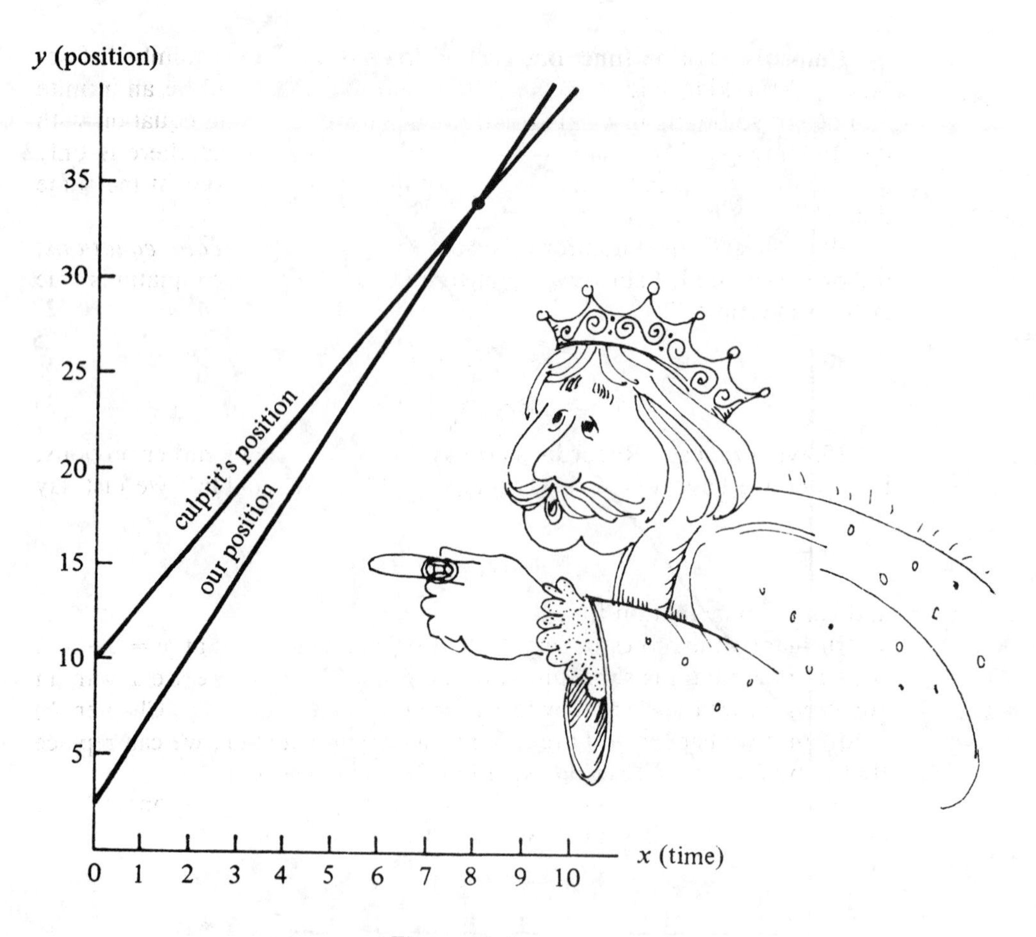

Figure 10–4

We were all struck by the beauty of the king's suggestion. However, Recordis still needed a little convincing. The professor explained. "At first the culprit is ahead of us. On the graph that means that the line representing the culprit's position is above our graph. Later on, we're ahead. The exact moment when we catch up with the culprit is shown on the graph where the two lines cross."

Now we had to look carefully at the diagram to determine the coordinates of the point where the two lines crossed. As far as we could tell, the lines crossed at the point (8, 34). In other words, we would catch the culprit 8 seconds after we started running, and at that moment we would both be 34 yards from the road. Since the river was exactly 34 yards from the road, that meant that we would catch the culprit at the river—but just barely.

"We must be sure," Recordis said. "We cannot take the chance that our graphical solution might be a little bit off. If we were back at the office and I had all my normal materials, then I could guarantee that the graph would be perfect; but in the woods it is harder to draw lines that are perfectly straight."

The king suggested that we look to see if we could find an algebraic way to find the solution. We stared at the two equations again.

our position: $y = f(x) = 2 + 4x$

culprit's position: $y = g(x) = 10 + 3x$

"We know that each of these equations has an infinite number of solutions," the king said. "In fact, we found that there will be an infinite number of solutions in most situations when we have one equation with two unknowns. (See Chapter 2.) However, we know that there is only one solution for x and y that solves both of these equations at the same time."

"We can call this situation a system of two *simultaneous equations*," the professor said, "since we want to make both of these equations true at the same time."

$$y = 2 + 4x$$

$$y = 10 + 3x$$

"I have an idea," Recordis said timidly. "This seems rather obvious, but since we have $y = 2 + 4x$ and $y = 10 + 3x$, couldn't we just say that

$$2 + 4x = 10 + 3x$$

and solve this equation for x?"

"Brilliant!" the professor said. "The first equation says that $y = 2 + 4x$. So we know from the substitution property that any time we see a y in an equation we can replace it with the expression $2 + 4x$. (See Chapter 2.) And, since we happen to have a y in the second equation, we can replace that y by $2 + 4x$ and get the equation Recordis suggested."

"Now we just have the familiar situation with one equation and one unknown," Recordis said. "We can handle that easily."

$$2 + 4x - \mathbf{3x} = 10 + 3x - \mathbf{3x}$$

$$2 + x = 10$$

$$2 + x - \mathbf{2} = 10 - \mathbf{2}$$

$$x = 8$$

"So our graphical solution was correct!" the professor said triumphantly.

"I hate to tell you this," Recordis said, "but it looks as if we've lost all information about y. We have found the correct value for x—but now how do we find the correct value for y?"

"No problem," the professor said confidently. "Once we know the value of x we can use either equation to pin down the value of y. For example, we can use the first equation to tell us that when $x = 8$ then $y = 2 + 4 \times 8 = 34$. And we can also use the second equation to see that when $x = 8$, then $y = 10 + 3 \times 8$, which also is 34. Therefore, the ordered pair $x = 8$, $y = 34$ is the solution to the two-equation simultaneous-equation system."

"We're lucky the two equations gave us the same value for y," Recordis said.

"It wasn't luck!" the professor said. "If the two equations had given different values for y it would mean that we had made a mistake in our calculation of the value of x."

"Now is the time to make our move," the detective said. "We know that we will catch the culprit."

We started running. At the same instant we left our bush the culprit left the thicket, and we realized to our total astonishment that the culprit we were chasing was not human—but instead a fierce fire-breathing dragon! However, in the cloud of dust we were not able to see it too clearly. We ran for 8 seconds, coming closer and closer to the dragon all the time. Suddenly we caught up with the culprit, but in the confusion and the cloud of dust the slippery dragon slipped out of our hands and splashed into the river.

"It's gone!" Recordis exclaimed in dismay. "Our calculations were correct, but it slipped away from us."

"You can't win them all," the detective said. "However, now that we know what the culprit looks like we will be sure to catch it sooner or later."

Crestfallen, we returned to the Main Conference Room. "We have made significant progress," the professor said, trying to cheer us up. "I'm sure our ability to solve simultaneous equations will help a lot in our quest to bring that dragon to justice."

We worked on some more examples of simultaneous equations to keep in practice.

equations $\begin{cases} y = 3 + 4x \\ y = 5 + 6x \end{cases}$

$$3 + 4x = 5 + 6x$$
$$-2 = 2x$$

solution: $x = -1, \; y = -1$

---

equations $\begin{cases} y = 20x \\ y = 2 + 30x \end{cases}$

$$20x = 2 + 30x$$
$$-2 = 10x$$

solution: $x = -1/5, \; y = -4$

---

equations $\begin{cases} y = 100 + 300x \\ y = 50 + 100x \end{cases}$

solution: $x = -0.25, \; y = 25$

---

equations $\begin{cases} y = 0.5 + 0.8x \\ y = 0.6 + 0.7x \end{cases}$

solution: $x = 1, \; y = 1.3$

"We can also solve the problem involving the demand and supply of potatoes," the king said. "We have found that farmers will grow more potatoes when the price is higher, and the quantity of potatoes that they grow (q) can be expressed as a function of the price (p):

$$\text{supply function:} \quad q = f(p) = -5 + 30p$$

However, if the price is higher, then people will buy fewer potatoes, and we have found that the quantity of potatoes that people demand can also be expressed as a function of the price:

$$\text{demand function:} \quad q = g(p) = 200 - 15p$$

There are lots of problems if the quantity that the sellers are supplying does not equal the quantity that the buyers are demanding. For example, we don't want surplus potatoes to rot away unsold, and we don't want people to be frustrated because the sellers are out of potatoes. We want to determine a price and a quantity such that the amount demanded equals the amount supplied, so we need to find the solution to this simultaneous two-equation system."

$$q = -5 + 30p$$

$$q = 200 - 15p$$

We could solve this equation in the same fashion as we had solved the others, and the solution was

$$p = 4.556, \quad q = 131.7$$

Figure 10–5 illustrates this solution.

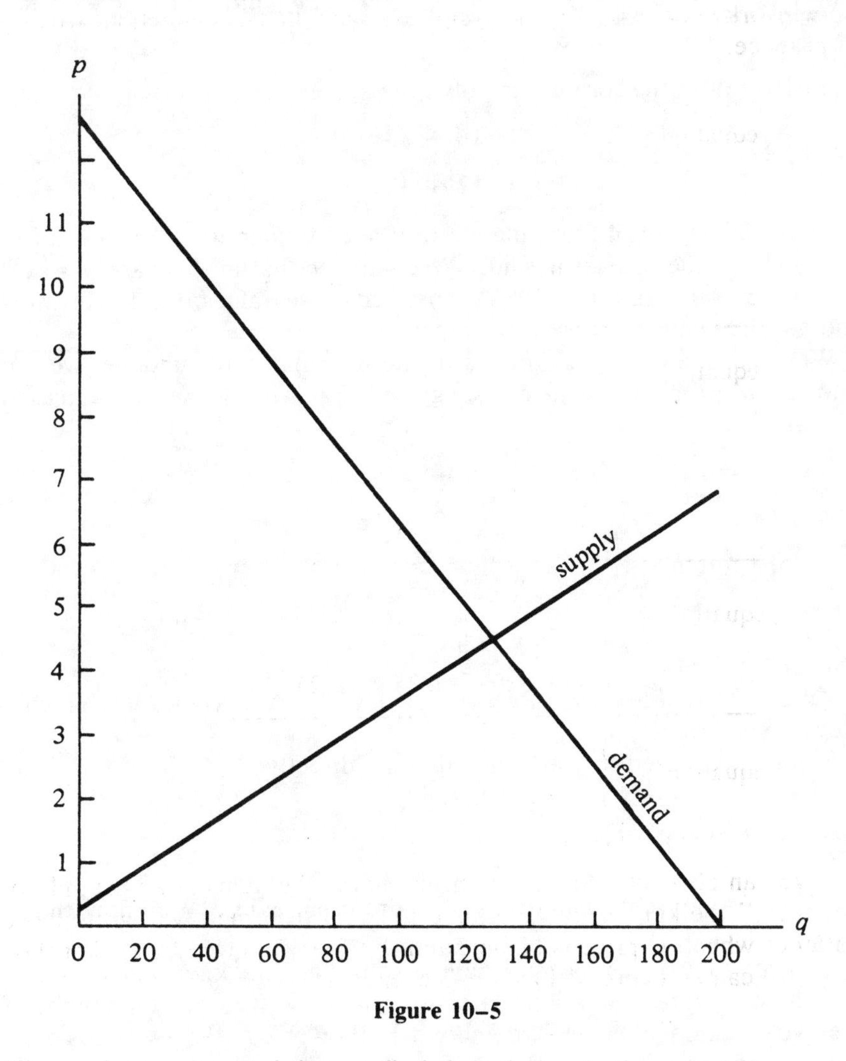

Figure 10–5

The professor suggested that we find the solution to the system:

$$y = a + bx$$

$$y = c + dx$$

"Remember that in this system we're pretending that we don't know the values of x and y but that we do know the values of a, b, c, and d."

We substituted the expression for y from the first equation into the second equation:

$$a + bx = c + dx$$

Now all we had to do was solve the equation for x.

$$a - c = dx - bx$$

$$a - c = (d - b)x$$

$$x = \frac{a - c}{d - b}$$

Now we inserted this formula for x into the first equation in order to find a formula for y:

$$y = a + b\left(\frac{a-c}{d-b}\right) = \frac{a(d-b)}{d-b} + \frac{b(a-c)}{d-b} = \frac{ad - ab + ab - bc}{d-b} = \frac{ad-bc}{d-b}$$

We tried to use the formula to solve the system

$$y = 16 + 18x$$

$$y = 12 + 10x$$

"In order to use the formula we first need to identify the values of a, b, c and d," the professor said. We could see that in this case $a = 16$, $b = 18$, $c = 12$, and $d = 10$. We inserted these values into the formula and got the result $x = -1/2$, $y = 7$.

We tried a few more examples of the formula, until suddenly we ran into a real problem. Recordis suggested that we try the two-equation system

$$y = 4 + 3x$$

$$y = 2 + 3x$$

Using the formula,

$$x = \frac{4 - 2}{3 - 3} = \frac{2}{0}$$

"We're in trouble now!" Recordis said. "The expression 2/0 is a highly illegal expression!"

Stunned, we tried to solve the equation directly:

$$2 + 3x = 4 + 3x$$

$$2 = 4$$

"That's a nonsense equation," the professor said. "We know that 2 doesn't equal 4."

"Maybe this system doesn't have a solution," the king said.

"There *must* be a solution!" Recordis said. "If we draw the graphs of the two equations then we can find where they cross. Any two lines must cross somewhere."

We drew the graphs of the two equations. (See Figure 10–6.) Sure enough, they did not cross at all.

"Silly!" the professor said. "You forgot that two *parallel* lines never cross!"

"So that means it is possible to construct a two-equation system that has no solution," the king said.

After recovering from the shock of this discovery, Recordis asked us to help him with a new diet problem. "I have decided that I will have milk

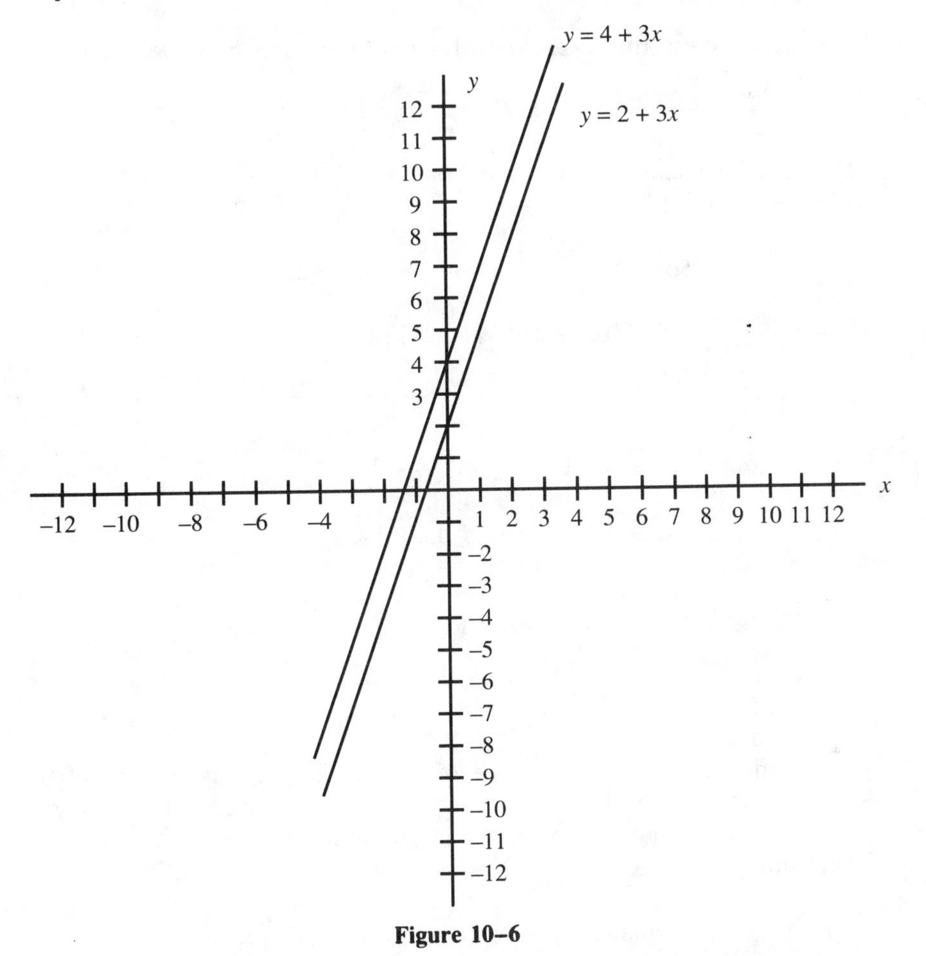

Figure 10–6

and soup for lunch," he said. "Each cup of milk provides exactly 8 grams of protein and 150 calories, whereas each bowl of soup provides exactly 3 grams of protein and 80 calories. I have scientifically calculated my diet so that I know I want exactly 41 grams of protein and 840 calories for lunch. Now all I need to do is calculate exactly how many bowls of soup and how many cups of milk I need to have for lunch."

"We cannot proceed until we turn the problem into a system of equations," the professor said. We decided to use x to stand for the number of cups of milk and y to stand for the number of bowls of soup. Then the total amount of protein in Recordis' lunch would be $8x + 3y$ grams and the total number of calories would be $150x + 80y$. We wrote out the two-equation system:

grams of protein: $8x + 3y = 41$

number of calories: $150x + 80y = 840$

"We have two equations with two unknowns," Recordis said, "but this equation is a little bit trickier than the others."

"We can use the substitution principle again," the king said. "We will have to write the first equation in a different form."

$$y = \frac{41 - 8x}{3}$$

Now, substituting this expression for y into the second equation,

$$150x + 80\left(\frac{41 - 8x}{3}\right) = 840$$

"Now we have simplified the situation so that we have only one equation with one unknown."

$$150x + \frac{80 \times 41}{3} - \frac{80 \times 8x}{3} = 840$$

$$150x + 1093\tfrac{1}{3} - 213\tfrac{1}{3}x = 840$$

$$-63\tfrac{1}{3}x = -253\tfrac{1}{3}$$

$$x = \frac{-253\tfrac{1}{3}}{-63\tfrac{1}{3}} = 4$$

We used the first equation to find the value of y:

$$y = 3$$

We tried another example of a two-equation system:

$$3x + 5y = 50$$

$$6x + 2y = 68$$

"The trick to solving this type of system is to focus on one of the equations first," the king observed. "Isolate one of the unknowns so it is all by itself on one side of the equation, and then substitute it into the other equation."

from the first equation: $y = 10 - \frac{3}{5}x$

substituting into the second equation: $6x + 2\left(10 - \frac{3}{5}x\right) = 68$

"Now that we've converted the system into one equation with one unknown we can use Standard Operating Procedures to proceed to the solution."

$$6x + 20 - \frac{6}{5}x \quad 68$$

$$\frac{24}{5}x = 48$$

$$x = 10$$

Now we can use either of the two original equations to find the value for y. Fortunately, both methods give the value $y = 4$. If the two equations gave different values of y, that would mean there must have been a mistake somewhere in our calculations.

We called this method the Substitution Method:

SUBSTITUTION METHOD FOR SOLVING A TWO-EQUATION SYSTEM WITH TWO UNKNOWNS (x AND y)

1. Use one of the equations to derive a formula for y in terms of x. (To do this, use the Golden Rule of Equations to isolate y on one side of the equation.)
2. Substitute the resulting formula into the other equation wherever y appears.
3. The result from step 2 is a single equation with a single unknown (x). Use this equation to solve for the value of x.
4. Insert the value of x into the expression for y from step 1 to determine the value of y.

We worked on another example.

$$4x + 3y = 38$$

$$6x - 3y = 12$$

We were about to proceed to solve this system when the king noticed something. "Let's just add the second equation to the first equation to form a new equation, like this:

$$(4x + 3y) + (6x - 3y) = 38 + 12$$

"Wait!" Recordis objected. "You can't do that! The Golden Rule of Equations says you must do exactly the same thing to both sides of the equation! You're adding $6x - 3y$ to the left-hand side, but you're adding 12 to the right-hand side!"

"But, according to the second equation, $6x - 3y$ is equal to 12," the professor said. "So what we're doing is perfectly legal."

"Oh," Recordis said. "I hadn't thought of that." He stared at the new equation,

$$4x + 3y + 6x - 3y = 38 + 12$$

and suddenly his eyes lit up.

"The $3y$ and the $-3y$ cancel out!" he said excitedly.

$$4x + 6x = 38 + 12$$

$$10x = 50$$

$$x = 5$$

We also found that $y = 6$.

"We have discovered a new solution method!" the professor said. "In some cases you can simplify a two-equation system merely by adding the two equations together."

We tried another example.

original system:

$$10x + \tfrac{1}{2}y = 8$$

$$-10x + 7\tfrac{1}{2}y = 40$$

new equation:

$$8y = 48$$

$$y = 6, \ x = \tfrac{1}{2}$$

The next system that the professor suggested was

$$6x + y = 44$$
$$3x + y = 23$$

"Adding those two equations together won't do any good," Recordis said. "But we can subtract the second equation from the first equation, like this:

$$(6x + y) - (3x + y) = 44 - 23$$

The y's cancelled each other out, leaving:

$$6x - 3x = 44 - 23$$
$$3x = 21$$
$$x = 7, \ y = 2$$

The professor's next suggestion was

$$6x + y = 32$$
$$2x - 2y = -8$$

"Adding those two equations together won't help—and neither will subtraction," Recordis said. The king had a new idea: He suggested multiplying the top equation by 2:

$$12x + 2y = 64$$
$$2x - 2y = -8$$

"This new equation system must be equivalent to the original system. And now we can add the two equations together to cancel out the y's."

$$14x = 56$$
$$x = 4,$$

Then we used the first equation to solve for y:

$$y = 8$$

We called this method the *Elimination Method.*

ELIMINATION METHOD FOR SOLVING A TWO-EQUATION SYSTEM WITH TWO UNKNOWNS (x AND y)

1. Multiply both sides of the first equation by a number chosen so that the coefficient of y in the first equation becomes the same as the coefficient of y in the second equation.
2. Subtract the second equation from the first equation. (This means to subtract the left-hand side of the second equation from the left-hand side of the first equation, and subtract the right-hand side of the second equation from the right-hand side of the first equation.)
3. The result from step 2 will be a new equation from which y has been eliminated. Solve that equation for x.
4. Insert the value for x into either one of the two original equations and then solve for the value of y.

Note: In some cases it will be more convenient to choose the multiplier in step 1 so that the coefficients of y in the two equations will be negatives of each other. In that case add the two equations instead of subtracting them. Also, in other cases it may be more convenient to eliminate x instead of y.

The following are examples of this procedure.

system:

$$2x - 3y = -7$$
$$x + 6y = 34$$

multiply the bottom equation by 2:

$$2x - 3y = -7$$
$$2x + 12y = 68$$

subtract:

$$-15y = -75$$
$$y = 5, x = 4$$

system:

$$100x + 16y = 148$$
$$-10x + y = -7$$

multiply the bottom equation by 10:

$$100x + 16y = 148$$
$$-100x + 10y = -70$$

add the two equations together:

$$26y = 78$$
$$y = 3, x = 1$$

The professor suggested that it would be useful to find a general formula for the two-equation system

$$a_1x + b_1y = c_1$$
$$a_2x + b_2y = c_2$$

Recordis and the king started to argue. The king liked the elimination method and Recordis liked the substitution method. They had a race to see who could come up with the general formula more quickly:

Elimination Method (the king):

Multiply the second equation by b_1/b_2:

$$a_1x + b_1y = c_1$$
$$b_1a_2x/b_2 + b_1y = b_1c_2/b_2$$

Subtract the second equation from the first equation:

$$(a_1 - b_1a_2/b_2)x = c_1 - b_1c_2/b_2$$

$$\left[\frac{a_1b_2 - a_2b_1}{b_2}\right]x = \frac{b_2c_1 - b_1c_2}{b_2}$$

$$x = \frac{b_2c_1 - b_1c_2}{a_1b_2 - a_2b_1}$$

Substitution Method (Recordis):

Find a formula for y from the second equation:

$$y = \frac{c_2 - a_2 x}{b_2}$$

Substitute that expression into the first equation:

$$a_1 x + b_1 \left[\frac{c_2 - a_2 x}{b_2} \right] = c_1$$

$$x \left[a_1 - \frac{a_2 b_1}{b_2} \right] = c_1 - \frac{b_1 c_2}{b_2}$$

$$x \left[\frac{a_1 b_2 - a_2 b_1}{b_2} \right] = \frac{b_2 c_1 - b_1 c_2}{b_2}$$

$$x = \frac{b_2 c_1 - b_1 c_2}{a_1 b_2 - a_2 b_1}$$

"The race is a tie," the professor declared. "That means that they are both good methods; sometimes it will be better to use one of them and other times it will be better to use the other."

Once we had found a formula for x, we were able to find a formula for y:

$$y = \frac{a_1 c_2 - a_2 c_1}{a_1 b_2 - a_2 b_1}$$

(See Exercise 27.)

We proceeded smoothly along using the formula until Recordis suggested that we try to solve the system

$$2x + 2y = 4$$

$$3x + 3y = 6$$

Putting these values into the formula we came up with the result $x = 0/0$.

"Oh, no!" Recordis cried. "0/0 is also a highly illegal expression, but it is not exactly the same as the other illegal expression we had."

We tried to solve the system directly:

$$2y = 4 - 2x$$

$$y = (4 - 2x)/2$$

$$y = 2 - x$$

$$3x + 3(2 - x) = 6$$

$$3x + 6 - 3x = 6$$

$$6 = 6$$

"There's nothing wrong with the equation $6 = 6$," Recordis said. "It is perfectly true. However, it doesn't tell us anything about the values of x and y."

We didn't know what was going on with this peculiar system, so we tried to draw the graph. And we found that both equations defined exactly the same line.

$$\text{first equation: } 2x + 2y = 4$$
$$2y = 4 - 2x$$
$$y = 2 - x$$
$$\text{slope} = -1, \ y\text{-intercept} = 2$$

$$\text{second equation: } 3x + 3y = 6$$
$$3y = 6 - 3x$$
$$y = 2 - x$$
$$\text{slope} = -1, \ y\text{-intercept} = 2$$

"But that means that *any* point along the line $y = 2 - x$ will be a solution to either equation," Recordis said in surprise. Sure enough, we tried several points along the line—(1, 1), (0, 2), (2, 0), (−1, 3)—and found that they were all solutions to both equations.

"So there is no one right answer," the king said. "There are as many right answers as we want."

"Of course that has to be right," the professor realized. "Suppose I gave you this two equation system:

$$x + y = 5$$
$$x + y = 5$$

"Those two equations are exactly the same!" Recordis said. "That means that you really only have on equation. You can't fool me into thinking that you can turn a one-equation system into a two-equation system just by writing the same equation down twice!"

"Exactly," the professor said. "And we know that in general a single equation with two unknowns will have an infinite number of solutions. (See Chapter 2.) Therefore, the system $x + y = 5$; $x + y = 5$ will also have an infinite number of solutions. Now suppose I gave you these two equations:

$$x + y = 5$$
$$2x + 2y = 10$$

"To get the second equation you merely multiplied both sides of the top equation by 2," Recordis observed. "So you haven't changed the solution to the equation at all, according to the Golden Rule of Equations.

Therefore, the two equations $x + y = 5$ and $2x + 2y = 10$ are equivalent equations. If the two equations in the system are equivalent equations, they will both define the same line and there will be an infinite number of points that will be solutions to both equations."

We were now able to write down the general procedure for two-equation systems:

SOLUTIONS FOR TWO-EQUATION SYSTEMS WITH TWO UNKNOWNS

Suppose we have an equation of the general form

$$a_1x + b_1y = c_1$$

$$a_2x + b_2y = c_2$$

In general, there will be one pair of values for x and y that will be a solution to both equations simultaneously. They can be found from the formulas

$$x = \frac{b_2c_1 - b_1c_2}{a_1b_2 - a_2b_1} \qquad y = \frac{a_1c_2 - a_2c_1}{a_1b_2 - a_2b_1}$$

However, it is possible that the equation system will have no solutions. This will occur if the graphs of the two equations are parallel lines.

It is also possible that the system will have an infinite number of solutions. This will occur if the two equations actually define the same line.

Note: in this chapter we have dealt only with linear equations, which means that the graph of the equation is a straight line. Therefore, there are no exponents for any of the variables, and no terms with two variables multiplied together. Some of these more complicated types of equations will be considered later.

Exercises

Solve the following systems of equations for x and y:

1. $y = 10 + 2x$
$y = 4.5x$

2. $y = 10 + 2x$
$y = 5x$

3. $y = 10 + 2x$
$y = 2x$

4. $y = 10 + 2x$
$y = 15 + 15x$

5. Suppose you face this equation system:

$2x + y = 3$
$4x - y = 3$

Form a new equation by adding these two equations together, and then solve the system.

Add or subtract these equations and then solve the two-equation systems:

6. $3x + 2y = 5$
$-3x + 5y = 16$

7. $2x + 6y = 10$
$2x + 10y = 6$

8. $x + 4y = 8$
$-2x + 3y = 8$
(Hint: multiply the first equation by 2 and then add the two equations.)

Solve these two-equation systems for x and y:

9. $3x + 3y = 10$
$2x + 2y = 8$

10. $1.5x - 3.2y = 4$
$6.8x + 1.9y = 10$

11. $x + y = 1$
$x - y = 1$

12. $x + y = 1$
$x - y = 0$

13. $xy = 1$
$x = 4y$

14. $xy = a$
$x = by$

15. $y = -2x + 10$
$y = -\frac{1}{2}x + 5$

16. $3x + 4y = 0$
$8x - 10y = 0$

17. $ax + by = 0$
$cx + dy = 0$

18. $y = m_1x + b_1$
$y = m_2x + b_2$

19. $5.1x - 4.3y = 3$
$10.2x - 8.6y = 6$

20. $11.1x - 12.4y = 16.3$
$10.878x - 12.152y = 15.974$

21. One day a Celsius thermometer and a Fahrenheit thermometer registered exactly the same numerical value for the temperature. What was the temperature that day?

22. The level of national income (Y) in a country can be found from the two-equation system

$$Y = C + I + G$$
$$C = a + bY$$

Solve for Y and C, assuming that I (investment spending), G (government spending), a, and b are known.

23. Calculate values for Y and C in the preceding problem, given that $I = 30$, $G = 30$, $a = 10$, and $b = 4/5$.

24. It has been discovered that national income (Y) and the interest rate (r) in a country satisfy these two equations:

$$Y - 20r = 0$$
$$Y + 20r = 200$$

What are the values of Y and r?

25. Suppose that the demand for a new good is given by the function $D_0 - dp$ (where p is the price of the good), and the supply of the good is given by the function $S_0 + sp$. Find the value of p such that demand and supply are equal.

26. Evaluate the formula in the preceding problem with $D_0 = 100$, $S_0 = 0$, $d = 1/2$, and $s = 1/2$.

27. Verify the formula in the chapter that gives the solution value for y for the system $a_1x + b_1y = c_1$, $a_2x + b_2y = c_2$.

28. When you use the formula discussed in the preceding exercise, does it matter which of the two equations you write first?

29. Farmer Floran has a total of 33 cows and chickens. Altogether the animals have 90 feet. How many cows and how many chickens are there?

❑ **30.** Use a computer spreadsheet program to illustrate the solution to a system of two linear equations. Design the graph so you can choose two values of x and have the screen show the appearance of the two lines between those two values. This way you can choose two values that are far apart to see the big picture; then you can choose two values of x that are close together to "zoom in" on the point of intersection.

11 Quadratic Equations

As Opening Day approached, the professor was trying to teach Pal how to catch fly balls so he could play outfield for the Royal Palace baseball team. "He can only catch the ball when it is exactly 10 units above the ground," the professor explained. "I need to know the exact time when the ball will be 10 units high." (Note: in this chapter the height above the ground is being measured in a very peculiar unit used only in Carmorra.)

The king suggested that we create a position function for the ball. "We need a function $y = f(x)$ such that if x is the time since the ball left the bat, then y is the height of the ball above the ground."

We had to hit a lot of fly balls before we could estimate the position function for a typical fly ball. As the professor (a baseball lover) said, "Sometimes scientific research is hard work." The result was

$$\text{If } x = \text{time and } f(x) = \text{height of ball,} \qquad f(x) = -x^2 + 7x$$

"Now we need to find the value of x such that $10 = f(x)$," the professor said.

$$10 = -x^2 + 7x$$

We decided to write all the terms of this equation on the left-hand side:

$$x^2 - 7x + 10 = 0$$

"The expression on the left-hand side is very familiar," the professor said. "That's just a multinomial with three terms—one involving x^2, one involving x, and one involving no letters. We should have a special name for a multinomial involving only powers of x." We decided to use the term *polynomial*.

> A polynomial in x is a multinomial in which each term contains x raised to a whole-number power.

Examples of polynomials:

$$x^2 - 7x + 10$$
$$3x^2 + 4x + 6$$
$$10x^2 - 4$$
$$2x^2 + 5x$$
$$x^3 - 1$$
$$10x^3 - 9x^2 - 8x - 7$$
$$ax^2 + bx + c$$
$$mx + b$$

Note that a polynomial can contain a term with no x, since $x^0 = 1$. However, the following multinomials are *not* polynomials:

$$\sqrt{x} + 5x + 4$$
$$x^{-2} + x^{-1} + 3$$
$$x^{-1/2}$$

"How many different kinds of polynomials are there?" Recordis asked.

"It seems that the important distinguishing feature is the highest power of x that appears," the professor said. "For example, in the polynomial $mx + b$, the highest power of x is 1; in the polynomial $ax^2 + bx + c$, the highest power is 2; and in the polynomial $x^3 - 1$, the highest power is 3." We decided to use the term *degree* to indicate the highest power of x appearing in the polynomial. For example:

$3x^2 + 4x + 6$	is a second-degree polynomial
$10x + 3$	is a first-degree polynomial
$x^3 - 5x^2 - 2$	is a third-degree polynomial

(A second-degree polynomial is also called a *quadratic polynomial,* so an equation of the form $ax^2 + bx + c = 0$ is called a *quadratic equation.*)

Igor wrote out the general form for a polynomial of degree n:

$$a_nx^n + a_{n-1}x^{n-1} + \cdots + a_3x^3 + a_2x^2 + a_1x + a_0$$

Igor used a_n to represent the coefficient of x^n, a_3 to represent the coefficient of x^3, and so on. He put three dots ($\cdots$) in the expression to indicate that some terms are probably missing (since we don't know the numerical value of n.) For variety, Igor showed that he could write a third degree polynomial in y:

$$5abcy^3 + 10qy^2 + 6y + 16$$

and a fourth degree polynomial in z:

$$z^4 + 2z^3 + 3z^2 + 4z + 5$$

(We investigated polynomials more in Chapter 13.)

"So what if we can classify our equation?" Recordis said. "That still doesn't help us solve it.

"We can't use the Golden Rule of Equations to isolate x, because then we are left with $x = (x^2 + 10)/7$, which doesn't work as a solution for x because it still contains x^2 on the right-hand side."

Further investigation revealed that Recordis was right—there was no easy way to solve this equation. "If only we had a linear equation, such as $-7x + 10 = 0$," Recordis moaned. "We know how to solve those."

After some more attempts that failed, the king had an idea. "In complicated situations such as this it helps to make a graph of the equation. Let's make a graph of the function

$$y = f(x) = x^2 - 7x + 10$$

and then find the values of x where the value of the function $f(x)$ is zero."

In order to make a graph of the function we first had to make a table showing the values of the function for several different values of x:

x	$f(x) = x^2 - 7x + 10$
−10	180
−8	130
−4	54
−1	18
0	10
1	4
3	−2
4	−2
6	4
8	18
10	40

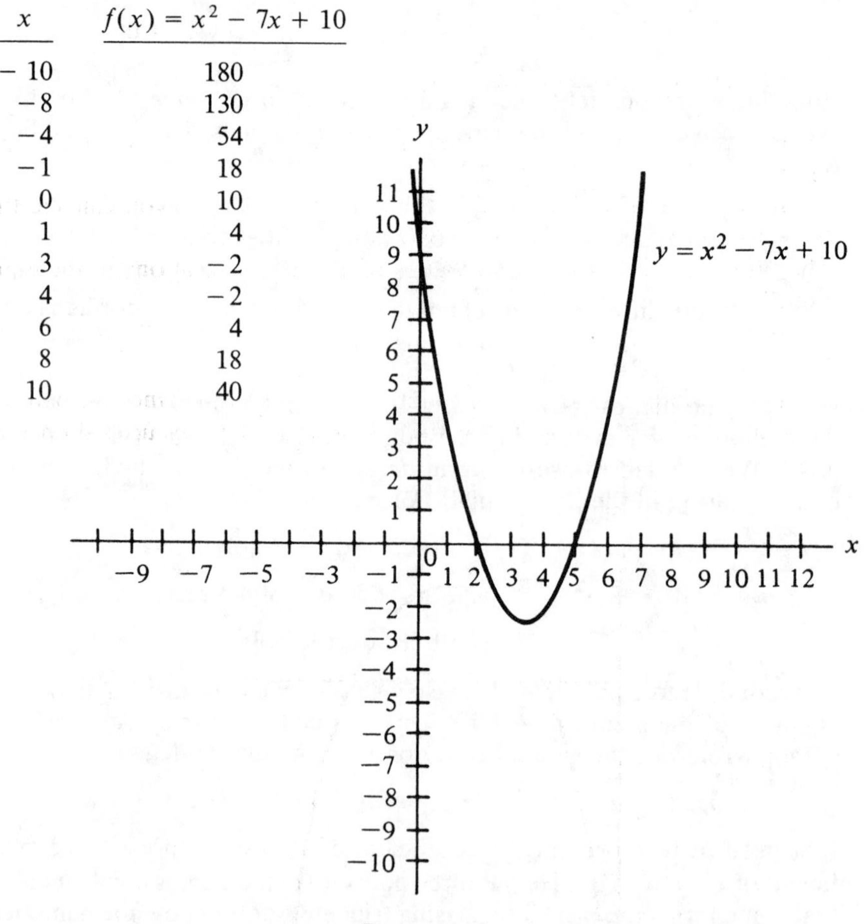

Figure 11–1

We made a graph of some of these points. (See Figure 11–1.)

We stared in awe as the picture slowly unfurled. "I've never seen anything like it!" the professor said. "The graph is a curve!" As we filled in the rest of the points it was clear that the graph of this equation was a smooth, graceful curve.

"I never knew that you could draw curves with algebraic equations!" Recordis said in wonder. "So far the only equations we have graphed have been represented by straight lines."

"Pretty! Pretty!" Pal said.

"Now the solution will occur at the point where the curve crosses the x axis—in other words, where the value of y is zero," Recordis said, looking closely at the diagram. "According to my eyes, that point occurs at approximately $x = 2$."

Recordis performed the calculation to see if $x = 2$ was indeed a solution to the equation $x^2 - 7x + 10 = 0$:

$$2^2 - 7 \cdot 2 + 10 = 4 - 14 + 10 = 0$$

"It does work!" Recordis said jubilantly. "$f(2)$ is zero, so $x = 2$ must be the solution! These quadratic equations are solvable after all!"

"Wait a minute!" the professor said. "The way I see it, the curve crosses the x axis at the point where $x = 5$, so I think the solution is $x = 5$."

"Ha!" Recordis said. "Let's just calculate $f(5)$ and see."

$$f(5) = 5^2 - 7 \cdot 5 + 10 = 25 - 35 + 10 = 0$$

Recordis was dumbstruck.

"See! The solution is $x = 5$," the professor said.

"But we already found that the solution was $x = 2$!" Recordis protested.

"If you look at the diagram," the king interjected, "you can see that the curve crosses the x axis at two points—at both $x = 2$ and $x = 5$. Therefore, there must be two values of x that are solutions to the equation."

"But which one is the right one?" Recordis demanded. "I think a problem should have only one right answer."

"There clearly must be two solutions," the king said, "since the ball will be 10 units high at two different times—once on the way up and once on the way down." (See Figure 11–2.)

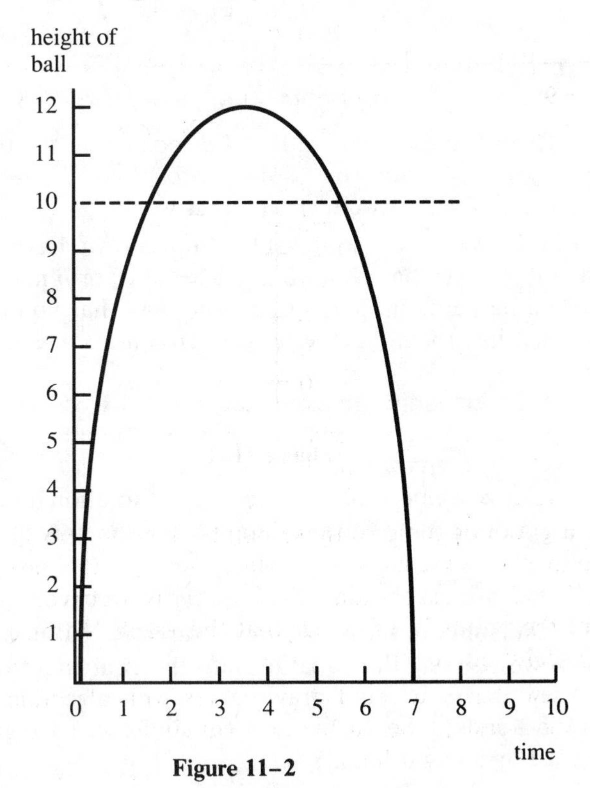

Figure 11–2

"We must find a better way to locate the solutions for quadratic equations," the professor said. "It is too much work to have to draw the curve every time."

However, we could not come up with anything. After we had lunch Igor tried to cheer us up by asking us to solve a curious problem he had come across:

$$(x + 5)(x - 6) = 0$$

"I know what to do in this situation," Recordis said. "We can use the FOIL method to multiply out the left-hand side." We rewrote the equation as

$$x^2 - x - 30 = 0$$

"Oh, no!" Recordis cried. "We ended up with a quadratic equation, which is the worst possible thing that could have happened to us at this moment, since we don't know how to solve those."

The professor hinted that maybe multiplying out the two factors was not an improvement, so we went back to trying to solve the equation written in its original form:

$$(x + 5)(x - 6) = 0$$

"If only we had two linear equations!" Recordis moaned. "If only we had these two equations:

equation: $x + 5 = 0$ solution: $x = -5$

equation: $x - 6 = 0$ solution: $x = 6$

Suddenly the king had a shrewd idea. "Let's try the two solutions $x = -5$ and $x = 6$," he said. Sure enough, we found that both of these suggested solutions worked:

$$(-5 + 5)(5 - 6) = 0$$

and

$$(6 + 5)(6 - 6) = 0$$

"Amazing!" the professor said. "It looks as if the equation $(x + 5)(x - 6) = 0$ can be split into two linear equations $x + 5 = 0$ and $x - 6 = 0$. I wonder why that works?"

"I think I know why," the king said. "Suppose we have the equation $ab = 0$. Then it is clear that either a must be zero, or b must be zero, or they both could be zero. In fact, there is no way that you can take two numbers and multiply them together to get zero unless one of the numbers is itself zero."

It took a little persuading to convince Recordis of that fact, but he eventually gave up trying to find two nonzero numbers that could be multiplied together to give zero.

"But now we have a clue about how to find the solution for the quadratic equation!" the professor realized. "Suppose we were given the equation

$$x^2 - x - 30 = 0$$

We can't find the solution to that equation directly, but it is easy to find the solution if we convert the equation into the equivalent equation

$$(x + 5)(x - 6) = 0$$

So, all we need to do is convert the second-degree polynomial into the product of two simple factors. For example, suppose the equation we're stuck with is

$$x^2 - 7x + 10 = 0$$

If we can factor that polynomial into two factors, then we can write the equation like this:

$$(x + m)(x + n) = 0$$

If we can find the correct values for m and n, then we can easily find the solution."

"How does that help?" Recordis asked. "We know that it is easy, using the FOIL method, to convert a product of two first-degree factors into a second-degree polynomial. However, we don't have the faintest clue about how to do that process in reverse—in other words, if we start with a second-degree polynomial, we don't know how to convert it into a product of two factors."

The king decided that we should assume that

$$x^2 - 7x + 10 = (x + m)(x + n)$$

and then see if we could find the correct values for m and n. Recordis suggested that we multiply out the right hand side using the FOIL method, since it was the only thing he could think of to do. (He was actually becoming fond of the FOIL method by now.) (See Chapter 7.)

$$x^2 - 7x + 10 = x^2 + (m + n)x + mn$$

"I see!" the professor said. "We need to choose m and n such that they add to give -7 and multiply to give 10. In other words, we want

$$m + n = -7$$

$$mn = 10$$

The professor tried to solve that two-equation system, but Recordis thought that guessing would be quicker. Since he correctly guessed $m = -2$ and $n = -5$, he turned out to be right.

Therefore, we could factor the polynomial $x^2 - 7x + 10$ into two factors, $(x - 2)$ and $(x - 5)$, so we could rewrite the equation

$$x^2 - 7x + 10 = 0$$

as

$$(x - 2)(x - 5) = 0$$

Now we could write the equation as two equations:

$$x - 2 = 0 \qquad \text{solution:} \quad x = 2$$

$$x - 5 = 0 \qquad \text{solution:} \quad x = 5$$

These two solutions were the same solutions that we had found on the graph.

We tried another example of a quadratic equation to see if we could solve it by factoring:

$$x^2 + 10x + 16 = 0$$

To factor the polynomial $x^2 + 10x + 16$, we needed to think of two numbers that added to give 10 and multiplied to give 16. Recordis suggested the first two numbers multiplying to give 16 that popped into his head: 4 and 4. However, these two didn't add up to 10, so we dropped that idea. The king suggested using 8 and 2, and we found that this expression worked:

$$x^2 + 10x + 16 = (x + 8)(x + 2)$$

Therefore, the solutions to the original equation were $x = -2$ and $x = -8$.

The next example we tried was

$$x^2 - 12x + 11 = 0$$

We needed two numbers that added to give -12 and multiplied to give 11, and we correctly guessed that -1 and -11 would work:

$$x^2 - 12x + 11 = (x - 1)(x - 11)$$

so the two solutions are $x = 1$ and $x = 11$.

The next example was

$$x^2 + 3x - 18 = 0$$

This one was harder. We needed two numbers that added to give 3 and multiplied to give -18. After some trial and error, we found that 6 and -3 worked, so

$$x^2 + 3x - 18 = (x + 6)(x - 3)$$

and the two solutions were $x = -6$ and $x = 3$.

"As I see it, factoring polynomials is just a guessing game," Recordis said.

We wrote a general rule for solving quadratic equations by the method of factoring:

SOLVING QUADRATIC EQUATIONS BY FACTORING

Suppose you are given an equation of the form

$$x^2 + bx + c = 0$$

(x is unknown; b and c are known.) To find the solution, think of two numbers m and n that when multiplied together give c and when added together give b. (In symbols: $mn = c$ and $m + n = b$.) Then the polynomial can be factored into the two factors $(x + m)$ and $(x + n)$, and the equation can be written

$$(x + m)(x + n) = 0$$

The two solutions are $x = -m$ and $x = -n$.

We needed a lot of practice before we became good at the factoring method, so the exercises list some of the examples that we used for practice.

"We should also add that every quadratic equation has two solutions," Recordis said. By now he was beginning to accept this strange fact, although he always said that he really preferred the old days when problems had only one right answer. (The two solutions are also called the two *roots* of the equation.)

"We don't know if every single possible quadratic equation has two solutions," the professor said.

"How could it not?" the king asked.

The professor suggested an example:

$$x^2 - 4x + 4 = 0$$

We solved this equation by factoring. We needed two numbers that multiplied to give 4 and added to give -4. We tried -2 and -2, and found

$$x^2 - 4x + 4 = (x - 2)(x - 2)$$

"So $x = 2$ is the only possible solution in this case." We decided that, although the polynomial could always be factored into two factors, in some exceptional circumstances the two factors would turn out to be identical, in which case there really was only one solution of the equation. In this situation the curve representing the equation just touches the x axis at one point. (See Figure 11–3.)

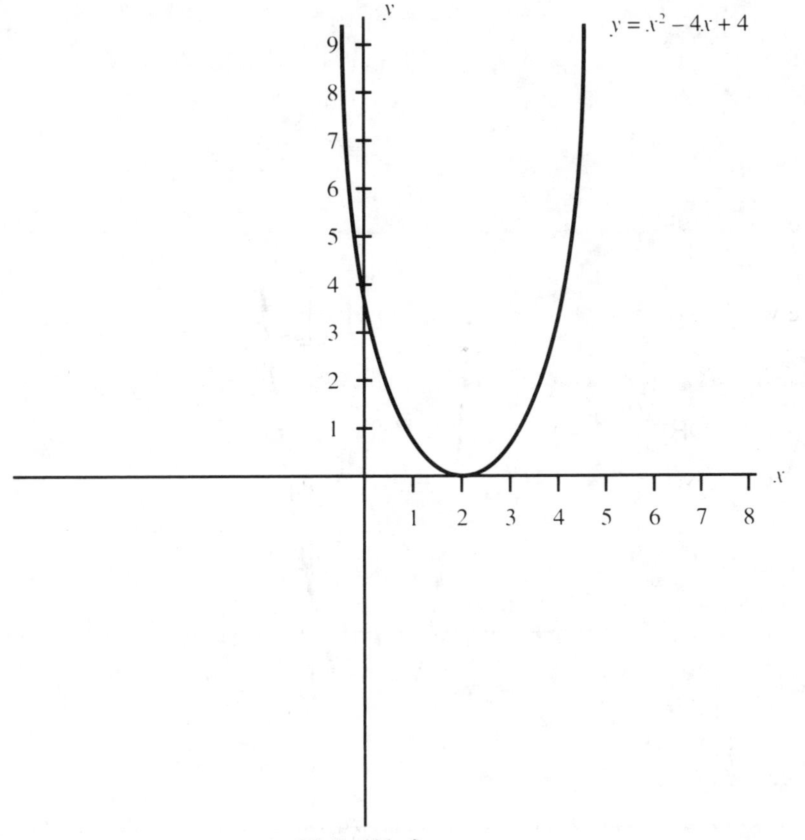

Figure 11–3

The next thing we knew a large arrow came sailing through the window of the Main Conference Room and lodged itself against the wall. Attached to the arrow was a note. With trembling hands the professor read:

"Ha! You will never be able to solve every quadratic equation this way! Your trial-and-error factoring method will soon become helpless when you meet the equations that I shall give you. Here is an example which you can try to solve (if you dare):

$$x^2 + 4x + 1 = 0$$

"I bet I can guess who that note is from," Recordis said in terror.

"I bet you can guess who this note is from," the professor continued to read the note. "Yours truly, the gremlin."

"Now is not the time to give up hope!" the king said valiantly. However, the equation $x^2 + 4x + 1$ turned out to be very difficult to solve. To use the factoring method we needed to think of two numbers that multiplied to give 1 and added to give 4, but there were no such numbers that we could think of.

In desperation we made a table of values and turned to the graphing method. (See Figure 11–4.)

x	$f(x) = x^2 + 4x + 1$
−8	33
−7	22
−6	13
−5	6
−4	1
−3	−2
−2	−3
−1	−2
0	1
2	13
3	22
4	33
5	46
6	61
7	78
8	97

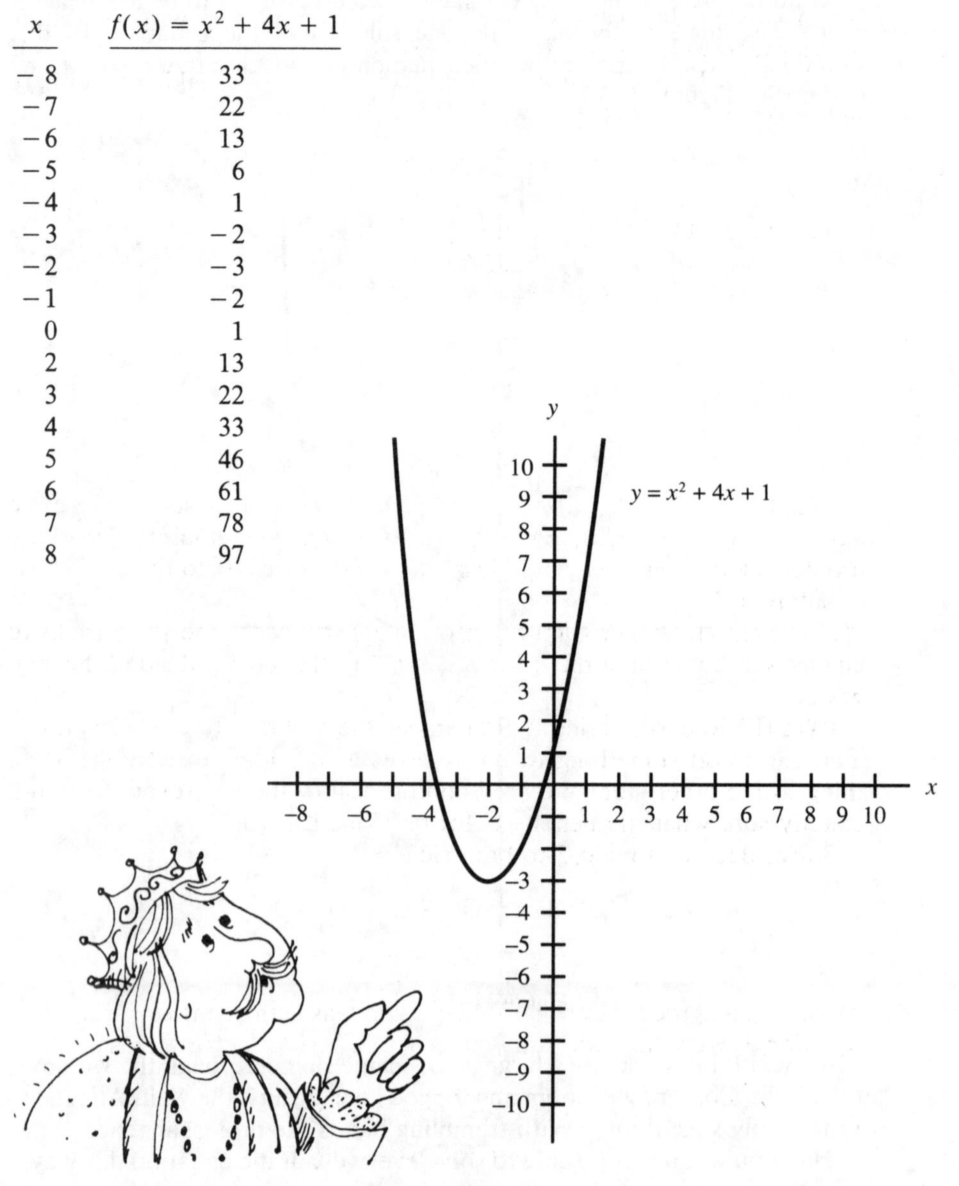

Figure 11–4

The graph made it clear that one of the solutions was between -3 and -4 and the other solution was between 0 and -1, but we could not tell from the graph what the exact values of the solutions were. "Just what I was afraid of!" Recordis moaned. "The solutions are weird numbers—not easy numbers like integers. With our luck, they will probably turn out to be irrational numbers."

The king stared thoughtfully at the equation:

$$x^2 + 4x + 1 = 0$$

"If only we could cancel out the effect of that x^2," he said. "Maybe it would help to take the square root of both sides of the equation."

$$\sqrt{x^2 + 4x + 1} = 0$$

But since we didn't know how to simplify the expression with the square root in it, this idea didn't seem to help much.

However, the professor had an idea for a case where taking the square root would help: "Suppose we had this equation," she said.

$$x^2 + 4x + 4 = 0$$

"Then we could take the square root."

$$\sqrt{x^2 + 4x + 4} = 0$$

$$\sqrt{(x + 2)^2} = 0$$

$$x + 2 = 0$$

$$x = -2$$

"That doesn't help us with our equation," Recordis said, "since we have $x^2 + 4x + 1 = 0$, not $x^2 + 4x + 4 = 0$. We wouldn't have any problems left at all if we could just adjust each problem to fit our method of solution."

However, the king thought that we might be able to use some tricks to change our equation a bit. "Let's add a 3 to the left hand side," he suggested.

"Wait!" Recordis cried. "Remember the Golden Rule of Equations! You can't add something to one side without adding exactly the same thing to the other side! We are bound to follow the law (even if I'm not exactly sure what the penalty is for breaking it.)"

So we decided to add 3 to both sides:

$$x^2 + 4x + 1 \mathbf{+ 3} = 0 \mathbf{+ 3}$$

$$x^2 + 4x + 4 = 3$$

We were all agreed that this new equation was equivalent to our original equation. "Now take the square root of both sides," the king said.

$$\sqrt{x^2 + 4x + 4} = \pm\sqrt{3}$$

$$\sqrt{(x + 2)^2} = \pm\sqrt{3}$$

$$x + 2 = \pm\sqrt{3}$$

"Now we can solve the equation!" Recordis said, dumfounded. The two solutions we found were

$$x = -2 + \sqrt{3}$$

$$x = -2 - \sqrt{3}$$

These two solutions can be written $x = -2 \pm \sqrt{3}$.

We checked to see that these two solutions both worked in the original equation. (See Exercise 53) We were able to use the square root function to come up with decimal approximations for the solutions:

$$x = -0.268 \quad \text{or} \quad x = -3.732$$

"I'm so happy that we outwitted the gremlin by finding a solution that I won't even complain too much about the fact that the solution is an irrational number." Recordis breathed a sigh of relief. (This method for solving a quadratic equation is called the method of *completing the square*.)

"Still, that method was a lot of work," the professor said. "We had better develop a more general method or else the gremlin might be able to overwhelm us by giving us too many problems to do at once."

We wrote out the most general form for a quadratic equation that we could imagine:

$$ax^2 + bx + c = 0$$

First, to make things more familiar, we divided both sides of the equation by a:

$$x^2 + \frac{bx}{a} + \frac{c}{a} = 0$$

We subtracted c/a from both sides to ease some of the clutter on the left hand side of the equation:

$$x^2 + \frac{bx}{a} = -\frac{c}{a}$$

"If only the left hand side was a perfect square," Recordis moaned. Wistfully, he imagined that the left hand side was written $(x + h)^2$, so he could take the square root: $\sqrt{(x + h)^2} = x + h$.

"It would be too much to hope for the answer to be that obvious," the professor said. Suddenly she had an idea: "Perhaps the left hand side might be

$$x^2 + 2hx + h^2$$

"You're out of luck," Recordis said. "We're stuck with this left hand side:

$$x^2 + \frac{b}{a}x$$

"They would almost be the same if $2h = b/a$," the professor noticed.

"Almost isn't good enough," Recordis replied. "We still don't have anything the equivalent of h^2."

"We could add $h^2 = (b/2a)^2 = b^2/4a^2$ to the left hand side."

"Remember the Golden Rule of Equations!" Recordis shouted. "You can't add whatever you like to one side of the equation—unless, of course, you intend to add the exact same thing to the other side."

"My plan exactly," the professor said.

original equation:

$$x^2 + \frac{bx}{a} = -\frac{c}{a}$$

new equation after adding $b^2/4a^2$ to both sides:

$$x^2 + \frac{bx}{a} + \frac{b^2}{4a^2} = \frac{-c}{a} + \frac{b^2}{4a^2}$$

We simplified the right-hand side by finding the common denominator and adding the two fractions:

$$x^2 + \frac{bx}{a} + \frac{b^2}{4a^2} = \frac{b^2 - 4ac}{4a^2}$$

We rewrote the left-hand side:

$$\left(x + \frac{b}{2a}\right)^2 = \frac{b^2 - 4ac}{4a^2}$$

Next we took the square root of both sides:

$$x + \frac{b}{2a} = \pm\sqrt{\frac{b^2 - 4ac}{4a^2}}$$

$$x = \frac{-b \pm \sqrt{b^2 - 4ac}}{2a}$$

"With this formula we can handle any quadratic equation easily!" Recordis said. We decided to call this formula the *quadratic formula*. Once we are given the values of a, b, and c, it becomes a simple matter of inserting the numbers in the formula and evaluating the resulting arithmetic expression.

SOLVING QUADRATIC EQUATIONS WITH THE QUADRATIC FORMULA

The general form of a quadratic equation can be written:

$$ax^2 + bx + c = 0 \qquad \text{(assuming } a \text{ does not equal 0)}$$

In order to solve for x, first identify the values of a, b, and c that apply to your particular situation. Then insert these values into this formula:

$$x = \frac{-b + \sqrt{b^2 - 4ac}}{2a} \qquad \text{or} \qquad x = \frac{-b - \sqrt{b^2 - 4ac}}{2a}$$

The two values of x that result from this formula are the two solutions to the equation. However, if $b^2 - 4ac = 0$, then there is only one solution; if $b^2 - 4ac$ is negative, then a new set of complications arise. (See Chapter 20.)

Exercises

Think of two numbers that:

1. add to give 2 and multiply to give 1.
2. add to give 0 and multiply to give -1.
3. add to give 5 and multiply to give 6.
4. add to give 9 and multiply to give 20.
5. add to give 9 and multiply to give 8.

Write these equations in the form $ax^2 + bx + c = 0$:

6. $3x^2 + 2x = -4$
7. $10x^2 - x = 6x^2 - 8x + 5$
8. $x(x - 4) = x(2x + 5) + 10$
9. $(x - 2)(x - 2) = -2x^2 + 10$
10. $9x^2 - 6x + 10 = 3x^2 - 4x + 4$

Solve these quadratic equations by factoring:

11. $x^2 - 9 = 0$
12. $x^2 = 16$
13. $x^2 - 4x = 0$
14. $10x^2 - 8x = 0$
15. $4x^2 + 6x + 5 = 3x + 5$
16. $x^2 + 2x + 1 = 0$
17. $x^2 - 11x + 28 = 0$
18. $x^2 - 21x + 110 = 0$
19. $x^2 - 20x + 19 = 0$
20. $x^2 + x - 6 = 0$
21. $2x^2 + 3x + 1 = 0$
22. $x^2 - x - 42 = 0$
23. $x^2 - x + 1/4 = 0$
24. $x^2 - \frac{3}{2}x + \frac{1}{2} = 0$
25. $y^2 + 3\frac{1}{3}y + 1 = 0$

Simplify these fractions:

26. $\dfrac{x^2 - 11x + 28}{x - 7}$
27. $\dfrac{x^2 + 2x - 15}{x + 5}$
28. $\dfrac{x^2 + 4x + 4}{x + 2}$
29. $\dfrac{x^2 + 8x + 12}{x + 6}$

Are these expressions polynomials?

30. $2x^2 - 6x - 4$
31. $x^3 + x^2 + x + 1 + x^{-1} + x^{-2} + x^{-3}$
32. $(x^2 + 3x + 4)/(x - 5)$
33. $\frac{1}{2}gt^2 + 40$
34. $(1/y)(y^3 + 3y^2 + 3y + y)$

Draw graphs of these equations:

35. $y = x^2$
36. $y = -x^2$
37. $y = x^2 + x + 1$
38. $y = -x^2 + x + 1$

Solve these equations with the quadratic formula to verify that the solutions in the chapter are correct:

39. $x^2 - x - 30 = 0$

40. $x^2 - 7x + 10 = 0$
41. $x^2 + 10x + 16 = 0$
42. $x^2 - 12x + 11 = 0$
43. $x^2 + 3x - 18 = 0$

Solve with the quadratic formula:

44. $2x^2 - 4x - 5 = 0$
45. $x^2 = x + 1$
46. $x^2 = 1 - x$
47. $3x^2 = x + 1$
48. $10x^2 + 10x + 1 = 0$
49. $10x^2 + 10x = 0$
50. $12x^2 + 12x + 3 = 0$
51. $x^2 + 49 = 14x$
52. $\frac{1}{2}x^2 + x + \frac{1}{2} = 0$
53. Verify that $-2 + \sqrt{3}$ and $-2 - \sqrt{3}$ are solutions to the equation $x^2 + 4x + 1 = 0$.
54. What is the quickest way to tell whether the solutions to a quadratic equation will be rational or irrational?
55. Can a quadratic equation have one rational solution and one irrational soiution?

Solve these equations for x:

56. $1 + 5/x + 6/x^2 = 0$ (Hint: multiply both sides byx^2.)
57. $2/x + 4/x^2 = 0$
58. $x - 1/x = 0$
59. $\sqrt{5x - 6} = x$
60. $\sqrt{20 - x} = x$
61. $\sqrt{x^2 + 2x + 1} = 2x - 5$
62. Solve for y: $y^4 - 13y^2 + 36 = 0$
63. Solve for y: $y^6 - 9y^3 + 8 = 0$

Create your own quadratic equations that have the solutions indicated:

64. $x = 3$ or 5
65. $x = 12$ or 11
66. $x = -5$ or 4
67. $x = -6$ or -7
68. $x = 2 + \sqrt{5}$ or $2 - \sqrt{5}$
69. If a ball is thrown straight up, then its height at time t will be given by the formula

$$h = -\tfrac{1}{2}gt^2 + v_0 t + h_0$$

The quantity g is a constant known as the *acceleration of gravity*; v_0 is the initial speed of the ball; and h_0 is the initial height of the ball. Write a formula that tells at what time the ball will reach a specified height h.
70. Evaluate the formula in the preceding problem using these values: $g = 9.8$, $v_0 = 20$, $h_0 = 2$, and $h = 10$.
71. What does the quadratic formula say is the solution to the equation $ax^2 + bx + c = 0$ if $b = 0$? What if $c = 0$? What if $a = 0$?
72. What are the dimensions of a rectangle if its perimeter is 40 and its area is 96?
73. Try to solve the equation

$$x^2 + 2x + 4 = 0$$

with the quadratic formula. What happens?

74. In the song "Twelve Days of Christmas," you receive one partridge in a pear tree every day for 12 days (total: 12), two French hens every day for 11 days (total: 22), and so on. For which gift do you receive a total of 42 items?

❑ **75.** Use a computer spreadsheet program to illustrate the solutions to a quadratic equation. Design the graph so you can choose two values of x, and have the computer illustrate the curve between those two values. You can choose values that are far apart to illustrate the big picture; you can choose two values that are close together to zoom in on one of the solutions.

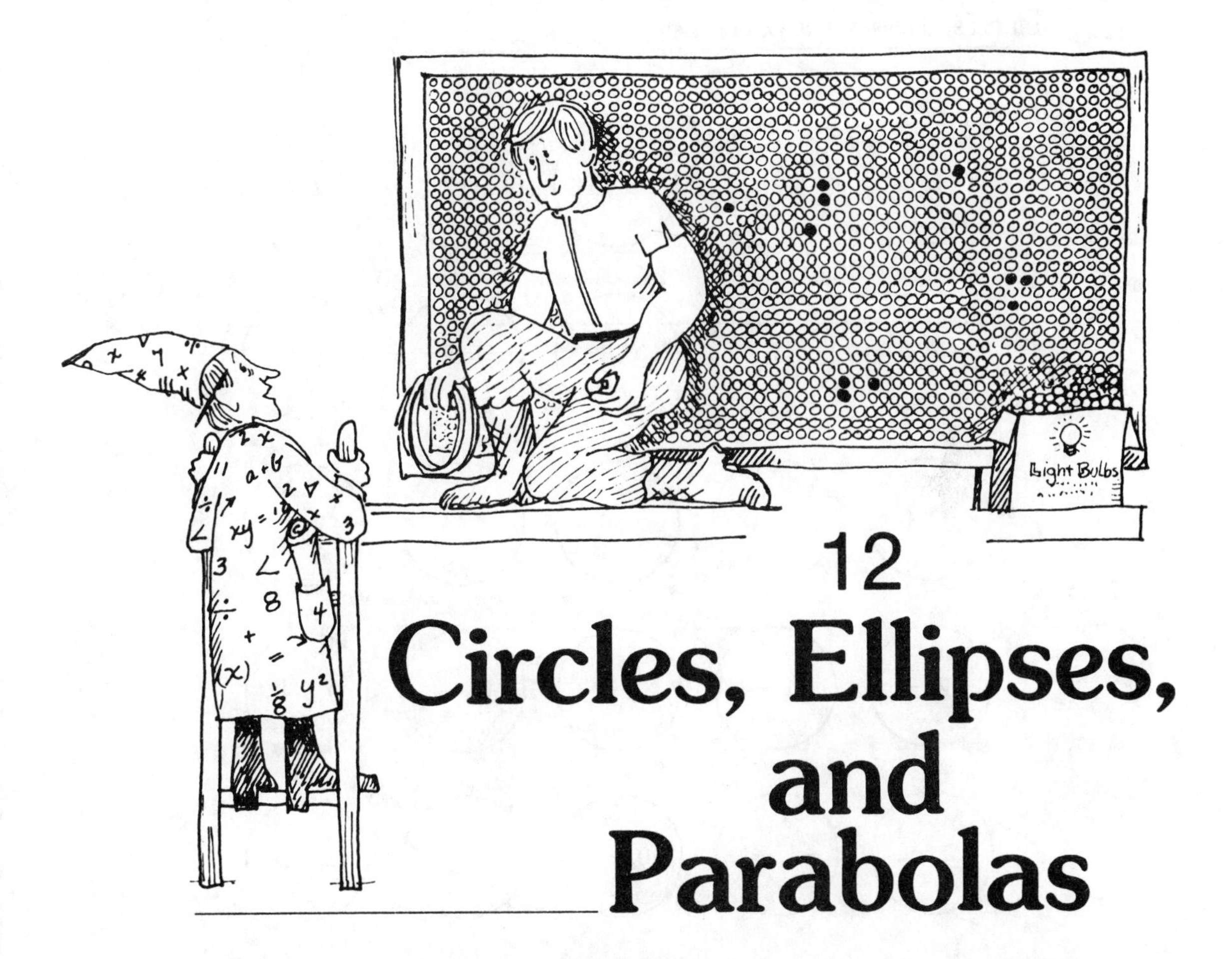

12 Circles, Ellipses, and Parabolas

Builder was working feverishly to finish the new stadium scoreboard so that it would be ready for opening day. "The scoreboard is made of a large rectangular grid of little light bulbs," he explained. "In order to make a particular number appear, we need to decide which light bulbs need to be lit to display that number. At the beginning of the game, both teams will have zero runs, so I am trying to figure out which lights need to be lit to form a circle to represent zero." (See Figure 12–1.)

"How do you tell which bulb is which?" Recordis asked.

"I labelled each bulb with two numbers," Builder explained. "The bulb in the center is (0, 0), and then every other bulb is labelled by a number

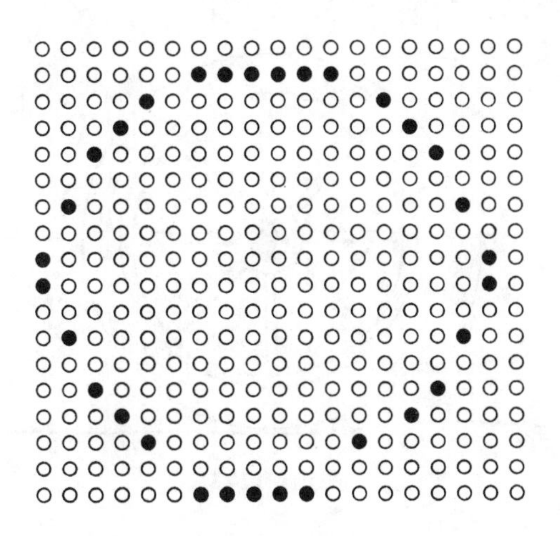

Figure 12–1

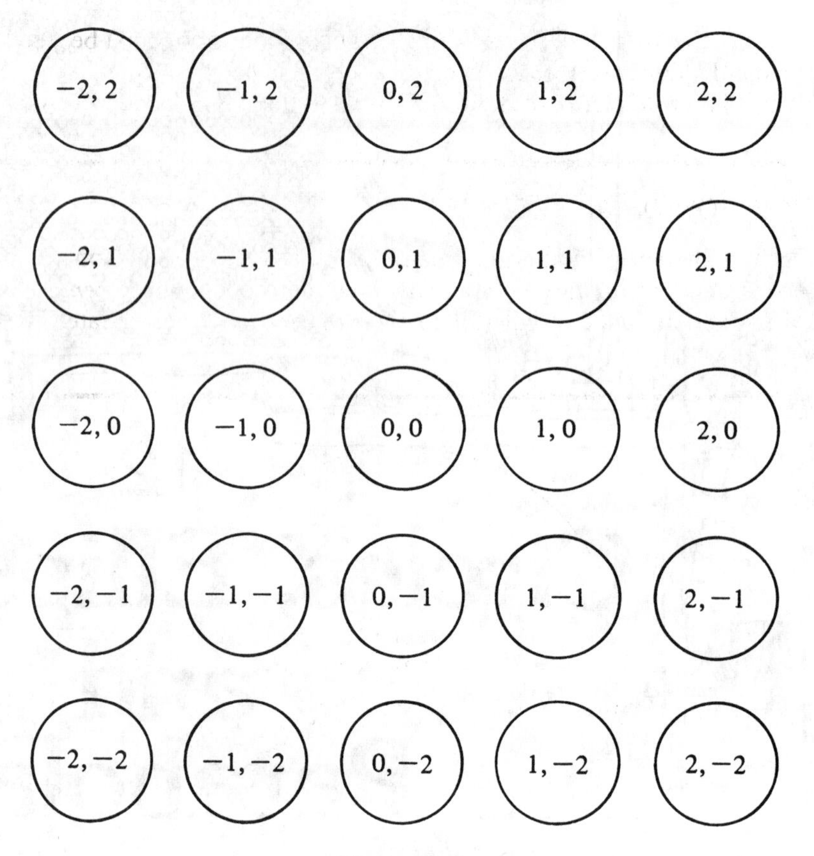

Figure 12–2

telling how far it is to the right of the middle bulb, and how far it is above it." (See Figure 12–2.)

"This is just like a system of rectangular coordinates," the professor said. "All we need to do is figure out the equation of a circle, and then the points that satisfy the equation will be the points on the circle."

"We must define exactly what a circle is," the king said.

"Everyone knows what a circle is!" Recordis said. "A circle is round."

However, we quickly realized that we needed a more precise definition. We watched a turning wheel, and noticed that the wheel gave a smooth ride if the center of the wheel was always at the same height above the ground. (See Figure 12–3.)

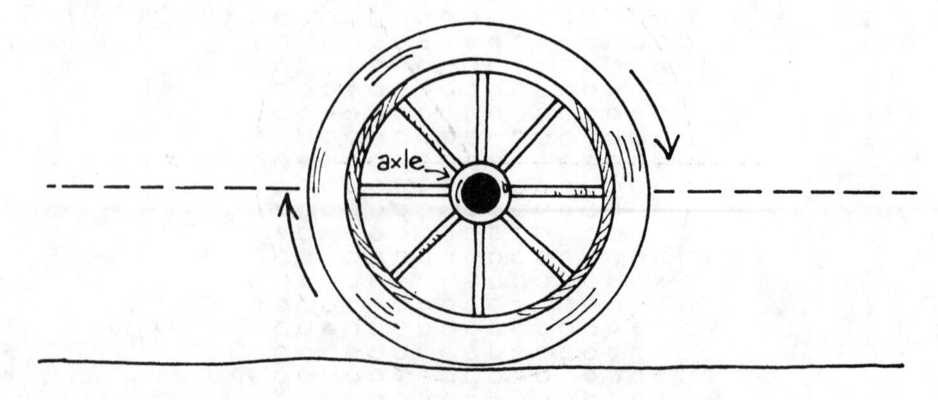

Figure 12–3

"I see!" the king said. "Every single point on the circle must be exactly the same distance from the center!"

Recordis wrote a formal definition in his book:

DEFINITION OF A CIRCLE

A circle is the set of points (in a plane) such that every point on the circle is the same distance from a fixed point. The fixed point is called the *center* of the circle, and the fixed distance is called the *radius* of the circle (abbreviated r.) See Figure 12–4.

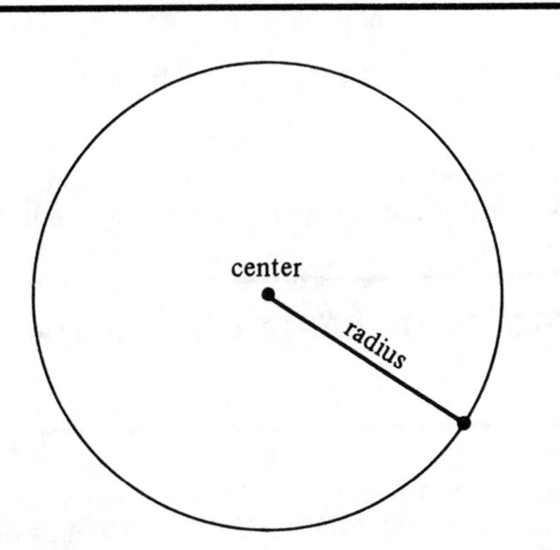

Figure 12–4

We decided to look for the equation of a circle with the center at the origin (0, 0). "Suppose that the point (x, y) is one of the points on our circle," the professor said. (See Figure 12–5.) "Then we need to think of an equation in two variables x and y such that every point that is on the circle will make the equation true, and every point that is not on the circle will make the equation false."

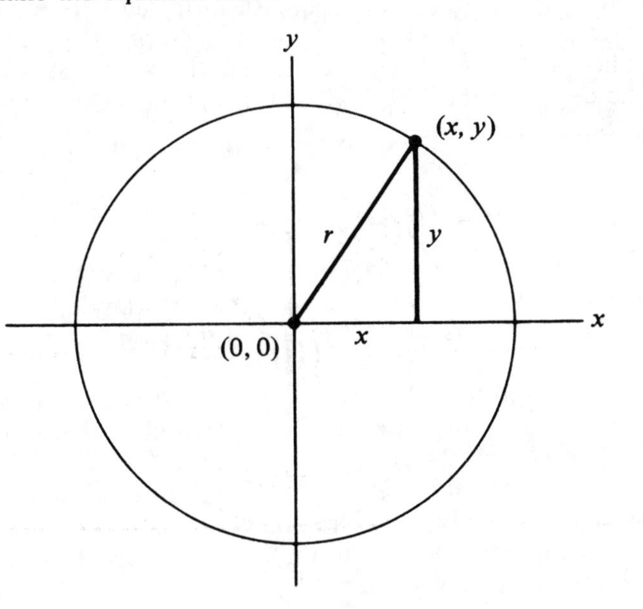

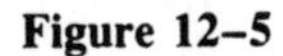

Figure 12–5

"If only we knew the distance from the point (x, y) to the point $(0, 0)$," Recordis sighed. "Then we could set up an equation to make sure that the distance was always equal to r."

"We do know the distance from (x, y) to $(0, 0)$!" the king suddenly remembered. "We found a formula that tells us the distance between any two points in a plane!" (See Chapter 9.) Recordis wrote it down: The distance between the two points (x_1, y_1) and (x_2, y_2) is

$$\sqrt{(x_1 - x_2)^2 + (y_1 - y_2)^2}$$

We used the distance formula to show that the distance from (x, y) to $(0, 0)$ was

$$\sqrt{(x - 0)^2 + (y - 0)^2} = \sqrt{x^2 + y^2}$$

Therefore, the equation of the circle could be written

$$r = \sqrt{x^2 + y^2}$$

Since Recordis still didn't like square-root signs very much, we decided to square both sides of the equation and write it like this:

EQUATION OF CIRCLE WITH CENTER AT ORIGIN AND RADIUS r

$$x^2 + y^2 = r^2$$

"We want to make the circle big, so everyone in the stadium will be able to see it," Builder said. So we decided on a radius of $r = 25$. Therefore, the equation of the circle became

$$x^2 + y^2 = 25^2$$

$$x^2 + y^2 = 625$$

"I bet I can guess four points that fit that equation before you can," Recordis said. He guessed the points $(0, 25)$, $(0, -25)$, $(25, 0)$, and $(-25, 0)$. Sure enough, we found that each of these points worked as a solution to the equation $x^2 + y^2 = 625$, so we plotted these points on a graph. (See Figure 12–6.)

The professor protested that we should start being systematic, instead of merely guessing, but Recordis quickly guessed another point: $(15, 20)$. We found

$$15^2 + 20^2 = 225 + 400 = 625 = 25^2$$

so the point $(15, 20)$ was indeed one of the points on the circle. "And if $(15, 20)$ is a point on the circle, then so must $(-15, 20)$, $(15, -20)$, and $(-15, -20)$," Recordis said, "because of the way the circle is symmetrical."

The king guessed that $(20, 15)$, $(-20, 15)$, $(20, -15)$, and $(-20, -15)$ must also be points on the circle. Igor drew all of these points on the diagram. (See Figure 12–7.)

"See! Just by guessing we were able to figure out a lot of points," Recordis said. However, we were unable to guess any more points, so we finally had to pay attention to the professor, who kept begging us to find a more systematic way.

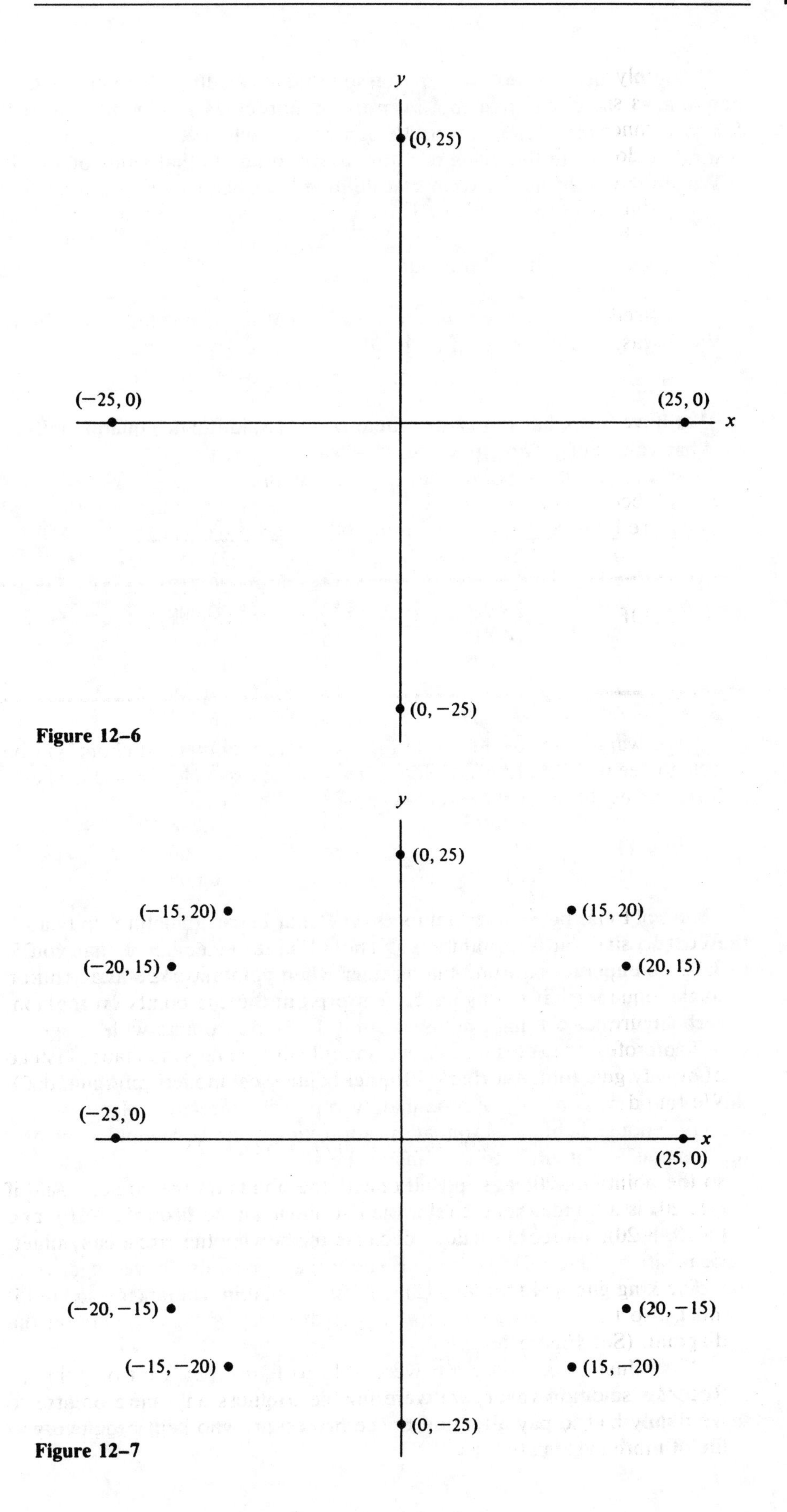

Figure 12–6

Figure 12–7

"What we need to do," the professor said shrewdly, "is rewrite the equation of the circle in such a form that it expresses a value for y as a function of a value for x. Then, we can choose whatever value of x we want, and calculate the value of y that corresponds to that value of x."

We started with the equation that defined the circle:

$$x^2 + y^2 = r^2$$

We subtracted x^2 from both sides:

$$y^2 = r^2 - x^2$$

We took the square root of both sides:

$$y = \sqrt{r^2 - x^2}$$

"We have just what we want!" the professor said. "The equation tells us what values of y correspond to what values of x."

We calculated some points that fit this equation:

x	$y = \sqrt{r^2 - x^2}$	x	$y = \sqrt{r^2 - x^2}$
0	25.000	13	21.354
1	24.980	14	20.712
2	24.920	15	20.000
3	24.819	16	19.209
4	24.678	17	18.330
5	24.495	18	17.349
6	24.269	19	16.248
7	24.000	20	15.000
8	23.685	21	13.565
9	23.324	22	11.874
10	22.913	23	9.798
11	22.450	24	7.000
12	21.932	25	0.000

"Most of these points aren't integers!" Recordis complained. "In fact, they're mostly irrational numbers. I should have remembered that you usually end up with an irrational number when you take the square root of something. So that means we can't represent the circle exactly on the scoreboard, since the light bulbs are only located at points with integer coordinates."

"I could have told you that." Builder said. "The pattern of lights on the scoreboard can only approximately represent the circle. However, we have enough light bulbs so that the approximate circle will look almost the same as a real circle to the fans in the stands."

"We have an even bigger problem with the equation $y = \sqrt{r^2 - x^2}$," the king said. "Since the $\sqrt{\ }$ symbol always means to take the positive value of the square root, this equation erroneously implies that y can only have positive values. However, we know that y can also have negative values, so we will have to put a plus or minus sign in front of the square root sign."

$$y = \pm\sqrt{r^2 - x^2}$$

"But this equation no longer defines y as a function of x!" the professor said in shock, "since it does not specify a unique value of y for every value of x."

"We can draw graphs of equations even if they do not define functions," Recordis said. (This type of equation defines a *relation* between x and y. See Chapter 8.)

Builder quickly completed the scoreboard design and moved on to a new problem. "Now we need to complete the plans for the giant circular race track that will be part of the new stadium complex," he said. "We have decided that the track will be a perfect circle one kilometer across. Now we need to figure out the distance that a runner must run in order to run all the way around the track." Igor drew a map. (See Figure 12–8.)

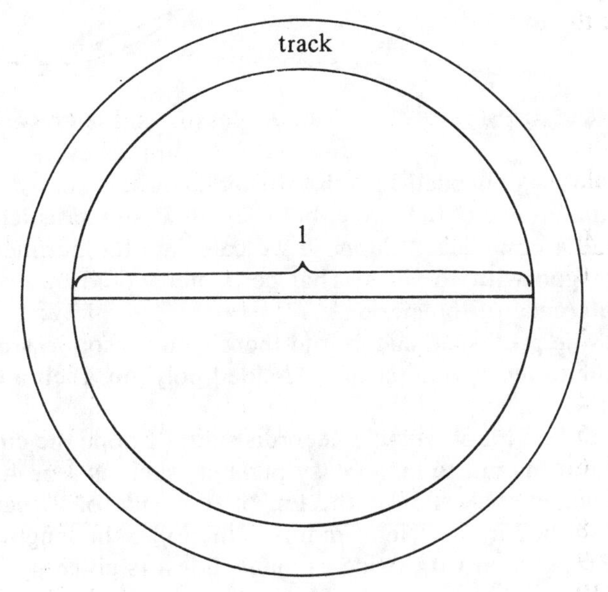

Figure 12–8

(We made up the name *circumference* to stand for the distance around the outside of a circle. We also made up the term *diameter* to stand for the distance across a circle. Note that the diameter is twice as long as the radius.)

"We've never measured distances around curves before," the professor said. "Our distance formula only tells us how to find the straight-line distance between two points."

"If only we had a square-shaped track instead of a circle!" Recordis moaned. (See Figure 12–9.) "Then it would be obvious from the Pythagorean theorem that the length of one side is $1/\sqrt{2}$, which is about 0.7071. That makes the total distance around the square equal to..." he calculated, "... 2.8284."

"That does not help us at all!" the professor said. "The square that you have drawn does not look at all like our circle."

The king looked thoughtfully at the map. "Suppose instead we had an octagon. The octagon looks a lot more like the circle." (See Figure 12–10.) We calculated that the length of one side of the octagon was 0.38268, and then we found that the total distance around the octagon was about 3.0615. (A geometric figure composed of a set of line segments joined end to end is called a *polygon,* and the total distance around the polygon is called the *perimeter*. When a polygon is placed inside a circle such that each corner, or *vertex*, of the polygon just touches the circle, then it is said that the polygon is *inscribed* in the circle.)

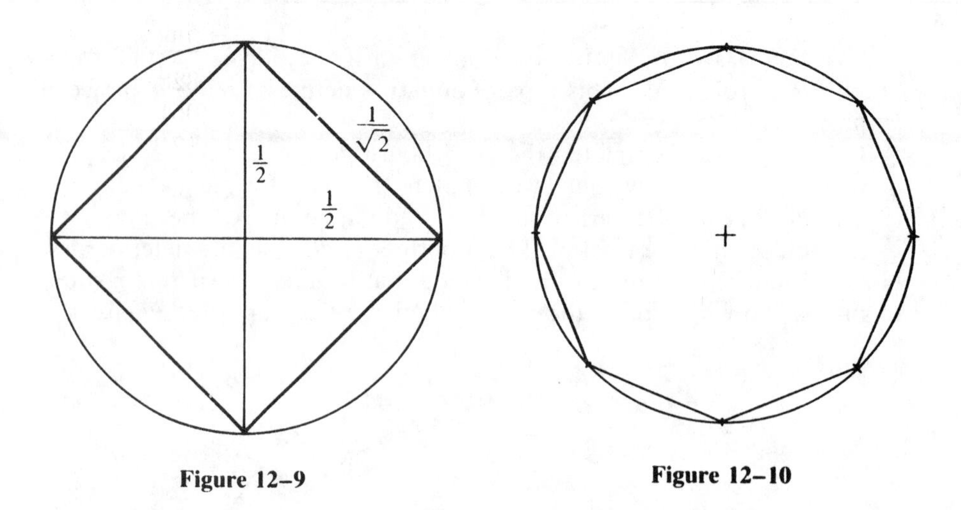

Figure 12–9 **Figure 12–10**

The professor conceded that the perimeter of the octagon was closer to the circumference of the circle, but she still wasn't satisfied. However, the king had a new idea. "Suppose we calculate the perimeter of an inscribed polygon with 16 sides! That perimeter would be even closer to the circumference of the circle."

"I see!" the professor said. "And there's no reason to stop at 16! We could calculate the perimeter of a 32-sided polygon, then a 64-sided polygon, and so on."

"That will be a lot of work!" Recordis said. "I hope the circumference of the circle turns out to be an easy number, such as 3 or 4."

We calculated that, if s_n is the length of a side of a regular n-sided polygon inscribed in a circle of radius r, then s_{2n}, the length of a side of an inscribed polygon with twice as many sides, is given by

$$s_{2n}^2 = 2r^2 - r\sqrt{4r^2 - s_n^2}$$

(See Exercise 30.)

In our case $r = 1/2$. Since we knew $s_4 = 1/\sqrt{2}$, we could find s_8 (the length of a side of the inscribed octagon):

$$s_8^2 = 2\left(\frac{1}{2}\right)^2 - \frac{1}{2}\sqrt{4\left(\frac{1}{2}\right)^2 - \left(\frac{1}{\sqrt{2}}\right)^2}$$

$$= \frac{1}{2} - \frac{1}{2}\sqrt{1 - \frac{1}{2}}$$

$$= \frac{1}{2} - \frac{1}{2}\sqrt{\frac{1}{2}}$$

$$s_8 = \sqrt{\frac{1}{2} - \frac{1}{2}\sqrt{\frac{1}{2}}}$$

We repeated the calculations, doubling the number of sides each time so the perimeter came closer and closer to the circumference:

number of sides	length of side	perimeter
4	0.70710678	2.82843
8	0.38268343	3.06147
16	0.19509032	3.12145
32	0.09801714	3.13655
64	0.04906767	3.14033
128	0.02454123	3.14128
256	0.01227154	3.14151
512	0.00613588	3.14157
1024	0.00306796	3.14159
2048	0.00153398	3.14159

"Just what I was afraid of," Recordis moaned. "The perimeter of the circle turns out to be a number about equal to 3.1416 . . . I suppose it will turn out to be one of these endlessly nonrepeating irrational numbers."

"Still, there must be a simple formula for that number," the professor said. "It must be the square root or the cube root of something." However, we could find no simple formula for the number representing the circumference of the circle.

The king was very distressed that there was a number that seemed to be singled out for such special treatment. He very much wanted to be an impartial ruler, and he hated to show favoritism to any one number.

"We will have to remember this number as being a fundamental irrational number," the professor said solemnly, "even if we can't understand the reason why it is such a special number. We will need a special symbol to represent this number."

"Let's call the number *pie*—since pies are round," Recordis said. "Remembering a pleasant concept as pie will help us to remember a difficult concept such as the circumference of a circle."

"I like your idea of calling the number *pi*," the professor said. "In fact, we can symbolize the number with the Greek letter pi (π):

FUNDAMENTAL NUMBER π

The symbol π, which is the Greek letter pi (pronounced *pie*) represents a fundamental irrational number which is approximately equal to 3.14159. . . . If the diameter of a circle is 1, then its circumference will be π.

"We can never find an exact decimal expression for pi," the professor said, "but if we wanted to we could calculate an approximate value containing thousands of decimal places."

"If you want me to do that you're going to have to pay me about a thousand times what you're paying me now!" Recordis protested.

We found that the circumference of any circle was equal to π times the diameter. And we found that the area of a circle could be found from the formula πr^2. (See Exercise 31.)

Formulas for circles

$\pi = 3.14159\ldots$		
$c = \pi d$	c = circumference,	d = diameter
$c = 2\pi r$	r = radius	$d = 2r$
$A = \pi r^2$	A = area	
$A = \frac{1}{4}\pi d^2$		

Builder needed some more help with the scoreboard design. "I have made the plans for the first circle," he said, "but now I also need to put a 0 on the scoreboard to represent the other team's score at the start of the game."

"And their score at the end of the game, I should add," the professor said.

Builder continued, "The first circle will be centered at the point (0, 0), and its equation will be $x^2 + y^2 = 25^2$. I want the second circle centered at the point (75, 0) That circle will be described by a different equation. I need you to tell me what that equation will be." (See Figure 12–11.)

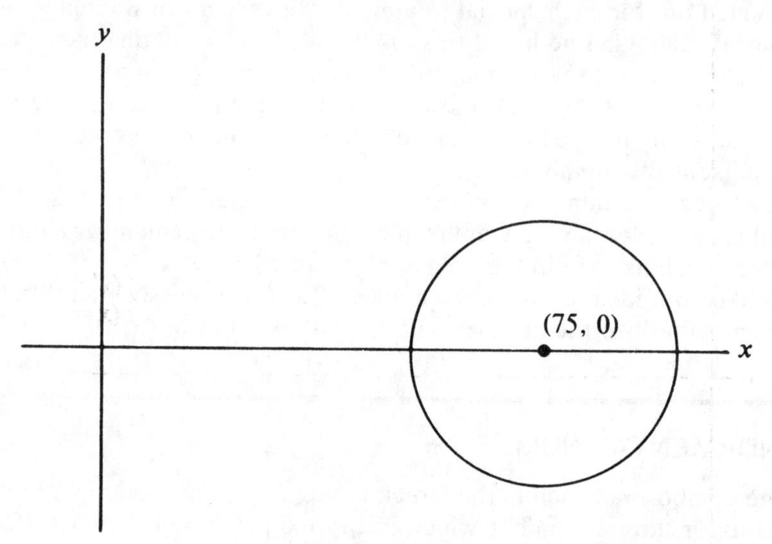

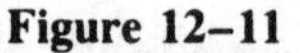
Figure 12–11

"That will be harder!" Recordis complained. "Our equation for a circle ($x^2 + y^2 = r^2$) only works when the center of the circle is at the origin. I don't have the slightest idea about how to write an equation for a circle that is not centered at the origin. If only we could just pick up the coordinate axes and move them so that the origin would be at the new point!" (See Figure 12–12.)

"That's it!" the professor said. "We have to set up a new coordinate system which has its origin at the point measured by (75, 0) in the old system." (See Figure 12–13.)

"What symbols shall we use for the new coordinate system?" the king asked. "We're using x and y to represent the old coordinate system."

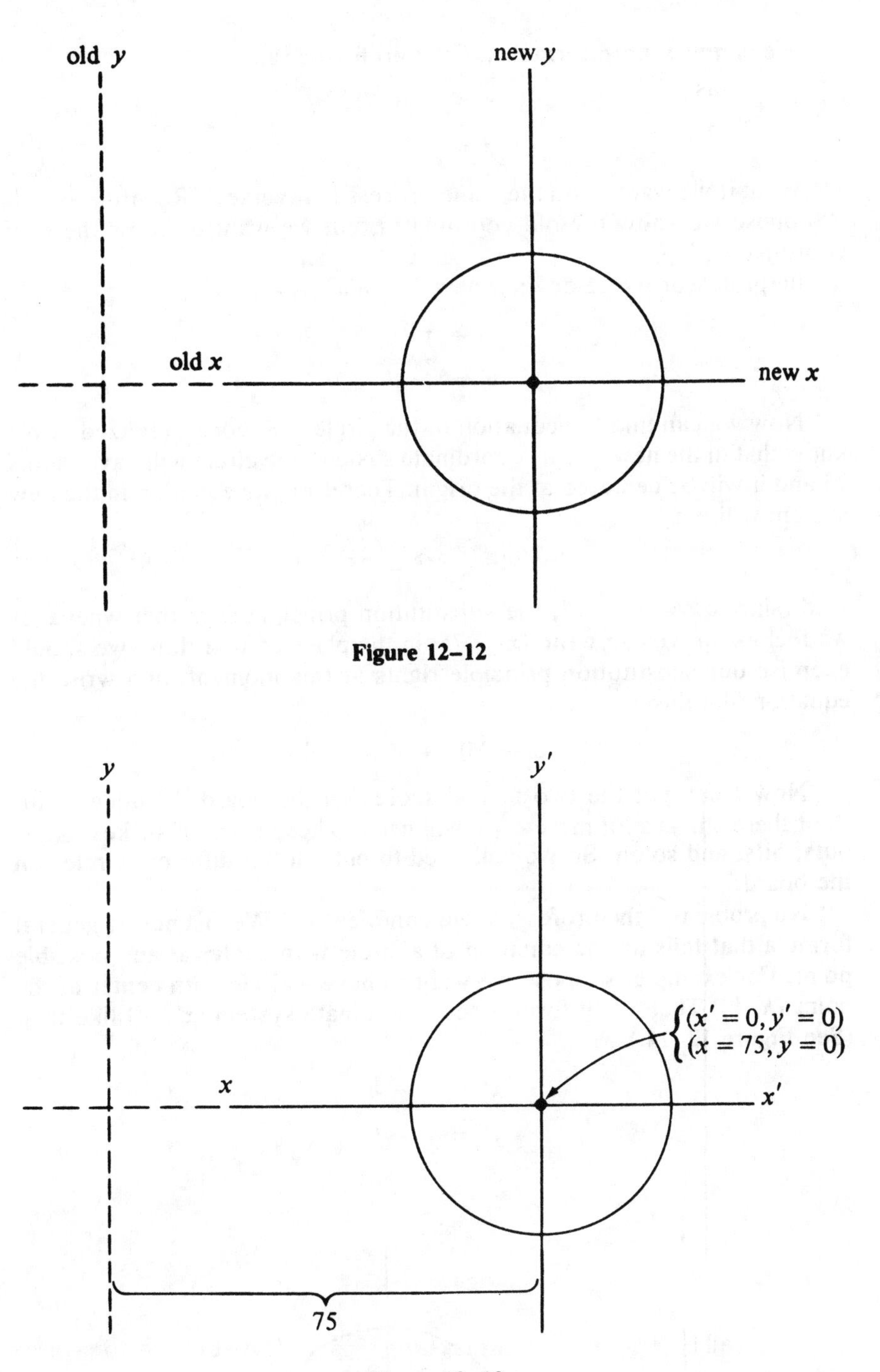

Figure 12–12

Figure 12–13

The professor suggested that we use x and y with little marks next to them, like this: x' and y'. She said that we should read these symbols as "x-prime" and "y-prime."

"Suppose we know the coordinates of a point measured under the new (x', y') system, but instead we really want to know the coordinates of the point in the old (x, y) system," Recordis demanded. "How do we do that?"

"We can use these formulas." the professor said.

$$x = x' + 75$$

$$y = y'$$

"What if we want to do the same process in reverse?" Recordis asked. "Suppose we know the old coordinates, but we want to know the new coordinates."

The professor wrote down a new formula:

$$x' = x - 75$$

$$y' = y$$

"Now we can find the equation of the circle!" Recordis realized. "We know that in the new (x', y') coordinate system, the circle will have radius 25 and it will be centered at the origin. Therefore, its equation in the new system will be:

$$x'^2 + y'^2 = 25^2$$

And, since $x' = x - 75$, the substitution principle says that whenever we feel like it we can write $(x - 75)$ in the place of x'. I think we should exercise our substitution principle rights at this moment, and write the equation like this:

$$(x - 75)^2 + y^2 = 25^2$$

"Now I can put the two teams' scores on the board," Builder said, "but there still is a lot more—we will need to keep track of strikes, balls, outs, hits, and so on. So we will need to put a lot of different circles on the board."

"No problem," the professor said confidently. "We just need a general formula that tells us the equation of a circle with center at any possible point. For example, suppose we want to have a circle with center at the point (h, k). Then we'll form a new coordinate system (x', y') like this. (See Figure 12–14.)

$$x' = x - h$$

$$y' = y - k$$

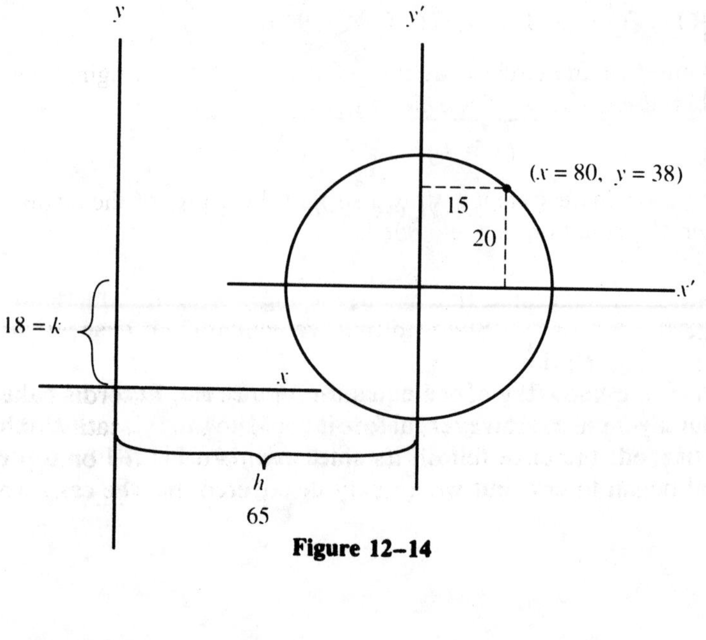

y
y'
(x = 80, y = 38)
15
20
x'
18 = k
x
h
65

Figure 12–14

"To get the new coordinate system, we need to shift the axes h units to the right and k units up. For example, suppose $h = 65$ and $k = 18$. If a point has coordinates $x = 80$, $y = 38$ in the old system, it will have coordinates $x' = 80 - 65 = 15$ and $y' = 38 - 18 = 20$ in the new system."

The technical term for shifting the axes in this way is *translation*. Therefore, these equations tell how to translate the coordinates from one system into the other.

TRANSLATION OF AXES

Suppose you are identifying the location of points in a plane by the two coordinates (x, y). In some cases it is convenient to translate the coordinates into a new system with the origin at a different location. Let (x', y') represent the coordinates of a point in the new system. Suppose that the origin of the new system $(x' = 0, y' = 0)$ is located at the point $(x = h, y = k)$ when measured in the original system. If you know the coordinates of a point (x, y) in the original system, then the new coordinates can be found from these formulas:

$$x' = x - h$$

$$y' = y - k$$

If you know the coordinates of the point (x', y') in the new system and want to know the coordinates of the point in the old system, then use these formulas:

$$x = x' + h$$

$$y = y' + k$$

(Note: It is assumed that the new x' axis is parallel to the old x axis, and the new y' axis is parallel to the old y axis. If the orientation of the axes is changed, then it is called a coordinate rotation. See a book on trigonometry.)

"Now we can write the general equation of a circle," the professor said.

EQUATION OF CIRCLE WITH RADIUS r

If the center of the circle is at the point (h, k) in the original coordinate system, then the equation of the circle is:

$$(x - h)^2 + (y - k)^2 = r^2$$

If a new coordinate system is drawn so that the center of the circle is at the origin, then the equation of the circle is:

$$x'^2 + y'^2 = r^2$$

To celebrate our discovery of the equation for a circle, Recordis baked a nice circular layer cake. However, before it could be served, catastrophe struck. Pal tripped, the cake fell on its side, and then Pal fell on top of the cake. Pal began to cry, but we quickly discovered that the cake was

fine except for the fact that it had been bent out of shape. (See Figure 12–15.)

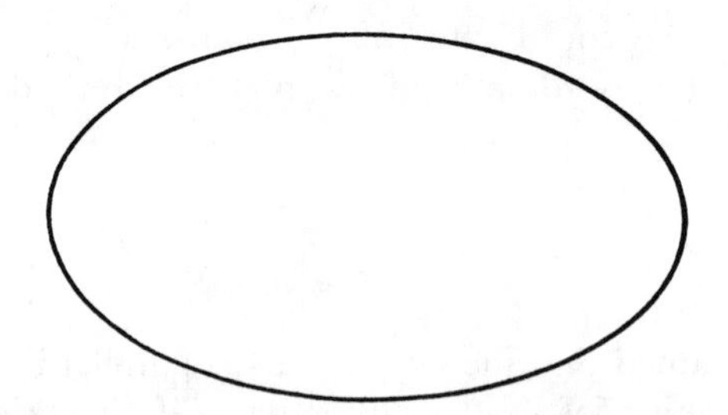

Figure 12–15

"It's not a circle any more!" Recordis complained.

"We need to think of a name for a squashed circle," the professor said. Igor suggested that we use the name *ellipse* to mean squashed circle.

"We can call the point in the exact middle of the ellipse the center," Recordis said.

"I see another important property," the professor said. "Suppose we cut the ellipse in half, like this. (See Figure 12–16.) Both halves are exactly the same. We can say that the ellipse is *symmetrical* about this line. We can call that line the *axis* of the ellipse."

"But you also could cut the ellipse in half like this." Recordis said. (See Figure 12–17.) "The ellipse is also symmetric about that line, so we could also call that line an axis."

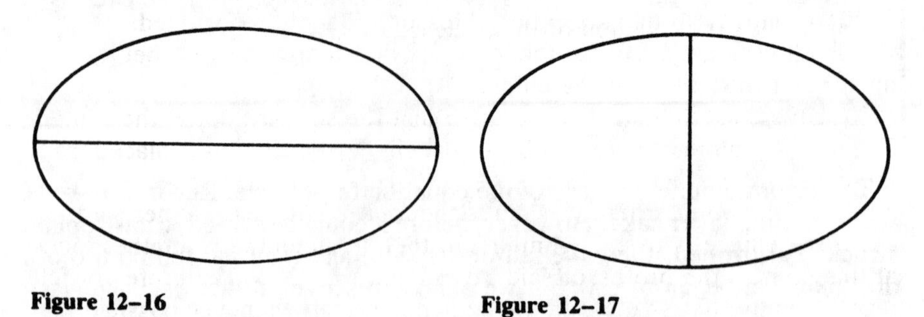

Figure 12–16 **Figure 12–17**

"It appears that an ellipse has two axes," the king said.

"My axis is longer than your axis!" the professor said.

"We'll call the long axis of the ellipse the *major axis,* and we'll call the short axis the *minor axis,*" the king said. (See Figure 12–18.)

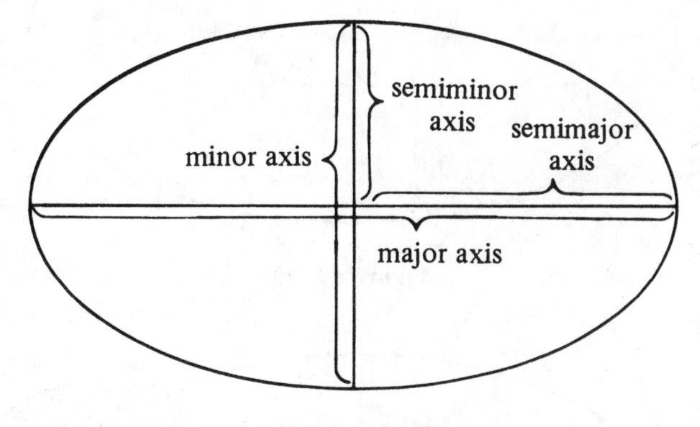

Figure 12–18

We also decided to call a distance equal to half of the major axis the semimajor axis, and a distance equal to half of the minor axis the semiminor axis.

"None of this helps us very much to determine exactly what an ellipse is," the professor said sadly.

At that moment the Royal Astronomer came running into the room triumphantly carrying a large scroll. "I have at last succeeded in making a diagram of the orbit of a comet!" he said excitedly. "Unfortunately, though, I have never seen this strange shape before." He unrolled the scroll. (See Figure 12–19.)

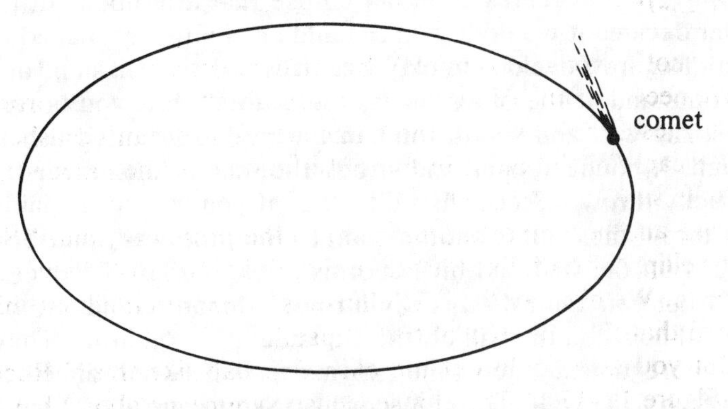

Figure 12–19

"That path is an ellipse!" the professor suddenly recognized.

"So ellipses are useful!" Recordis said in astonishment. "I bet the sun must be at the center of the ellipse."

However, we found to our surprise that the sun was not at the center. The sun was along the major axis of the ellipse, but it was displaced from the center. (See Figure 12–20.)

"I call the point where the sun is the *focus,*" the astronomer said.

"If the ellipse is to be symmetrical, then it should have another focus at this point." the professor said. (See Figure 12–21.) (The professor felt very strongly that shapes should be symmetrical whenever possible.)

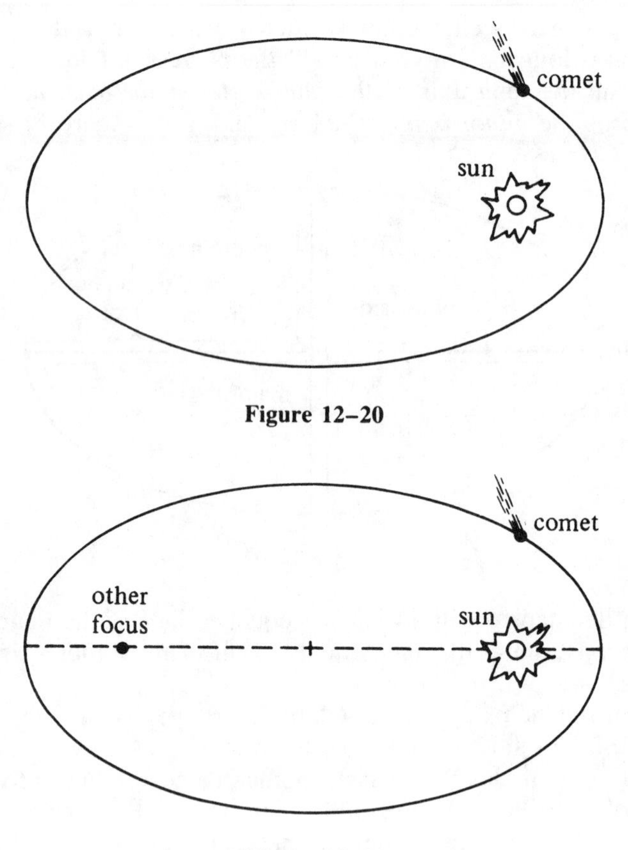

Figure 12–20

Figure 12–21

However, no matter how hard we tried, we were unable to come up with an equation representing an ellipse that afternoon. In the evening Builder decided it would be fun to build an elliptically shaped playground for Pal. Recordis decided to play a game with Pal. "I'll stand at one focus, and you stand at the other focus," he said. "Then you bounce the ball against the wall and see if you can hit me." Pal hurled the ball against the wall. It bounced back and struck the surprised Recordis.

"Lucky throw," Recordis said. "See if you can do it again."

Pal threw the ball at another point along the wall, and once again it bounced off the wall and hit Recordis.

"I bet you can't do three in a row!" Recordis said. However, once again Pal bounced the ball off the wall and it hit Recordis. They tried this many more times. Each time, no matter where Pal aimed the ball, it bounced off the wall and hit Recordis. (See Figure 12–22.)

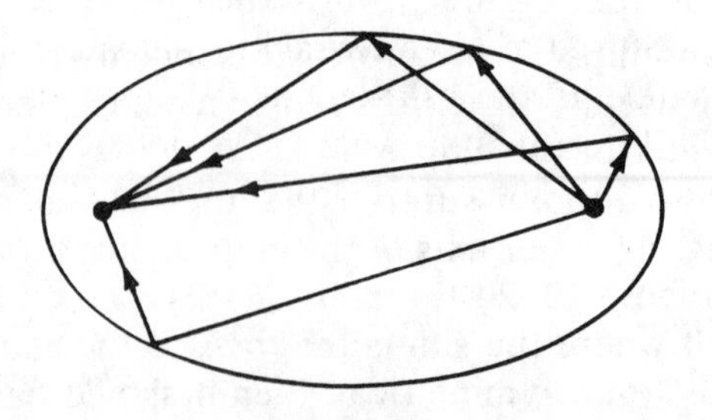

Figure 12–22

Finally, the battered record keeper realized that they had discovered a new property. "If you throw a ball from one focus of an ellipse, it will always bounce off the wall and hit the person standing at the other focus."

The professor noticed something else important. She had been watching the game (from a safe distance) with her stopwatch. "Each throw took exactly the same amount of time to hit you!" she said excitedly. "And, since Pal always throws the ball at exactly the same speed, that means the ball travelled precisely the same distance each time!"

"So what?" Recordis asked.

"Suppose we pick any point on the ellipse," the professor explained. "Call the distance from that point to the first focus d_1, and the distance to the second focus d_2. Then we know that $d_1 + d_2$ will be the same for all of the points on the ellipse!" (See Figure 12–23.)

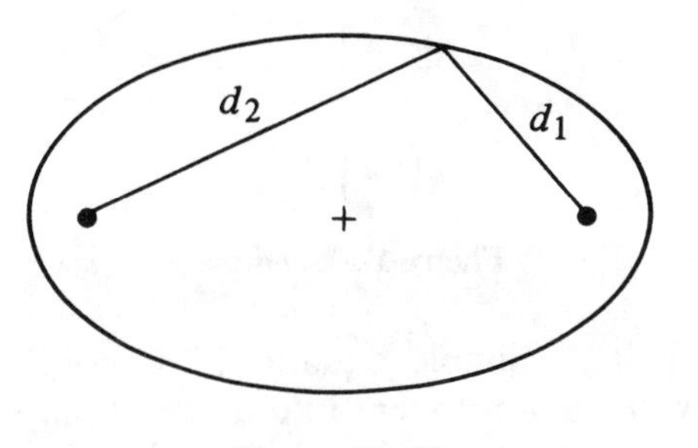

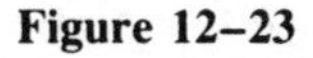

Figure 12–23

Recordis entered this fact in his book as the definition of an ellipse:

> **DEFINITION OF ELLIPSE**
>
> An ellipse is the set of all points in a plane such that the sum of the distances to two fixed points is a constant. The two fixed points are called the *foci* (plural of focus).

"Now we have to find the equation for an ellipse," the professor said.

"For convenience, we can put the center of the ellipse at the origin," Recordis said. "And we can put the major axis along the x axis."

We decided to use the letter a to stand for the length of the semimajor axis of the ellipse, and we let the letter b stand for the length of the semiminor axis. (See Figure 12–24.) Then we set to work to grind out the equation. That proved to be easier said than done. But, as the professor said during the middle of the calculations, "We're not doing anything that's fundamentally different from what we've done before—we're just doing a lot of it at once." After the dust cleared, we finally found that the equation for the ellipse could be written in this form:

$$\frac{x^2}{a^2} + \frac{y^2}{b^2} = 1$$

(See Exercise 62.)

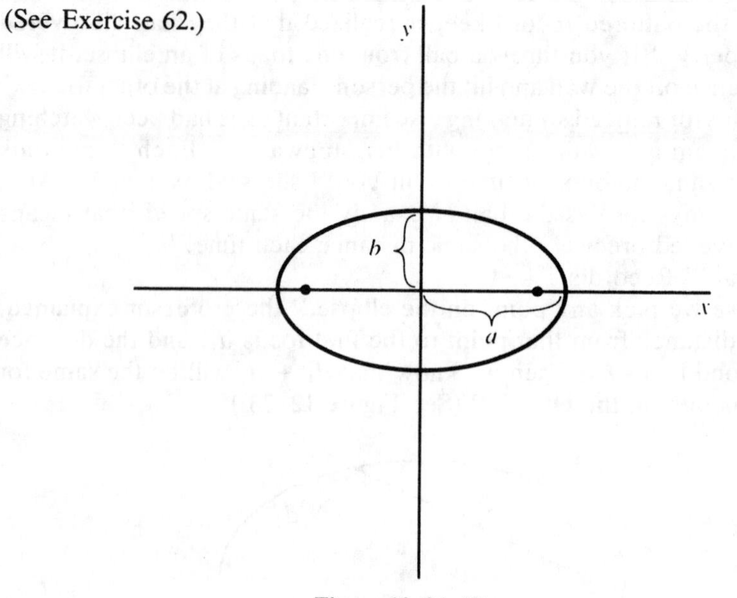

Figure 12–24 ellipse

"That's an amazingly simple equation!" Recordis said delightedly, "considering how much work went into its creation."

We quickly made a list of obvious solutions that satisfied that equation:

$$(a, 0), \quad (-a, 0), \quad (0, b), \quad (0, -b)$$

"And if we need to calculate more points, we can derive an equation for y in terms of x, Recordis said."

$$y = \pm b\sqrt{1 - \frac{x^2}{a^2}}$$

Igor drew some ellipses. (See Figure 12–25.)

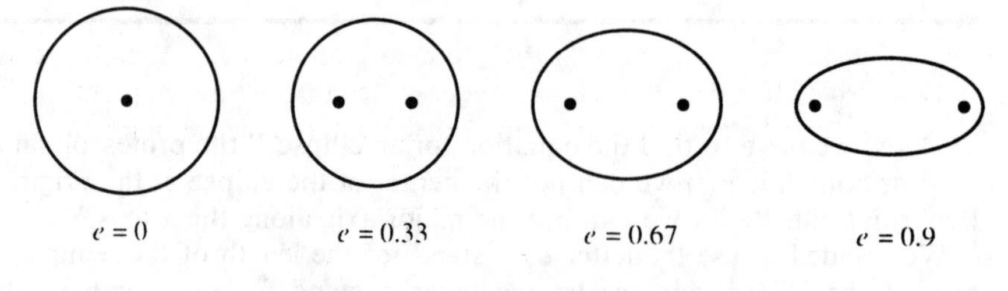

Figure 12–25

We noticed one significant fact: They were all different shapes. "There is only one possible shape for a circle," Recordis said, "but there seem to be many possible shapes for an ellipse. Sometimes the ellipses are almost the same as circles, but other times they are very flat. I think that the ones that are like circles are normal, and the flatter ones are more and more eccentric."

"We need some way to measure the shape of an ellipse," the professor said. "It seems that if the focal points are close together, then the ellipse looks almost the same as a circle."

"On the other hand, if the focal points are very far apart, then the ellipse becomes very flat," the king said. "The important quantity seems to be the distance between the focal points divided by the length of the major axis of the ellipse." We decided to call that quantity the *eccentricity* of the ellipse:

$$\text{eccentricity} = \frac{\text{distance between focal points}}{\text{major axis}}$$

We let e represent the eccentricity and f represent the distance from the center to one of the focal points. Then

$$e = \frac{f}{a}$$

(See Figure 12–26.)

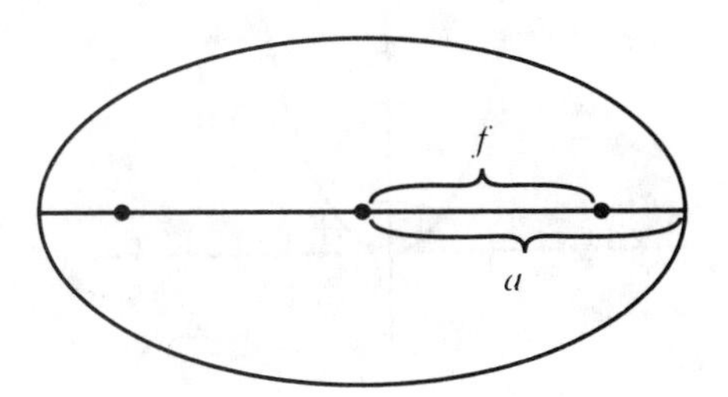

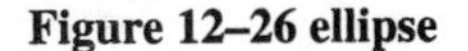
Figure 12–26 ellipse

We also found another expression for the eccentricity:

$$e = \sqrt{\frac{a^2 - b^2}{a^2}}$$

(See Exercise 33.)

If $e = 0$, then $a = b$, the two focal points coincide, and the ellipse is really a circle. If $e = 1$ the ellipse degenerates into a line segment.

We also found the equation of an ellipse centered at the point (h, k):

$$\frac{(x - h)^2}{a^2} + \frac{(y - k)^2}{b^2} = 1$$

(We found this equation by performing a translation of the coordinate axes.)

The astronomer was quite pleased with these results. "As it turns out, the orbits of most planets are very close to being circles—that is, they have very low eccentricities. On the other hand, the orbits of some comets that I have studied have very high eccentricities."

"Ellipses and circles are too hard," Recordis said. "Let's draw the graph of a simple curve—such as $y = x^2$." Igor drew the graph. (See Figure 12–27.)

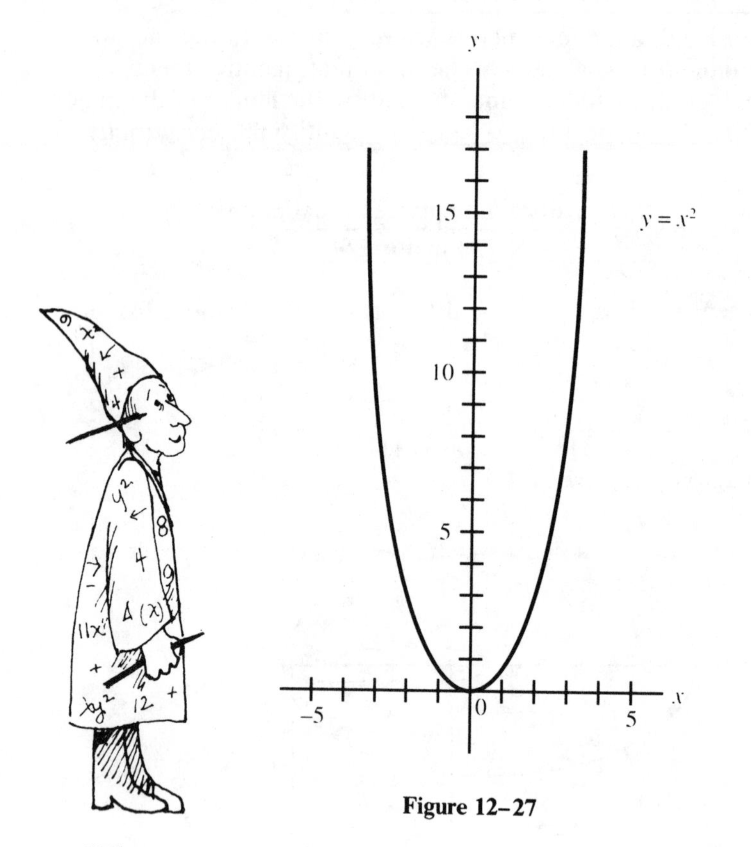

Figure 12–27

"That curve has a nice shape," the professor said. "We should give a special name to that type of curve."

Igor suggested the name *parabola*. The professor realized that there could be parabolas with many different sizes. "Any equation of the form $y = cx^2$, where c is a real number, will be a parabola," she said. Igor drew some examples. (See Figure 12–28.)

We noticed that a parabola was symmetrical about a line drawn through its middle, so we called that line the *axis* of the parabola. (See Figure 12–29.) We called the point at the tip of the parabola the *vertex*.

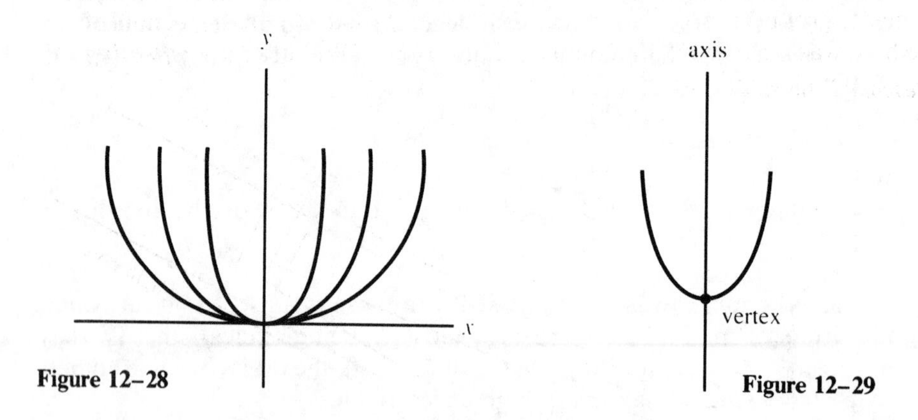

Figure 12–28

Figure 12–29

"We have already found one use for parabolas," the professor said. "If you throw a ball, then the path it follows will be shaped like a parabola." (See Chapter 11.)

Builder built a play area for Pal that had a parabolically shaped wall at one end. The equation of the parabola he used was $y = \frac{1}{4}x^2$. Recordis decided to play a new game. He stood at the point (0, 1) and told Pal to go far away from the parabola. "Now throw the ball and see if you can bounce it off the parabola and hit me!" he said.

Pal hurled the ball straight at the parabola. It bounced off the wall and struck Recordis. Once again Pal tried—and once again the ball struck the record keeper. However, when Pal threw the ball along a path that was not parallel to the axis of the parabola, it did not hit Recordis.

"How can that be?" Recordis asked. "No matter where Pal stands, whenever he throws the ball parallel to the axis it bounces off the parabola and hits me." (See Figure 12–30.)

"We will call the point you're standing at the *focus* of the parabola," the professor said, "since it seems to be such a special point."

The Royal Astronomer suddenly realized that this was a very important result. "Let's make a mirror whose cross-section is a parabola!" he said excitedly. "Then, let's point the parabola at a star. When the light rays from the star hit the parabola, they will all be reflected back to the focal point. Then we can see faint objects better, because we will be able to collect many different light rays from the star and focus them in the same place."

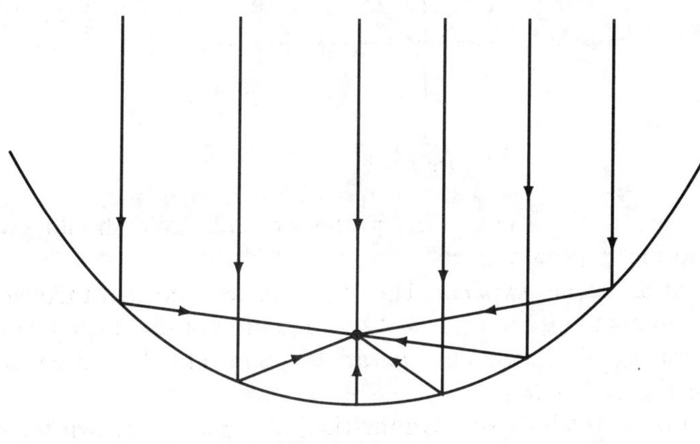

Figure 12–30

The Astronomer quickly demonstrated his idea. He took a circular piece of glass, ground a hole out of the middle such that the cross-section of the hole was a parabola, and then put it in a tube. "We'll call it a *reflecting telescope*," he said. (See Figure 12–31.)

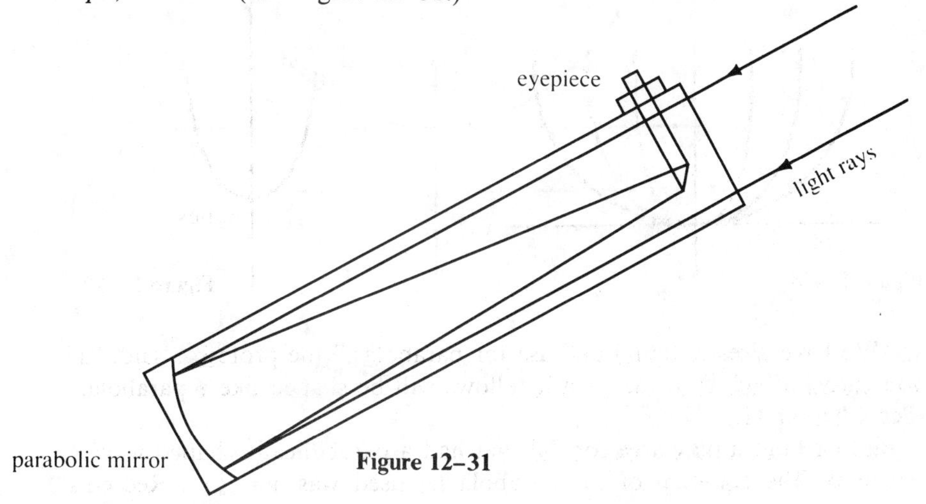

Figure 12–31

(If you take a parabola and spin it about its axis, then the three-dimensional surface that the parabola traces out is called a *paraboloid.* The surface of the telescope mirror is an example of a paraboloid. Notice that you need to add a small flat diagonal mirror to reflect the light out of the side of the tube so you can look at it, and then you also need to add an eyepiece to magnify the image.)

The astronomer pointed the telescope at the moon and some planets, and we found that we could see them much better.

"We still have a problem," Recordis said. "We don't know how to calculate the location of the focus if we know that the equation of the parabola is $y = cx^2$."

We worked on that problem for a while. In the course of our investigations, we discovered another interesting fact about parabolas: Any point on the parabola was exactly the same distance away from the focus as it was from a particular line that we called the *directrix.* (See Figure 12–32).

We decided to find the equation for a parabola with the focus at the point $(0, a)$ and the directrix at the line $y = -a$. If (x, y) represents a point on the parabola, then the distance to the focus point is $\sqrt{(y-a)^2 + x^2}$, and the distance to the directrix line is $y + a$. We set these two quantities equal:

$$(y + a) = \sqrt{(y - a)^2 + x^2}$$

$$(y + a)^2 = (y - a)^2 + x^2$$

$$y^2 + 2ay + a^2 = y^2 - 2ay + a^2 + x^2$$

$$2ay = -2ay + x^2$$

$$4ay = x^2$$

$$y = \frac{1}{4a}x^2$$

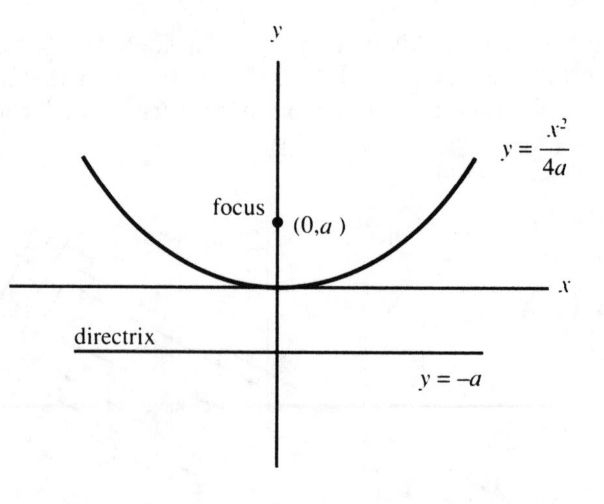

Figure 12–32 parabola

GENERAL FORM FOR A PARABOLA

A parabola is the set of all points such that the distance to a fixed point (called the focus) is the same as the distance to a fixed line (called the directrix). If the vertex of the parabola is at the point (0, 0), the focus is at the point $(0, a)$, and the directrix is the line $y = -a$, then the equation of the parabola is

$$y = \frac{1}{4a}x^2$$

If the vertex is at the point (h, k), the focus is the point $(h, k + a)$, and the directrix is the line $y = k - a$, then the equation of the parabola is

$$(y - k) = \frac{1}{4a}(x - h)^2$$

Just before we were ready to go home, the professor suddenly felt in a playful mood. "An ellipse is the set of points such that the *sum* of the distances to two fixed points is a constant," she said. "I wonder what kind of curve we would get if we found the set of points such that the *difference* between the distances to two fixed points is a constant."

The professor ground out some algebra, and found that her proposed curve must satisfy the equation

$$\frac{x^2}{a^2} - \frac{y^2}{b^2} = 1$$

Notice that this equation is the same as the equation for the ellipse, except for the minus sign. However, we found that the graph of this equation did not look at all like an ellipse. (See Figure 12–33.)

"That's a two-branch curve!" the king said in surprise. Igor suggested that we call this curve a *hyperbola*. We noticed that each branch of a hyperbola looks like a misshapen parabola.

"It's fun drawing pictures like this!" Recordis said. "After all the work we've put into algebra, it's about time that we're finally getting some enjoyment from it."

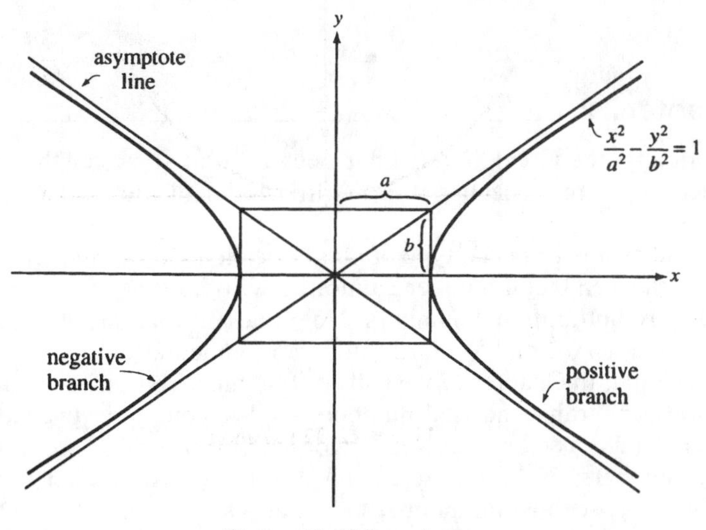

Figure 12–33 hyperbola

SUMMARY OF CONIC SECTIONS

A **circle** is the set of points in a plane that are all at the same distance (the radius) from a fixed point (the center).

$$\text{Equation: } x^2 + y^2 = r^2$$

r is the radius; the center is at (0, 0)

An **ellipse** is the set of points in a plane such that the sum of the distances to two fixed points (the foci) is constant.

$$\text{Equation: } \frac{x^2}{a^2} + \frac{y^2}{b^2} = 1$$

a is the semimajor axis; b is the semiminor axis; the center is at (0, 0); the foci are at $(0, ea)$ and $(0, -ea)$, where the eccentricity e comes from the formula

$$e = \sqrt{\frac{a^2 - b^2}{a^2}}$$

A **parabola** is the set of points in a plane such that the distance to a fixed point (the focus) equals the distance to a fixed line (the directrix).

$$\text{Equation: } y = \frac{1}{4a} x^2$$

with focus at $(0, a)$ and directrix $y = -a$

A **hyperbola** is the set of points in a plane such that the difference in distances to two fixed points is constant

$$\text{Equation: } \frac{x^2}{a^2} - \frac{y^2}{b^2} = 1$$

An equation of the form

$$Ax^2 + Bx + Cy^2 + Dy + E = 0$$

can be graphed as a circle, ellipse, or hyperbola; form a new coordinate system with $x' = x + B/2A$ and $y' = y + D/2C$ so the center will be at the origin. (See Exercise 73.)

Notes to Chapter 12

- Historically, the Greek letter π has been used to stand for the circumference of a circle because it is the first letter of the Greek word for perimeter.
- The number π is a special type of number called a *transcendental number.* A transcendental number cannot be written in the form a^b where a and b are both rational numbers. Note that all square roots and other roots, such as $\sqrt{2}$ and $\sqrt[4]{15}$, are not transcendental even though they are irrational. In Chapter 17 we will see that most values for logarithmic functions are transcendental numbers, and if you study trigonometry you will find that most values of trigonometric functions are transcendental numbers. If you study calculus you will discover another fundamental transcendental number called e, which is about 2.71828 . . .
- The four curves circle, ellipse, parabola, and hyperbola are called *conic sections* because they can be formed by the intersection of a plane with a

right circular cone. (See Figures 12–34 and 12–35.) If the plane is perpendicular to the axis of the cone, the intersection will be a circle. If the plane is slightly tilted, the result will be an ellipse. If the plane is parallel to one element of the cone, the result will be a parabola. If the plane intersects both nappes of the cone, the result will be a hyperbola. (Note that a hyperbola has two branches.)

- It is possible to define ellipses, parabolas, and hyperbolas by one equation. A conic section can be defined as the set of points such that the distance from a fixed point divided by the distance from a fixed line is a constant. The fixed point is called the focus, the fixed line is called the directrix, and the constant ratio is called the eccentricity of the conic section (abbreviated e.) When $e = 1$, this definition exactly matches the definition of a parabola. If e is less than one, then the conic section is an ellipse. If e is greater than one, then the conic section is a hyperbola. (See Exercise 61.)

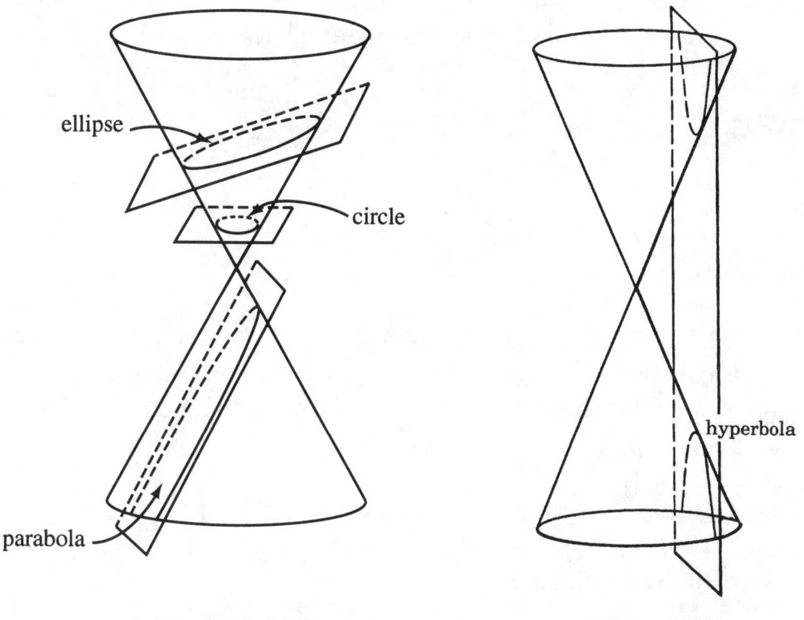

Figure 12–34 **Figure 12–35**

- If you have a computer with graphics capabilities, then you are quite familiar with Builder's problem in deciding which light bulbs to light to form patterns on the scoreboard. The computer divides the screen up into an array of little dots, and you have to tell the computer which dots to light up in order to draw the pictures that you want. (See Exercise 73.)
- There are many applications for elliptical and paraboloid-shaped objects in addition to the ones described in the chapter. The microphones that pick up crowd noises at football games are shaped like paraboloids, as are the reflectors of automobile headlights. If you put an object at the focus of a large paraboloidal reflector aimed at the sun you can observe a very effective solar oven at work.
- If the ceiling of a building is shaped like an ellipse, then a person standing at one focus can hear another person whispering from the other focus—even if there is a considerable distance between the two focal points. The capitol in Washington, D.C., St. Peter's in Rome, and Grand Central Station in New York contain rooms where you can observe this "whispering gallery" effect.

Exercises

Calculate approximate values for the area and circumference for these circles:

1. $r = 5$
2. $r = 1$
3. $r = 10$
4. $r = 5.64$
5. $r = 3{,}960$ miles
6. $r = 93$ million miles
7. $r = 100$ million light years

Calculate the radius of each circle:

8. circumference $= 6.28$
9. circumference $= 1$
10. area $= 1/4$
11. circumference $= 12.56$
12. circumference $= 10$
13. The fraction 22/7 is sometimes used as an approximation for the value of π. How accurate is that approximation? (*I.e.*, what is the difference between 22/7 and the true value of π?)

Sketch the circles defined by these equations:

14. $(x - 1)^2 + (y - 1)^2 = 1$
15. $(x - 3)^2 + (y - 1)^2 = 1$
16. $(x - 2)^2 + (y - 1)^2 = 4$
17. $(x - 5)^2 + (y - 1)^2 = 16$

Determine whether or not the pairs of circles defined by these equations intersect. (In other words, do they cross each other?) If they do intersect, determine whether they intersect at one point or two points.

18. $x^2 + y^2 = 100$; $x^2 + (y - 20)^2 = 100$
19. $(x - 1)^2 + (y - 1)^2 = 16$; $(x - 3)^2 + (y - 3)^2 = 16$
20. $(x - 2)^2 + (y - 3)^2 = 25$; $(x - 9)^2 + (y - 11)^2 = 25$
21. $(x - 4)^2 + (y - 5)^2 = 100$; $(x - 32)^2 + (y - 15)^2 = 4$
22. Rewrite the equation of the circle

$$(x - h)^2 + (y - k)^2 = r^2$$

in this form:

$$Ax^2 + Bx + Cy^2 + Dy + E = 0$$

Each of these equations defines a circle. Identify the center and radius of each circle.

23. $x^2 - 2x + y^2 - 2y + 1 = 0$
24. $x^2 + y^2 - 10y = 0$
25. $x^2 + 4x + y^2 - 2y - 4 = 0$
26. $x^2 - 20x + y^2 + 6y + 107 = 0$
27. $x^2 + 4x + y^2 + 2y + 2 = 0$
28. $x^2 + y^2 - 49 = 0$
29. Show that if (x_1, y_1) is a point on a circle centered at the origin, then $(-x_1, y_1)$, $(x_1, -y_1)$, and $(-x_1, -y_1)$ are also points on the same circle.
30. Show that, if s_n is the length of one side of a regular polygon inscribed in a circle of radius r, then the length of one side of a regular polygon with twice as many sides inscribed in the same circle is given by the formula $s_{2n}^{\,2} = 2r^2 - r\sqrt{4r^2 - s_n^{\,2}}$.

31. Suppose we fill a circle with radius r with n isosceles triangles. (See Figure 12–36.) Each triangle has an altitude of approximately r and a base of approximately $2\pi r/n$. What is the approximate total area of all of the triangles?

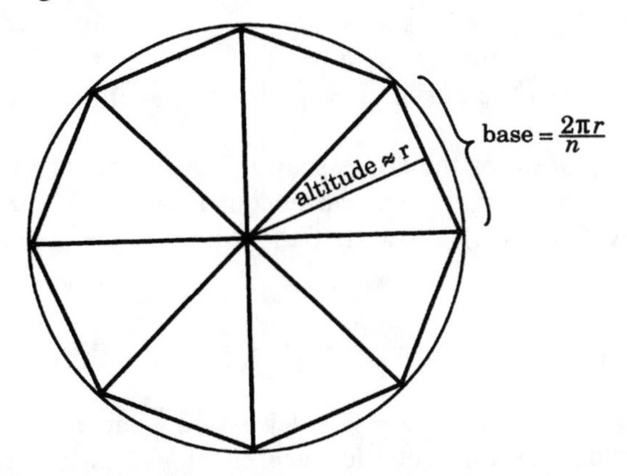

Figure 12–26

32. Show that if x is the x-coordinate of a point on an ellipse, then y can be found from the formula $y = \pm b\sqrt{1 - x^2/a^2}$. (The center of the ellipse is at the origin, the semimajor axis is a, and the semiminor axis is b.)

33. Show that the eccentricity of an ellipse can be found from the formula $e = \sqrt{a^2 - b^2}/a$.

The area of an ellipse with semimajor axis of length a and semiminor axis of length b can be found from the formula $A = \pi ab$. Calculate the areas and eccentricities of these ellipses:

34. $a = 5$, $b = 3$ **35.** $a = 10$, $b = 1$ **36.** $a = 100$, $b = 95$

In each exercise below, you are given the length of the semimajor axis and the semiminor axis for the orbit of a planet. (These lengths are measured in a unit called the *astronomical unit*, equal to the distance from the earth to the sun.) Calculate the eccentricity of the orbit for each planet.

37. Mars: $a = 1.5237$ $b = 1.5170$
38. Jupiter: $a = 5.2028$ $b = 5.1967$
39. Saturn: $a = 9.5388$ $b = 9.5240$
40. Pluto: $a = 39.439$ $b = 38.184$

Each of these equations defines an ellipse. Determine the semimajor axis and the semiminor axis for each ellipse.

41. $x^2/100 + y^2/81 = 1$
42. $x^2 + 4y^2 = 4$
43. $(x - 16)^2/36 + (y - 11)^2/121 = 1$
44. $(x - 7)^2/11 + (y - 34)^2/28 = 1$
45. What are the coordinates of the two focal points of the ellipse defined by the equation

$$\frac{(x - h)^2}{a^2} + \frac{(y - k)^2}{b^2} = 1?$$

Identify the focus point and the equation of the directrix for the parabolas defined by these equations:

46. $y = x^2$
47. $y = 10x^2$
48. $y = cx^2$
49. $y = x^2/16$
50. $y - 5 = (1/4)(x + 3)^2$
51. $y - 8 = (x - 3)^2$
52. $x = y^2$
53. $x = (1/4)y^2$
54. $y - k = (x - h)^2$
55. $y - k = b(x - h)^2$

Determine the equations of the parabolas with these characteristics:

56. focus (0, 8) directrix $y = 0$
57. focus (10, 6) directrix $y = 4$
58. focus (17, 9) directrix $x = 12$
59. focus (0, 10) directrix $y = -10$
60. Draw a graph of the curve $xy = 1$. What does the shape of that curve look like?
61. Find the equation of the conic section with focus at the point $(0, p)$, directrix at the line $x = 0$, and eccentricity e. (See the note at the end of the Chapter.) Show that the equation can be written in the form

$$\frac{(x - h)^2}{A} + \frac{y^2}{B} = 1$$

where $h = p/(1 - e^2)$, $A = e^2p^2/(1 - e^2)^2$, and $B = e^2p^2/(1 - e^2)$. What curve does this equation represent if e is less than 1? What curve does it represent if e is greater than 1?
62. Starting from the definition of an ellipse, show that the equation of the ellipse is $x^2/a^2 + y^2/b^2 = 1$. (Warning: This is hard.)

Here's how to sketch the graph of the hyperbola $x^2/a^2 - y^2/b^2 = 1$. First, plot the two points $(a, 0)$ and $(-a, 0)$. Then draw a rectangle centered at the origin with dimensions $2a$ by $2b$. (See Figure 12–33). Draw the diagonals of that rectangle, and extend them off to infinity in all four directions. These lines are called *asymptote lines*. As the value of x becomes very large or very small, the hyperbola itself becomes almost indistinguishable from the asymptotes. However, the hyperbola never actually touches the asymptote lines.

Sketch these hyperbolas:

63. $x^2 - y^2 = 1$
64. $x^2/25 - y^2/16 = 1$
65. $x^2/9 - y^2/4 = 1$
66. $x^2 - 9y^2 = 36$

Sketch these curves. (Are they all hyperbolas?)

67. $x^2 - y^2 = 10$
68. $x^2 - y^2 = 5$
69. $x^2 - y^2 = 4$
70. $x^2 - y^2 = 2$
71. $x^2 - y^2 = 0$

72. You have observed a planet at two different times and you are trying to figure out the equation of its orbit. Suppose you know that (x_1, y_1) and (x_2, y_2) are two points on the ellipse of the planet's orbit, which has its center at the origin and major axis along the x axis. However, you do not know the value of a, the semimajor axis. Derive a formula for a in terms of x_1, y_1, x_2, and y_2.
73. The equation $Ax^2 + Bx + Cy^2 + Dy + E = 0$ represents one of the conic sections not centered at the origin. Determine what the equation will look like if you use a new coordinate system given by these translation formulas:

$$x' = x + \frac{B}{2A} \quad \text{and} \quad y' = y + \frac{D}{2C}$$

13 Polynomials

The next day the detective led us to a high bluff overlooking Scenic Lake. "The dragon is playing in the lake!" the detective pointed out. The dragon's long, snaky body was forming undulating curves that went up and down. The dragon looked up and saw us, but it made no move to escape. Instead, it continued to form wavy curves in the water.

"I bet you can't find an algebraic equation that describes this shape!" the dragon called to us defiantly. Recordis quickly sketched the curve formed by the dragon (See Figure 13–1.)

"We've never seen a curve like that before!" Recordis cried.

"How about this one?" the dragon asked, forming a curve which now changed directions three times. (See Figure 13–2.)

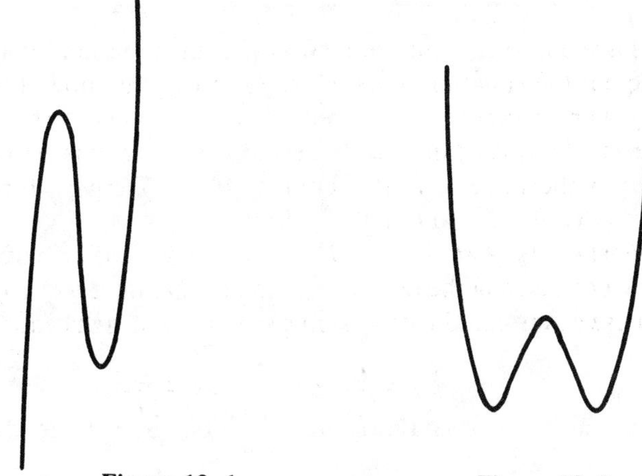

Figure 13–1

Figure 13–2

"What a horrible, vile monster!" Recordis exclaimed.

"I hope you can come out and play another day," said the dragon as it quickly swam off.

"The dragon seems to think that this is all a game." Recordis fumed. "It is refusing to take the matter seriously."

"Let's take the dragon up on his challenge," the professor said in a spirit of adventure. "Let's think of a curve that changes its direction twice."

"We have found no such curve!" Recordis said heatedly. "We know that the graph of a first-degree equation, such as $y = ax + b$, does not change direction at all. And we know that the graph of a second-degree equation, such as $y = x^2$, changes direction once. But I don't know any curve that changes direction twice."

"If the same pattern is to be maintained, then perhaps a third-degree curve will change direction twice, and perhaps a fourth-degree curve will change direction three times," the king suggested.

"It's worth a try," the professor said. We made up an arbitrary third-degree polynomial in x:

$$x^3 - 9x^2 + 18x + 1$$

Then, by setting this polynomial equal to y, we formed an equation in two variables:

$$y = x^3 - 9x^2 + 18x + 1$$

We calculated a table of values:

x	y
−5	−439
−4	−279
−3	−161
−2	−79
−1	−27
0	1
1	11
2	9
3	1
4	−7
5	−9
6	1
7	29
8	81
9	163

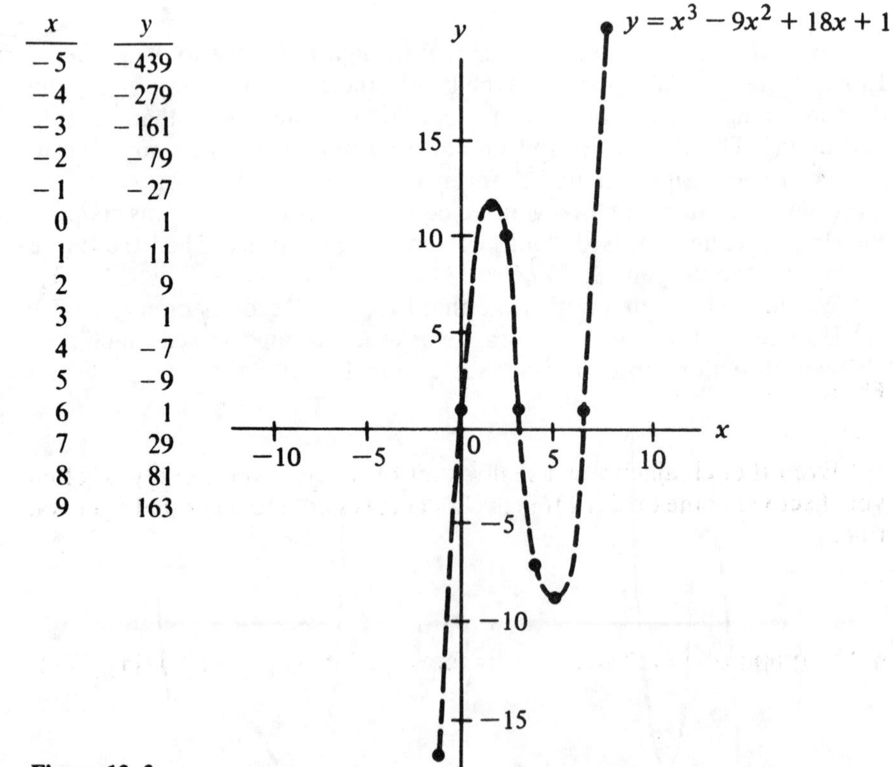

Figure 13–3

We then drew a graph of some of these points. (See Figure 13–3.)

"The curve does start out going up, then it turns down, and finally it turns up again!" Recordis said in awe.

We also found that we could draw a fourth-degree polynomial that had three turning points. (See Figure 13–4.)

$$y = x^4 - 20x^3 + 124x^2 - 240x + 5$$

x	y
−3	2,462
−2	1,157
−1	390
0	5
1	−130
2	−123
3	−58
4	5
5	30
6	5
7	−58
8	−123
9	−130
10	5
11	390
12	1,157

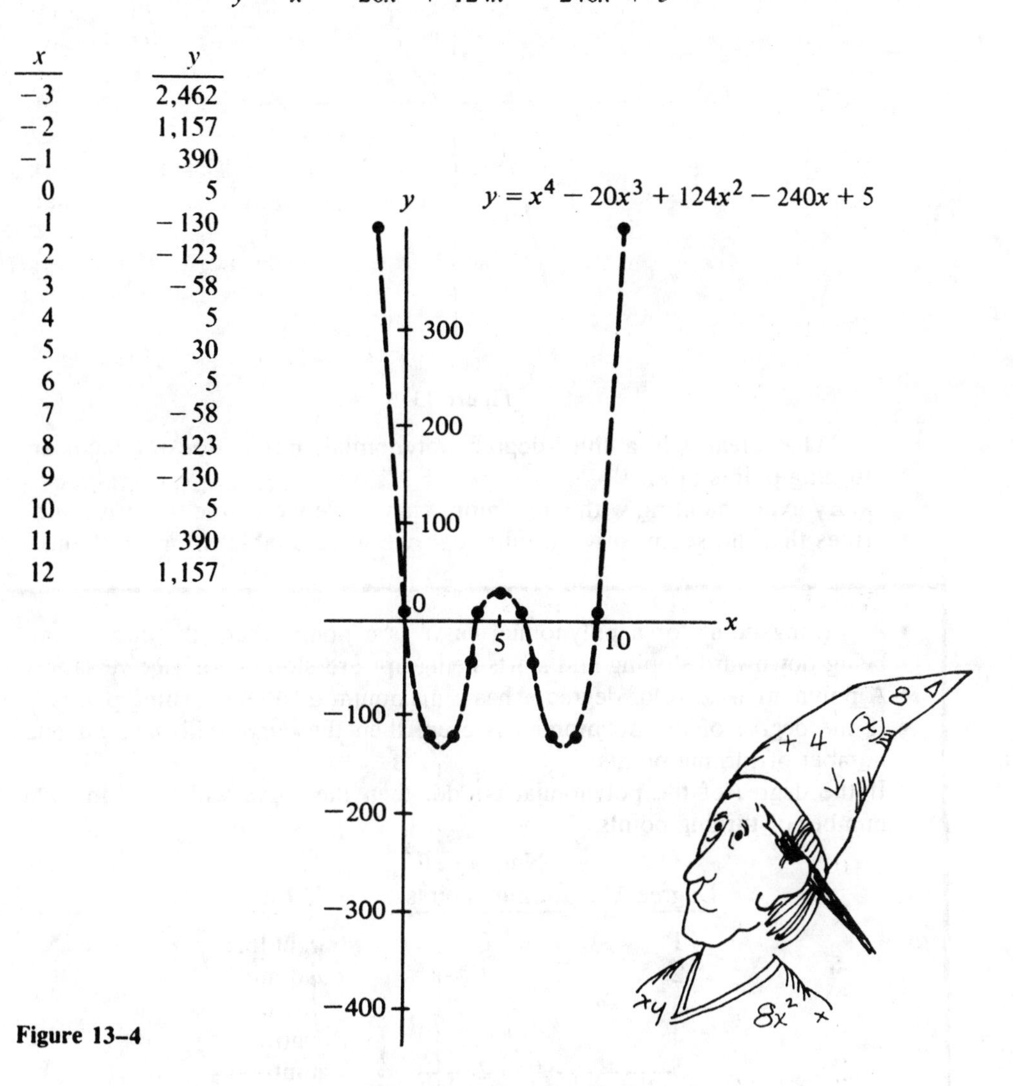

Figure 13–4

"Even though algebra is usually a lot of tedious work, every so often you discover some exciting things," Recordis said. He suggested a general rule:

- The graph of an nth-degree polynomial will have $n - 1$ turning points.

"Not always!" the professor said. She suggested that we draw a graph of the third-degree curve $y = x^3$. (See Figure 13–5.)

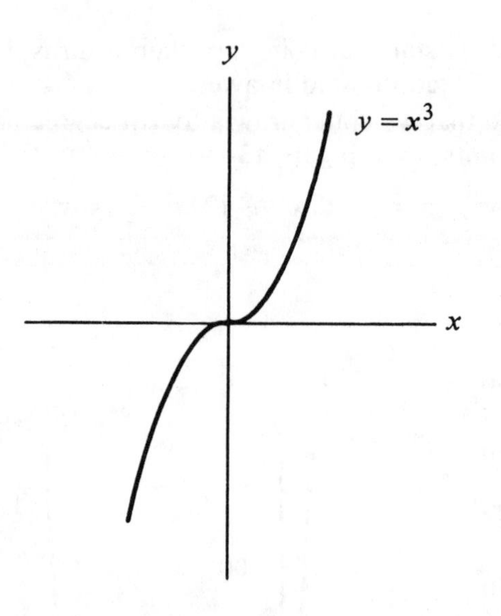

Figure 13–5

"That clearly is a third-degree polynomial, but it doesn't have any turning points at all."

By experimenting with some more curves we were able to derive some rules that did seem to work (although we were unable to prove them).

- A turning point for a polynomial curve is a point where the curve stops being downward sloping and starts being upward sloping, or vice versa.
- A polynomial curve of degree n has a maximum of $n - 1$ turning points.
- If the degree of the polynomial is even, then the curve will have an odd number of turning points.
- If the degree of the polynomial is odd, then the curve will have an even number of turning points.

Degree	Number of turning points	Term
1	0	straight line
2	1	quadratic
3	0 or 2	cubic
4	1 or 3	quartic
5	0 or 2 or 4	quintic

"Now we must investigate polynomial equations of the form $f(x) = 0$, where $f(x)$ is a polynomial in x," the professor said. "I always say that the best defense is a good offense, and since the gremlin might show up some day and ask us to solve this type of equation, we should study them first and be prepared."

"I can guess how many solutions these equations will have," Recordis said. "A third-degree curve like this (see Figure 13–6) crosses the x axis three times, so the corresponding equation will have three solutions; and a fourth-degree curve like this (see Figure 13–7) crosses the x axis four times, so the corresponding equation will have four solutions."

Using the turning point rules, we were able to show that a polynomial equation of degree n can have as many as n solutions. (Note that we have

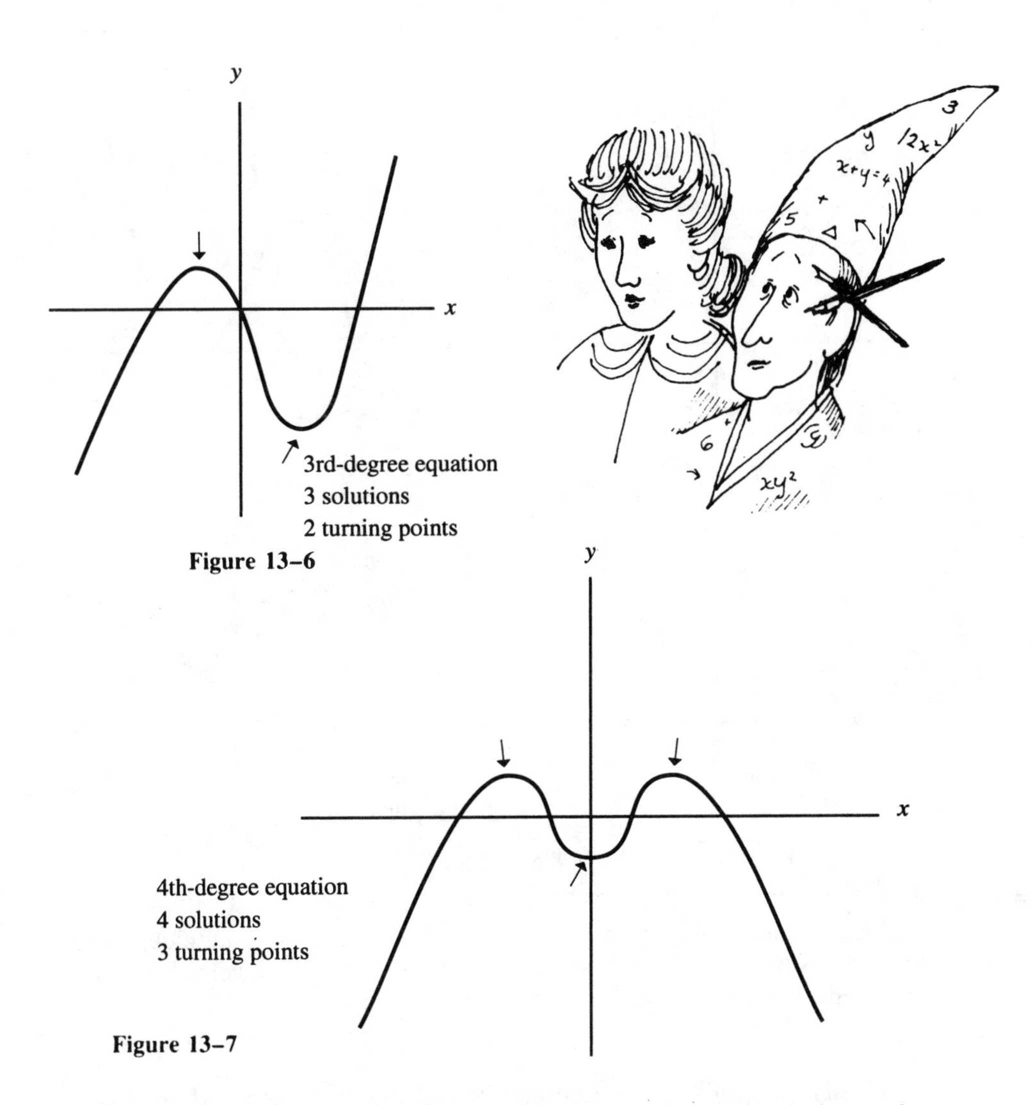

Figure 13–6

Figure 13–7

already found that a first-degree (linear) equation has one solution and a second-degree (quadratic) equation can have as many as two solutions.)

"However, an equation of degree *n* doesn't have to have *n* solutions," the professor said. She drew graphs of a second-degree equation that had no solutions, a third-degree equation that had two solutions, a fourth-degree equation that had two solutions, and a fourth-degree equation that had no solutions. (See Figure 13–8.)

Using the turning-point rules, the king suggested two more rules:

- If the degree of the polynomial is *odd*, then the equation must have at least one solution.
- If the degree of the polynomial is *even*, then the equation might have no solutions.

(When we talk about solutions here we are talking about solutions that are real numbers. You might be wondering: How can there be such a thing as a solution that is not a real number? However, we made some amazing discoveries in Chapter 20.)

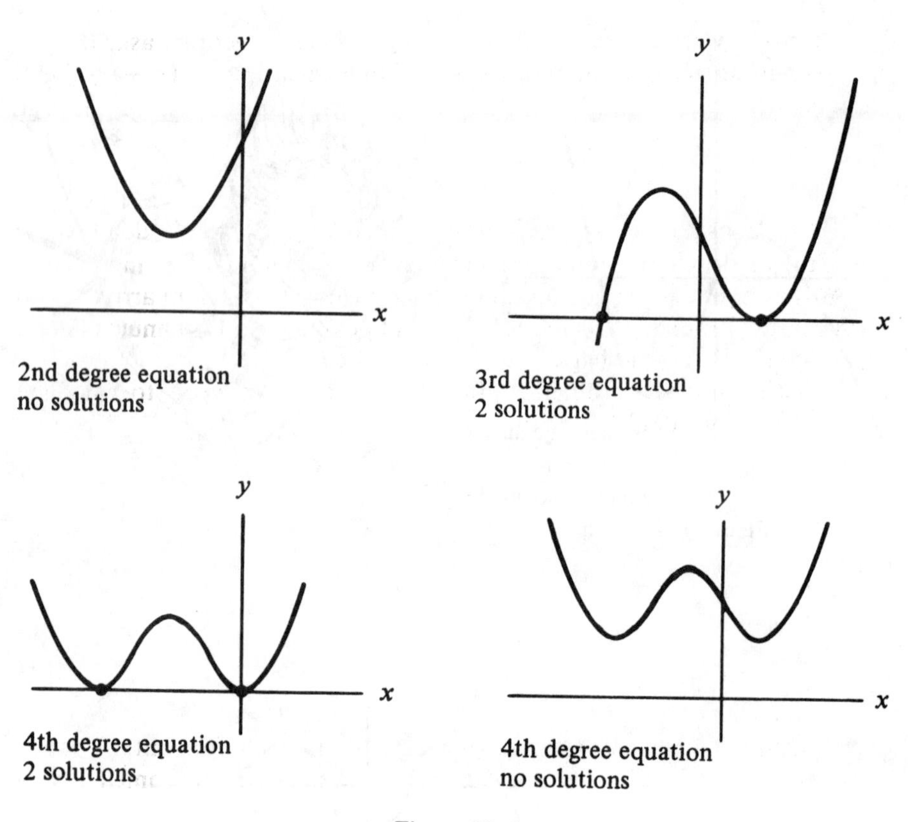

Figure 13–8

"I hate to disillusion you, but none of these rules give any clues as to what the solutions actually are," Recordis said. Igor suggested that we try to find the solution to the third-degree equation

$$x^3 + 4x^2 + 4x + 1 = 0$$

"If only we could *factor* that equation!" the professor said wistfully. "Suppose we could write that third-degree polynomial as the product of three factors, like this:

$$(x - r_1)(x - r_2)(x - r_3) = 0$$

Then it would be clear that x could be either r_1, r_2, or r_3 and still make the equation true."

"But we don't have the faintest idea how to factor the polynomial $x^3 + 4x^2 + 4x + 1 = 0$," Recordis pointed out. "In desperate situations such as this, my rule is: When in doubt, guess!"

Recordis guessed that $x = 1$ was a solution, but when that didn't work he did correctly guess that $x = -1$ was a solution, since $(-1)^3 + 4(-1)^2 + 4(-1) + 1 = 0$.

"However, we know that a third-degree equation can have as many as three solutions," the professor cautioned, "so we still need to look for two more." She suggested that we try to factor the polynomial into this form:

$$x^3 + 4x^2 + 4x + 1 = (x + 1)(x - r_2)(x - r_3)$$

"Since $x = -1$ is one of the solutions, I'll bet you a hundred dollars that $x - (-1) = (x + 1)$ is one of the factors."

"But how do we find the values for r_2 and r_3?" Recordis asked.

"We could try dividing both sides of that equation by $(x + 1)$," the king said.

$$\frac{x^3 + 4x^2 + 4x + 1}{x + 1} = (x - r_2)(x - r_3)$$

"How does that help?" Recordis wanted to know. "We don't know how to perform a division problem when two polynomials are involved."

At that moment the concession manager from the stadium arrived with a desperate problem. "I have 1,441 bags of peanuts and 11 peanut venders. How can I divide the bags evenly among the venders?"

"That's an easy problem," Recordis said. "We need to calculate 1,441/11, using regular division."

$$\begin{array}{r} 131 \\ 11\overline{)1441} \\ \underline{11} \\ 34 \\ \underline{33} \\ 11 \\ \underline{11} \end{array}$$

"The answer is 131."

Suddenly the professor fell out of her chair with astonishment. "Look!" she gasped. "The division problem 1,441/11 is *exactly* the same as

$$\frac{x^3 + 4x^2 + 4x + 1}{x + 1}$$

"Too much algebra has scorched your wits!" Recordis said. "1,441/11 is a regular arithmetic problem—not at all like a difficult algebra problem involving polynomials."

"But what does the number 1,441 mean?" the professor asked rhetorically. "It means 1,000 + 400 + 40 + 1, which is

$$(1 \times 1{,}000) + (4 \times 100) + (4 \times 10) + 1$$
$$= (1 \times 10^3) + (4 \times 10^2) + (4 \times 10^1) + 1$$

So, 1,441 is the same as the polynomial

$$(1 \times x^3) + (4 \times x^2) + (4 \times x) + 1$$

The only difference is that it contains a 10 instead of an x. So let's set up the division problem as if we were dividing two regular numbers.

$$x + 1\overline{)x^3 + 4x^2 + 4x + 1}$$

First, we ask how many times does $(x + 1)$ go into x^3? It will go in x^2 times, so we write x^2 on the top line, like this:

$$\begin{array}{r} x^2 \\ x + 1\overline{)x^3 + 4x^2 + 4x + 1} \end{array}$$

Then we multiply x^2 times $(x + 1)$, with the result $x^3 + x^2$, and then we write that result down on the second line:

$$\begin{array}{r} x^2 \\ x + 1\overline{)x^3 + 4x^2 + 4x + 1} \\ x^3 + x^2 \end{array}$$

Then we subtract:

$$\begin{array}{r} x^2 \\ x + 1 \overline{)\, x^3 + 4x^2 + 4x + 1} \\ \underline{x^3 + x^2} \\ \mathbf{0 + 3x^2} \end{array}$$

Then we bring down the $4x$:

$$\begin{array}{r} x^2 \\ x + 1 \overline{)\, x^3 + 4x^2 + 4x + 1} \\ \underline{x^3 + x^2} \\ 3x^2 + \mathbf{4x} \end{array}$$

Then we ask how many times $x + 1$ goes into $3x^2$, so we write $3x$ on the top line, and then multiply again:

$$\begin{array}{r} x^2 + \mathbf{3x} \\ x + 1 \overline{)\, x^3 + 4x^2 + 4x + 1} \\ \underline{x^3 + x^2} \\ 3x^2 + 4x \\ \underline{\mathbf{3x^2 + 3x}} \end{array}$$

and subtract again:

$$\begin{array}{r} x^2 + 3x \\ x + 1 \overline{)\, x^3 + 4x^2 + 4x + 1} \\ \underline{x^3 + x^2} \\ 3x^2 + 4x \\ \underline{3x^2 + 3x} \\ \mathbf{0 + x} \end{array}$$

Then, bring down the 1:

$$\begin{array}{r} x^2 + 3x \\ x + 1 \overline{)\, x^3 + 4x^2 + 4x + 1} \\ \underline{x^3 + x^2} \\ 3x^2 + 4x \\ \underline{3x^2 + 3x} \\ x + \mathbf{1} \end{array}$$

Then, since $x + 1$ goes into $x + 1$ exactly once, we get the answer

$$\begin{array}{r} x^2 + 3x + 1 \\ x + 1 \overline{)\, x^3 + 4x^2 + 4x + 1} \\ \underline{x^3 + x^2} \\ 3x^2 + 4x \\ \underline{3x^2 + 3x} \\ x + 1 \\ \underline{x + 1} \\ 0 \end{array}$$

"Since the result is 0 after the last subtraction, that means there is no remainder. Therefore, the algebraic division problem

$$\frac{x^3 + 4x^2 + 4x + 1}{x + 1} = x^2 + 3x + 1$$

is just the same as the arithmetic division problem 1,441/11 = 131."

"How do we know that the result is right?" Recordis asked. "Not that I don't trust you, but . . ."

"We can multiply $x + 1$ by $x^2 + 3x + 1$ to see if the result really is $x^3 + 4x^2 + 4x + 1$," the king suggested.

$$\begin{aligned}(x + 1)(x^2 + 3x + 1) &= x^3 + 3x^2 + x + x^2 + 3x + 1 \\ &= x^3 + 4x^2 + 4x + 1\end{aligned}$$

"Therefore, the equation

$$x^3 + 4x^2 + 4x + 1 = 0$$

and the equation

$$(x + 1)(x^2 + 3x + 1) = 0$$

are the same thing," the king said. "The equation will be true if either $x + 1 = 0$ or $(x^2 + 3x + 1) = 0$."

"We can easily find the two solutions to the equation $x^2 + 3x + 1 = 0$ using the quadratic formula," Recordis said.

$$x = \frac{-3 \pm \sqrt{9 - 4}}{2}$$

$$x = -0.3820 \quad \text{or} \quad x = -2.6180.$$

To satisfy Recordis' skepticism, we verified that $x = -1$, $x = -0.3820$, and $x = -2.6180$ were indeed the solutions to the equation

$$x^3 + 4x^2 + 4x + 1 = 0$$

(Note that when you write an irrational solution as a decimal approximation the equation will not be exactly true.)

"Since this is a third-degree equation, we know that it can have at most three solutions, so we have found all of the possible solutions," the professor said, satisfied.

"Therefore, whenever we find one solution for a polynomial equation, we can use algebraic division to simplify the equation," the king said.

"Algebraic division is a lot of work!" Recordis said. However, he did write down some clues for how to work algebraic division:

- Write the terms in both the dividend and the divisor in order of decreasing power of x. If the coefficient of one power of x is zero, then mark its place with a zero. Throughout the operation, keep terms with the same power of x aligned in the same column.

Here are some examples:

$$\begin{array}{r|l} & \quad\quad\quad\quad\quad 5x^2 + x + 3 \\ \hline 2x^2 + 4x - 3 & 10x^4 + 22x^3 - 5x^2 + 9x - 9 \\ & \underline{10x^4 + 20x^3 - 15x^2} \\ & \quad\quad 2x^3 + 10x^2 + 9x \\ & \quad\quad \underline{2x^3 + 4x^2 - 3x} \\ & \quad\quad\quad\quad 6x^2 + 12x - 9 \\ & \quad\quad\quad\quad \underline{6x^2 + 12x - 9} \end{array}$$

Therefore,

$$\frac{10x^4 + 22x^3 - 5x^2 + 9x - 9}{2x^2 + 4x - 3} = 5x^2 + x + 3$$

with no remainder.

$$\begin{array}{r} 2x^2 + 3x + 5 \\ x - 4\overline{)2x^3 - 5x^2 - 7x - 15} \\ \underline{2x^3 - 8x^2} \\ 3x^2 - 7x \\ \underline{3x^2 - 12x} \\ 5x - 15 \\ \underline{5x - 20} \\ 5 \end{array}$$

The remainder is 5, so

$$\frac{2x^3 - 5x^2 - 7x - 15}{x - 4} = 2x^2 + 3x + 5 + \frac{5}{x - 4}$$

"None of this information tells us how to find one solution in the first place!" Recordis said. "We'll still have to guess to find the first solution." (It turns out that there is no general way to solve polynomial equations of degree 3 or greater. Guessing is sometimes the best way. If you study math further then you will learn some more techniques that provide clues about the solutions.)

"We still must find a way to catch the dragon," the detective said. He noticed a pile of loose logs lying on the bluff. "If we could somehow build those logs into a high lookout tower, then we could observe the dragon's movements more easily."

Notes to Chapter 13

- Synthetic division is a short way of dividing a polynomial by a binomial of the form $x - b$. For example, to find

$$\frac{4x^3 - 17x^2 + 21x - 18}{x - 3}$$

by algebraic division, we would have to write:

$$\begin{array}{r} 4x^2 - 5x + 6 \\ x - 3\overline{)4x^3 - 17x^2 + 21x - 18} \\ \underline{4x^3 - 12x^2} \\ - 5x^2 + 21x \\ \underline{- 5x^2 + 15x} \\ 6x - 18 \\ \underline{6x - 18} \end{array}$$

In order to make the division shorter, we leave out all the x's and just write the coefficients. Also, we reverse the sign of the divisor ($x - 3$ in this case) so as to make every intermediate subtraction into an addition. Finally, we condense everything into three lines.

First, write the coefficients on a line:

4	−17	21	−18	3

The 4 is the coefficient of x^3; -17 is the coefficient of x^2; 21 is the coefficient of x; -18 is the constant term; the 3 comes from reversing the sign of the -3 in the divisor.

Second, bring down the first coefficient into the answer line:

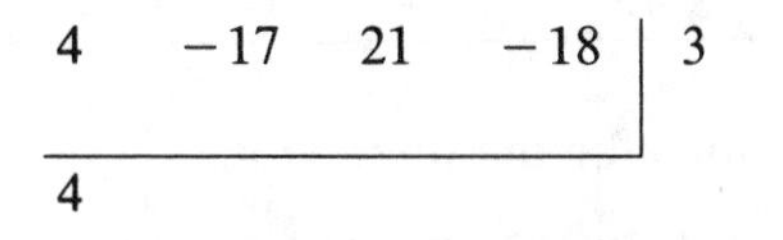

4	−17	21	−18	3
4				

Third, multiply the 4 in the answer by the 3 in the divisor, write the result (12) in the next column of the middle line, and then add -17 to 12:

4	−17	21	−18	3
	12			
4	−5			

Repeat the process for the next column: multiply -5×3, write the result on the middle line, and then add:

4	−17	21	−18	3
	12	−15		
4	−5	6		

Multiply 6×3, write the result on the middle line, and then add again:

4	−17	21	−18	3
	12	−15	18	
4	−5	6	0	

The farthest right entry in the answer line is the remainder (in this case 0). The remaining elements in the answer line are (from right to left) the coefficients of x^0, x^1, and x^2. Therefore, in this case the result is $4x^2 - 5x + 6$. Compare this result with the result found by algebraic division.

- The *rational root theorem* says that if the polynomial equation

$$a_n x^n + a_{n-1}x^{n-1} + a_{n-2}x^{n-2} + \cdots a_2x^2 + a_1x + a_0 = 0$$

(where $a_0, a_1, \ldots, a_n$ are all integers) has any rational roots, then each rational root can be expressed as a fraction in which the numerator is a factor of a_0 and the denominator is a factor of a_n. This theorem sometimes makes it easier to find the roots of complicated polynomial equations, but it provides no help if there are no rational roots to begin with.

For example, suppose we are looking for the rational roots of the equation

$$x^3 - 9x^2 + 26x - 24 = 0$$

In this case $a_n = 1$ and $a_0 = -24$. Therefore, the rational roots, if any, must have a factor of 24 in the numerator and 1 in the denominator. The factors of 24 are 1, 2, 3, 4, 6, 8, 12, and 24. If we test all of these possibilities, it turns out that the three roots are 2, 3, and 4.

- It is possible to identify the locations of the turning points for a polynomial curve with the use of calculus.

Exercises

Solve these equations for the unknown:

1. $x^3 - x = 0$ (Hint: factor an x out of each term.)
2. $x^3 - 2x = 0$
3. $2x^3 - 3x^2 - 5x = 0$
4. $3x^5 + 4x^4 - 10x^3 = 0$
5. $y^2 - 13y + 36 = 0$
6. $x^4 - 13x^2 + 36 = 0$ (Hint: See previous exercise.)
7. $x^4 - 2x^2 - 35 = 0$
8. $x^6 + 4x^3 - 96 = 0$
9. $x^{10} - 33x^5 + 32 = 0$
10. $x^3 - 9x^2 + 20x = 0$
11. $x^3 - 2x^2 - 35x = 0$

In each exercise below, make up a polynomial equation that has the solutions indicated:

12. $x = 1, x = 2$, or $x = 3$
13. $x = -3, x = 4$, or $x = 6$
14. $x = 11, x = 1$, or $x = -2$
15. $x = 2 + \sqrt{3}, x = 2 - \sqrt{3}$, or $x = 4$
16. $x = 0, x = 1$, or $x = -3$

Perform these divisions:

17. $(x^2 + 2x + 2)/(x - 1)$
18. $(2x^4 + 5x^3 + 7x^2 + 5x)/(2x^2 + 3x + 4)$
19. $(15x^4 - 15x + 1)/(3x^3 - 3)$
20. $(5x^4 + 4x^3 - 5x^2 - 4x + 1)/(x^2 - 1)$
21. $(3x^4 - x^3 - 33x^2 - 25x - 61)/(x^2 - x - 12)$

Perform these divisions using synthetic division. For example, to find $(2x^2 + 7x + 5)/(x + 8)$:

$$\begin{array}{rrr|r} 2 & 7 & 5 & -8 \\ & -16 & 72 & \\ \hline 2 & -9 & 77 & \end{array}$$

so the answer is $2x - 9$ with a remainder of 77.

22. $(x^2 + 2x + 2)/(x + 2)$
23. $(x^2 - 3x - 1)/(x - 1)$
24. $(x^2 + x + 1)/(x - 4)$
25. $(3x^2 + 5x + 1)/(x + 3)$
26. $(6x^3 - 6x^2 + 5x + 10)/(x - 3)$
27. $(3x^3 + 2x^2 + 5x + 4)/(x - 6)$
28. $(2x^3 + 6x^2 - 8x - 9)/(x - 1)$
29. $(5x^3 + 8x^2 - 6x + 6)/(x - 4)$
30. $(x^3 + x^2 - 13x - 4)/(x + 4)$
31. $(3x^3 - 13x^2 - 29x - 6)/(x - 6)$
32. $(6x^3 + 10x^2 - 33x - 27)/(x + 3)$
33. $(8x^3 - 2x^2 + 11x + 3)/(x + 1/4)$

Perform these divisions:

34. $(x^2 - 1)/(x - 1)$
35. $(x^3 - 1)/(x - 1)$
36. $(x^4 - 1)/(x - 1)$
37. $(x^5 - 1)/(x - 1)$
38. Can you guess what $(x^n - 1)/(x - 1)$ is?

Find the solutions to these equations. (Hint: Use the rational root theorem. Once you have found one solution, use synthetic division to simplify the equation before you look for the rest of the solutions.)

39. $x^3 - 16x^2 + 65x - 50 = 0$
40. $x^3 - 3x^2 - 4x + 12 = 0$
41. $x^3 + 6x^2 + 12x + 8 = 0$
42. $x^3 - 5x^2 + 3x + 9 = 0$
43. $x^3 - 24x^2 + 176x - 384 = 0$
44. $4x^3 + 4x^2 - 36x - 36 = 0$
45. $12x^3 + 4x^2 - 3x - 1 = 0$
46. $x^3 - 5x^2 - 49x + 245 = 0$

In each exercise below, one solution of the equation is given. Find the remaining solutions if there are others.

47. $x^3 + 3x^2 - 21x - 63 = 0 \quad (x = -3)$
48. $2x^3 + 8x^2 - 14x - 56 = 0 \quad (x = -4)$
49. $x^4 - 2x^3 - 7x^2 + 12x = 0 \quad (x = 3)$
50. $x^3 - 5x^2 - 3x + 15 = 0 \quad (x = \sqrt{3})$
51. $x^3 + 10x^2 + 20x - 24 = 0 \quad (x = \sqrt{8} - 2)$

Suppose $f(x)$ is a polynomial in x. Suppose we perform the division $f(x)/(x - r)$ and observe the remainder. It turns out that the remainder is $f(r)$. (This result is called the *remainder theorem*.) For example, we'll calculate $(x^3 + 3x^2 - 4x + 5)/(x - 1)$ by synthetic division:

1	3	−4	5	1
	1	4	0	
1	4	0	5	

The quotient is $(x^2 + 4x)$ with a remainder of 5. Sure enough, $f(1) = 5$. Perform the following divisions to show that the remainder theorem holds:

52. $(5x + 6)/(x - 2)$
53. $(x^2 + x + 5)/(x + 5)$
54. $(3x^2 - 16x + 7)/(x - 7)$
55. $(x^2 - x - 12)/(x - 4)$
❑ **56.** Write a computer program to draw a graph of a polynomial. Have the program read in the coefficients of the polynomial first, and then put the graph on the screen. Design the program so you can adjust the location on the screen where the origin of the coordinate axes appear, and make it possible to either magnify your view of a portion of the curve or else obtain a wide-angle view of a larger section of the curve.
❑ **57.** Write a computer program that reads in the coefficients of two polynomials $f_1(x)$ and $f_2(x)$ and then performs the algebraic division $f_1(x)/f_2(x)$.
❑ **58.** Write a computer program that reads in the coefficients of a fourth-degree polynomial $f(x) = a_4x^4 + a_3x^3 + a_2x^2 + a_1x + a_0$ and then finds a solution to the equation $f(x) = 0$. Have the computer read in an initial guess for the solution (x_1), and then calculate a closer guess (x_2) from this formula:

$$x_2 = x_1 - \frac{a_4x^4 + a_3x^3 + a_2x^2 + a_1x + a_0}{4a_4x^3 + 3a_3x^2 + 2a_2x + a_1}$$

Keep repeating the process of using this formula to find guesses that become closer to the solution. When you find a value of x such that $|f(x)| < 0.0001$, then treat that as the solution. (This method is known as *Newton's method*. See a book on calculus for more details.)

For Exercises 59 to 68, use a spreadsheet, graphing calculator, or the program from Exercise 56 to graph the polynomial. Then determine the values of x that make the polynomial zero, either by looking at the graph or by using the program from Exercise 58.

59. $x^3 - x^2 - 44x + 84$
60. $x^3 + 2x^2 - 15x - 36$
61. $x^3 + x^2 - 6x - 18$
62. $x^3 + 2x^2 + 3x + 4$
63. $x^4 + 2x^3 - 47x^2 + 48x + 252$
64. $x^4 + 14x^3 + 67x^2 + 144x + 126$
65. $x^4 - 4x^3 - 44x^2 + 96x + 576$
66. $x^4 - 6x^3 - 12x^2 + 88x - 96$
67. $x^4 + 8x^3 + 26x^2 + 40x + 24$
68. $x^4 + 2x^3 - 47x^2 - 48x + 900$

14 Series

Builder quickly designed a triangle-shaped lookout tower using the logs. "The top row will have one log, the second row will have two logs, and so on," he explained. (See Figure 14–1.) "There will be a total of 20 rows, but so far I've only numbered the logs for the first seven rows."

"How many logs will we need?" the king asked.

"We can find the answer by addition," the professor said.

(number of logs) =

$$1 + 2 + 3 + 4 + 5 + 6 + 7 + 8 + 9 + 10$$
$$+ 11 + 12 + 13 + 14 + 15 + 16 + 17 + 18 + 19 + 20$$

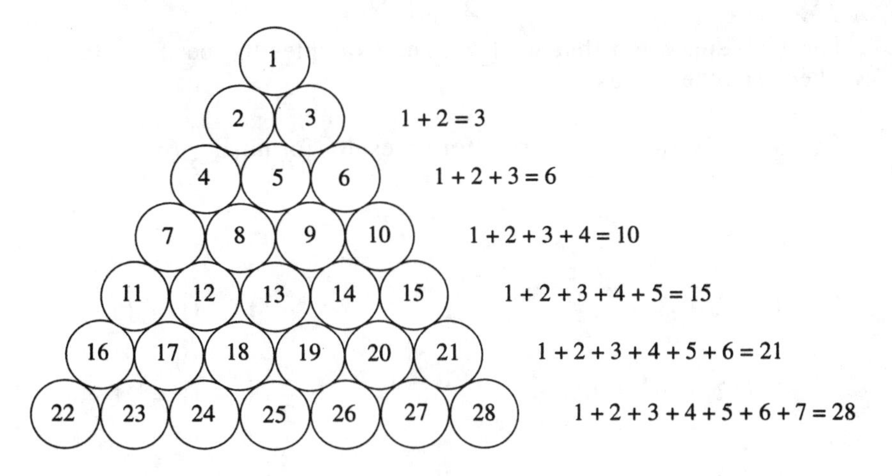

Figure 14–1

"It will take a long time to add together 20 numbers!" Recordis protested.

"That's *your* job!" the professor responded.

"There must be an easier way!" Recordis stared at the long sum. However, at that moment Igor seemed to develop a slight malfunction. The numbers floated aimlessly around the screen. Recordis had to hit the visiomatic picture chalkboard machine several times until finally a semblance of order was restored. Still, things were not quite correct. "You've got the bottom row backwards!" he screamed.

$$\begin{array}{l} 1 + \ 2 + \ 3 + \ 4 + \ 5 + \ 6 + \ 7 + \ 8 + \ 9 + 10 \\ + 20 + 19 + 18 + 17 + 16 + 15 + 14 + 13 + 12 + 11 \end{array}$$

However, Recordis suddenly noticed something. "Let's add together the numbers in each column," he suggested.

$$21 + 21 + 21 + 21 + 21 + 21 + 21 + 21 + 21 + 21$$

Each column summed to 21, and since there were 10 columns the total sum was $10 \times 21 = 210$.

"Therefore, we need 210 logs!" Recordis said triumphantly.

The professor did not believe that the answer could be found from so simple a formula, but since Recordis refused to calculate the sum the long way the professor had to do it herself. Sure enough, the result was 210.

"We can make a general rule for the sum of all the whole numbers from 1 up to a particular number n," Recordis said. "Write the sum like this:

$$1 + 2 + 3 + 4 + \cdots + n$$

$$\begin{array}{l} = 1 + 2 + 3 + 4 + \cdots + \frac{n}{2} \\ + n + (n - 1) + (n - 2) + (n - 3) + \cdots + \left(\frac{n}{2} + 1\right) \end{array}$$

(The three dots $\cdots$ indicate that not all of the terms have been listed.) Then, since the sum of each column is $(n + 1)$, and since there are $n/2$ columns, the total sum is

$$(\text{sum of all of the integers from 1 to } n)$$

$$= \frac{n}{2}(n + 1)$$

The king suggested that we try some examples to see if the formula worked for other cases.

series	brute-force result	formula: $\frac{n}{2}(n + 1)$
$1 + 2$	3	$\frac{2}{2}(2 + 1) = 3$
$1 + 2 + 3 + 4$	10	$\frac{4}{2}(4 + 1) = 10$
$1 + 2 + \cdots + 6$	21	$\frac{6}{2}(6 + 1) = 21$
$1 + 2 + \cdots + 8$	36	$\frac{8}{2}(8 + 1) = 36$

"Ha!" the professor said. "We have only shown that this formula works if n is an even number!" However, we tried some examples where n was an odd number, and the professor was forced to concede that the formula seemed to work in that case as well.

$$1 + 2 + 3 = 6; \quad \frac{3}{2}(3 + 1) = 6$$

$$1 + 2 + \cdots + 5 = 15; \quad \frac{5}{2}(5 + 1) = 15$$

$$1 + 2 + \cdots + 7 = 28; \quad \frac{7}{2}(7 + 1) = 28$$

"We still have not *proved* that this formula is true," the professor said suspiciously, but she soon gave up trying to think of any counterexamples.

"Now we can find the sum of all of the numbers from 1 to 1,000," the king said. "I always wondered what that sum was, but I never had time to do the calculations myself."

$$\begin{aligned} 1 + 2 + 3 + \cdots + 1{,}000 &= \left(\frac{1{,}000}{2}\right)(1{,}000 + 1) \\ &= 500 \times 1{,}001 \\ &= 500{,}500 \end{aligned}$$

The professor suggested that we use the name *series* to mean a long sum like this. She also suggested that we use a crooked symbol, Σ, to mean summation.

"Why?" Recordis asked in puzzlement.

"The crooked symbol is a Greek capital letter called *sigma*, which is a bit like an *S*," the professor explained. "So it seems natural to use it for summation. If we want to add up all the possible values of a letter i that can stand for integers, we can write it like this:

$$\Sigma\, i$$

"How do we know what values of i you want to add?" Recordis demanded. "Nothing you have written there tells us where the values of i are supposed to start, or where they are supposed to stop."

"We can put the starting number at the bottom of the Σ and we can put the stopping number at the top," the king suggested.

$$\sum_{i=1}^{20} i$$

"This symbol means: Add up all of the values of i, starting at $i = 1$ and continuing until $i = 20$. For example,

$$\sum_{i=1}^{20} i = 1 + 2 + \cdots + 20 = \left(\frac{20}{2}\right)(20 + 1) = 210$$

as we already know."

We called this notation *summation notation*. We wrote some more examples of summation notation:

$$\sum_{i=1}^{5} i = 1 + 2 + 3 + 4 + 5 = 15$$

$$\sum_{j=1}^{10} j^2 = 1 + 4 + 9 + 16 + 25 + 36 + 49 + 64 + 81 + 100 = 385$$

"I don't like this notation very much," Recordis said. "It's too hard to understand."

"I'm sure you'll get used to it with practice," the professor said encouragingly. "Another amazingly versatile feature of this new notation is that we don't always have to start at one." She wrote an example:

$$\sum_{i=11}^{20} i = 11 + 12 + 13 + \cdots + 20$$

"We don't have a simple formula for that sum," Recordis complained.

"We'll have to find one," the professor said. "It's boring to have the variable in the sum only go up by 1 each time, so let's derive a general formula for a series with the first term a, the next term $(a + d)$, then $(a + 2d)$, and so on. We can think of d as the *difference* between the successive terms of the series."

We let S_n represent the sum of the first n terms of this series:

$$S_n = a + (a + d) + (a + 2d) + (a + 3d) + \cdots + [a + (n - 1)d]$$

(A series such as this, in which the difference between any two successive terms is always the same, is called an *arithmetic series*.)

"We can also write the expression for the series in terms of summation notation," the professor said.

$$S_n = \sum_{i=0}^{n-1} (a + id)$$

We combined all of the a's into one term and wrote the series like this:

$$S_n = na + [0 + d + 2d + 3d + \cdots + (n - 1)d]$$

Then we factored a d out of the expression in brackets:

$$S_n = na + d[1 + 2 + 3 + \cdots + (n - 1)]$$

We realized that we could use our formula to simplify the sum in brackets:

$$1 + 2 + 3 + \cdots + (n - 1) = \frac{1}{2}(n - 1)n$$

Therefore,

$$S_n = na + \frac{1}{2}n(n - 1)d$$

The professor suggested that we try an example:

$$S_n = 12 + 16 + 20 + 24 + 28 + 32$$

We identified the first term in the series: $a = 12$; the difference between adjacent terms in the series: $d = 4$; and the number of terms in the series: $n = 6$. Then we inserted these values into the formula:

$$S_n = 6 \times 12 + \frac{1}{2} \times 6 \times (6 - 1) \times 4$$

$$= 132$$

We calculated some more examples:

$$2 + 4 + 6 + 8 + 10 + 12 = 6 \times 2 + \frac{1}{2} \times 6 \times 5 \times 2 = 42$$

$$2 + 4 + 6 + \cdots + 100 = 50 \times 2 + \frac{1}{2} \times 50 \times 49 \times 2 = 2{,}550$$

$$35 + 42 + 49 + \cdots + 84 = 8 \times 35 + \frac{1}{2} \times 8 \times 7 \times 7 = 476$$

$$6 + 6\frac{1}{3} + 6\frac{2}{3} + \cdots + 8\frac{2}{3} = 9 \times 6 + \frac{1}{2} \times 9 \times 8 \times \frac{1}{3} = 66$$

"You had better be glad that I thought of a short way to calculate the sums of these series," Recordis told the professor. "Where would you be without me?"

"I will be eternally in your debt," the professor told him.

"I therefore hope you will be willing to pay me back through all eternity," Recordis said.

"All right," the professor slyly agreed. I will pay you back a little bit each day from now to forever. Unfortunately, I don't have very much—but I will give you one dollar today, 1/2 dollar tomorrow, then 1/4 dollar the next day, then 1/8, and so on forever."

Recordis agreed to this plan. "I won't be getting very much each day, but since I will be paid for an infinite number of days, I will have an infinite amount of money!" He wrote the expression for the amount of money he would have:

$$1 + \frac{1}{2} + \frac{1}{4} + \frac{1}{8} + \frac{1}{16} + \frac{1}{32} + \cdots$$

"This is a new kind of series," Recordis continued. "In an arithmetic series, we found that the difference between any two consecutive terms was a constant. In this series, it seems that the ratio between any two consecutive terms is always the same—in this case, one-half."

(A series in which the ratio between two consecutive terms is always the same is called a *geometric* series.)

"I hate to disillusion you," the king said, "but I don't think you will end up with an infinite amount of money."

"Well, at least it will be close to infinity," Recordis said.

"Actually, I think you will only end up with two dollars," the king said.

$$1 + \frac{1}{2} + \frac{1}{4} + \frac{1}{8} + \cdots = 2$$

"How can the sum of an infinite number of terms be a finite number?" Recordis screamed. However, the king calculated the decimal approximations for the sum of these terms and illustrated with a graph:

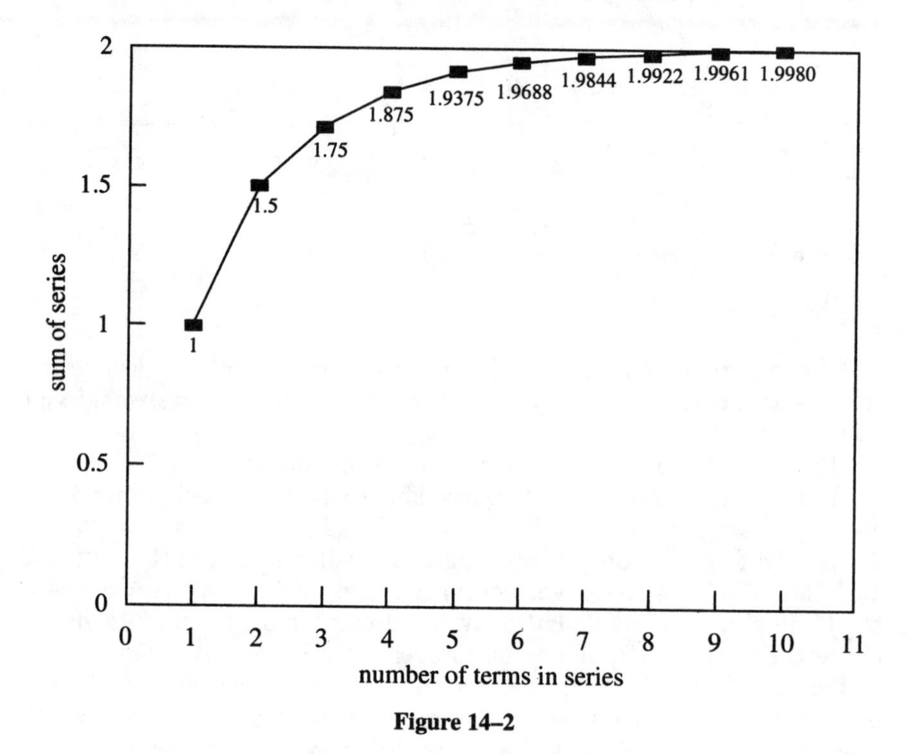

Figure 14–2

With a sinking feeling, Recordis realized that, no matter how many terms were added to this series, the result would still be less than 2.

"You tricked me!" Recordis accused the professor. "We must make sure that this never happens again. How can we tell what the sum of a geometric series will be?"

That problem proved to be very difficult, and we had not made any progress before we were interrupted by a visit from Mrs. O'Reilly. "I will be saving 100 dollars each year for the next ten years," she explained. "I need you to tell me how much I will have in the bank at the start of the tenth year."

"That's easy," Recordis said. "You will have $10 \times 100 = 1{,}000$ dollars!"

"But you forgot that the bank pays five percent interest!" Mrs. O'Reilly reminded him.

"That's right," Recordis said. "After ten years you will have 100×1.05^{10}."

"But now you're forgetting that each year I'm putting an additional 100 dollars in the bank," she said. "At the start of the first year I will have 100; at the start of the second year I will have $100 + 100 \times 1.05$; at the start of the third year I will have

$$100 + (100 \times 1.05) + (100 \times 1.05^2)$$

At the start of the tenth year I will have

$$100 + (100 \times 1.05) + (100 \times 1.05^2) + \cdots + (100 \times 1.05^9)$$

and so on."

"This is another geometric series," the professor said. "In this case it is clear that the ratio between any two consecutive terms is 1.05. However, this series has a finite number of terms, whereas the previous series had an infinite number of terms."

"Oh no!" Recordis complained. "This could not have occurred at a worse time! We don't know a formula for the sum of a geometric series!"

"Don't panic!" the professor said. "We just need to be systematic. Let's develop a general formula for a geometric series in which a is the first term and r is the constant ratio between consecutive terms." Again we used S_n to stand for the sum of the first n terms of the series.

$$S_n = a + ar + ar^2 + ar^3 + \cdots + ar^{n-1}$$

"In our case, we have $a = 100$ and $r = 1.05$."

We wrote the same expression using summation notation:

$$S_n = \sum_{k=0}^{n-1} ar^k$$

"We can factor an a out of every term," Recordis said helpfully.

$$S_n = a(1 + r + r^2 + r^3 + \cdots + r^{n-1})$$

"So the only hard part is figuring out a simple formula for this expression," the professor said.

$$1 + r + r^2 + r^3 + \cdots + r^{n-1}$$

"That expression is as simple as it can get!" Recordis insisted. "There are no like terms to combine. The expression can't be factored, and it can't be simplified by taking a power or a root of it. What else is there?"

We were becoming desperate now. "In times such as this," the king said, "there seems to be a paradoxical truth: The only way to make the expression simpler is first to make it more complicated. We must think of something to multiply this expression by such that the result turns out to be simpler than the original formula." We tried several guesses. We quickly found that multiplying the long expression by r did not help. Neither did multiplying by $1/r$, or multiplying by r^2, or $1/r^2$, or $(r + 1)$. However, something exciting happened when we multiplied the long expression by $(r - 1)$:

$$(r - 1)(1 + r + r^2 + r^3 + \cdots + r^{n-1}) =$$

$$\begin{array}{rl} & r + r^2 + r^3 + \cdots + r^{n-1} + r^n \\ -1 & - r - r^2 - r^3 - \cdots - r^{n-1} \end{array}$$

"When we add those together, almost all of the terms will cancel out!" Recordis said gleefully. The only terms left were -1 and r^n, so the result was:

$$(r - 1)(1 + r + r^2 + r^3 + \cdots + r^{n-1}) = (r^n - 1)$$

Next we divided both sides by $(r - 1)$, to make up for the fact that we multiplied by $(r - 1)$ earlier:

$$1 + r + r^2 + r^3 + \cdots + r^{n-1} = \frac{r^n - 1}{r - 1}$$

We checked to make sure that the formula worked for $n = 1$, $n = 2$, and $n = 3$:

$$(r^1 - 1)/(r - 1) = 1$$

$$(r^2 - 1)/(r - 1) = 1 + r$$

$$(r^3 - 1)/(r - 1) = 1 + r + r^2$$

(We used synthetic division to calculate these results. See Chapter 13.)

We wrote out the general formula for the geometric series:

$$S_n = a + ar + ar^2 + \cdots + ar^{n-1} = a\frac{r^n - 1}{r - 1}$$

"We can also write that another way if we want to," the professor said.

$$S_n = a\frac{1 - r^n}{1 - r}$$

"And now we can solve the problem of the bank balance," the king said.

$$\text{balance} = 100 + (100 \times 1.05) + (100 \times 1.05^2) + \cdots + (100 \times 1.05^9)$$

Using the formula with $a = 100$, $r = 1.05$, and $n = 10$,

$$100\left(\frac{1.05^{10} - 1}{1.05 - 1}\right) = 1257.79$$

We calculated some more examples of geometric series:

$$1 + 2 + 4 + 8 = 1 \times (2^4 - 1)/(2 - 1) = 15$$

$$1 + 2 + 4 + \cdots + 128 = 1 \times (2^8 - 1)/(2 - 1) = 255$$

$$10 + 100 + 1{,}000 + 10{,}000 + 100{,}000 = 10(10^5 - 1)/(10 - 1)$$

$$= 111{,}110$$

$$3 + 1 + 1/3 + 1/9 + 1/27 + 1/81 = 3\left[\tfrac{1}{3}^6 - 1\right]/\left(\tfrac{1}{3} - 1\right) = 4.4938$$

"Now we must investigate the question of when a geometric series will have an infinite sum," Recordis said.

"Clearly the series can have an infinite sum only if it has an infinite number of terms," the professor said. We also realized that if the ratio between successive terms was greater than 1, then the sum of an infinite series would have to be infinity, since each term would be larger than the last one. For example, the sum of the series

$$1 + 2 + 4 + 8 + 16 + 32 + 64 + \cdots$$

is clearly infinity.

The professor continued, "An infinite geometric series can only have a finite sum if r is less than 1. Now, let's look at the formula for the sum, in the case where r is less than 1 and n is very large.

$$a + ar + ar^2 + \cdots + ar^{n-1} = a\frac{1 - r^n}{1 - r}$$

"If r is less than 1, then r^n will become very small as n approaches infinity," the king realized. He calculated some examples: $(1/2)^{100} = 7.89 \times 10^{-31}$; $(9/10)^{1000} = 1.75 \times 10^{-46}$; and $(1/10)^{50} = 1 \times 10^{-50}$.

"In fact, in the limit where n goes to infinity, we can say that r^n is zero!" the professor said. "Then we have the result we need

$$a(1 + r + r^2 + r^3 + r^4 + \cdots) = a\frac{1}{1 - r}$$

For example,

$$1 + 1/2 + 1/4 + 1/8 + \cdots = \frac{1}{1 - 1/2} = 2$$

$$1 + 0.8 + (0.8)^2 + (0.8)^3 + \cdots = \frac{1}{1 - 0.8} = 5$$

$$\frac{1}{1.05} + \frac{1}{(1.05)^2} + \frac{1}{(1.05)^3} + \cdots = \left(\frac{1}{1.05}\right)\left(\frac{1}{1 - 1/1.05}\right) = \frac{1}{0.05} = 20$$

We returned to the Main Conference Room while Mrs. O'Reilly returned to the hotel and Builder set to work to construct the lookout tower. "It will take a while to put all 210 logs in place," he said.

SUMMARY OF ARITHMETIC AND GEOMETRIC SERIES

An arithmetic series is the sum of n terms that all have a common difference between them:

$$S_n = a + (a + d) + (a + 2d) + (a + 3d) + \cdots + [a + (n - 1)d]$$

where a is the initial term and d is the common difference. The sum is given by this formula:

$$S_n = na + \frac{1}{2}n\,(n - 1)d$$

In the special case where $a = 1$ and $d = 1$:

$$S_n = 1 + 2 + 3 + 4 + \cdots + n = \frac{1}{2}n\,(n + 1)$$

A geometric series is the sum of n terms that all have a common ratio between them:

$$S_n = a + ar + ar^2 + ar^3 + \cdots + ar^{n-1}$$

where a is the initial term and r is the common ratio. The sum is given by this formula:

$$S_n = a\frac{r^n - 1}{r - 1} = a\frac{1 - r^n}{1 - r}$$

If there are an infinite number of terms there will be a finite sum if $0 < r < 1$:

$$S_n = a + ar + ar^2 + ar^3 + \cdots = a\,\frac{1}{1 - r}$$

Exercises

Write these series using summation notation:

1. The sum of all integers from 10 to 30.
2. The sum of all odd numbers from 1 to 49.

3. The sum of all even numbers from 20 to 60.
4. The sum of all numbers that are the squares of integers that are less than 100.
5. Show that $\sum_{i=1}^{10} af(x_i) = a \sum_{i=1}^{10} f(x_i)$.

Use the property from the previous exercise to calculate these sums:

6. $\sum_{i=1}^{10} 3i$ **7.** $\sum_{i=1}^{10} 4i$ **8.** $\sum_{i=1}^{10} 10i$

Calculate the sums of these series:

9. $1 + 2 + 3 + 4 + \cdots + 1{,}000{,}000$
10. $5 + 10 + 15 + 20 + \cdots + 100$
11. $16 + 22 + 28 + \cdots + 76$
12. $1 + 1\frac{1}{2} + 2 + 2\frac{1}{2} + \cdots + 10$
13. $-3 - 2 - 1 + 0 + 1 + 2 + 3 + 4 + 5$
14. $-10 - 7 - 4 - 1 + 2 + 5 + 8 + 11$
15. $3/10 + 3/100 + 3/1{,}000 + 3/10{,}000 + \cdots$
16. $2/10 + 2/100 + 2/1{,}000 + 2/10{,}000 + \cdots$
17. $18/100 + 18/10{,}000 + 18/1{,}000{,}000 + \cdots$
18. $1 + 4/5 + (4/5)^2 + (4/5)^3 + (4/5)^4 + \cdots$
19. $1 + 2/3 + (2/3)^2 + (2/3)^3 + (2/3)^4 + \cdots$
20. $10 + 100 + 1{,}000 + 10{,}000 + \cdots + 100{,}000{,}000$
21. $1.05 + 1.05^2 + 1.05^3 + 1.05^4 + 1.05^5$
22. $1.05 + 1.05^2 + 1.05^3 + \cdots + 1.05^{20}$
23. $2 + 4 + 8 + 16 + 32 + 64 + 128$
24. $3 + 9 + 27 + 81 + 243 + 729$
25. $1 - 1 + 1 - 1 + 1 - 1$
26. $1 - 1 + 1 - 1 + 1 - 1 + 1$
27. $1 - 3 + 9 - 27 + 81 - 243$
28. $-3 + 6 - 12 + 24 - 48 + 96 - 192$

What is the nth term in each series?

29. $2 + 4 + 6 + 8 + \cdots$
30. $25 + 30 + 35 + 40 + \cdots$
31. $12 + 24 + 36 + 48 + \cdots$
32. $56 + 59 + 62 + 65 + \cdots$
33. $32 + 16 + 8 + 4 + 2 + 1 + 1/2 + \cdots$
34. $3 + 9 + 27 + 81 + 243 + \cdots$
35. $100 + 100(1.05) + 100(1.05)^2 + \cdots$

Determine the common difference d between the terms in the arithmetic series $S = a + (a + d) + (a + 2d) + \cdots + [a + (n - 1)d]$ if:

36. $S = 55, n = 10, a = 1$
37. $S = 610, n = 20, a = 2$
38. $S = 760, n = 16, a = 10$
39. $S = 800, n = 5, a = 100$

Determine the first term in the geometric series $S = a + ar + ar^2 + ar^3 + \cdots + ar^{n-1}$ if:

40. $S = 1726.285 \quad r = 1.05 \quad n = 15$
41. $S = 310 \quad r = 2 \quad n = 5$

42. $S = 40\ 4/9 \qquad r = 3 \qquad n = 6$
43. $S = 3.9514 \qquad r = 4/5 \qquad n = 7$
44. The *present value* of an amount of money X that you receive n years in the future is

$$\text{(present value)} = \frac{X}{(1 + R)^n}$$

where R is the interest rate. Note that money you will receive in the future is worth less than money you have now. If you receive a stream of income in which X_i is the amount that you will receive in year i, then the total present value of the income stream is

$$\text{(present value)} = \sum_{i=0}^{n} \frac{X_i}{(1 + R)^i}$$

What is the present value of a stream of income in which you receive \$5 one year from now, and then \$5 every year for the next ten years after that? (Let $R = 0.05$.)
45. What is the present value of an income stream in which you receive \$5 every year forever? (Let $R = 0.05$.)
46. The formula for the first n terms of the arithmetic series $a + (a + d) + (a + 2d) + \cdots$ is $na + (1/2)n(n - 1)d$. What is the formula for the first $n + 1$ terms of this series?
47. The formula for the first n terms of the geometric series

$$a + ar + ar^2 + ar^3 + \cdots \quad \text{is} \quad a(r^n - 1)/(r - 1)$$

What is the formula for the sum of the first $n + 1$ terms of this series?
48. Suppose you are starting out on a 4-mile hike. You walk at a constant speed of 5 miles per hour. However, your enthusiastic dog travels at a constant speed of 9 miles per hour. Your dog keeps running ahead to the finish line, then running back to you, then running ahead to the finish again, and so on. Set up a geometric series that represents the total distance that your dog runs, and then calculate the sum of that series. After you have done it the hard way, can you think of an easier way to calculate the total distance that the dog travels?
49. Suppose everyone always spends 60 cents and saves 40 cents every time he or she receives an additional dollar of income. Suppose someone spontaneously increases spending by 100. Then the person who receives that 100 will increase spending by 60. The person receiving the 60 will increase spending by $(0.6)^2(100)$; the person receiving that amount will increase spending by $(0.6)^3(100)$; and so on. How much will total spending go up?
50. Suppose you know the first term, the last term, and the common difference d for an arithmetic series. Derive a formula that will tell you the sum of the series.
51. How many gifts do you receive in total in the song "Twelve Days of Christmas?" (See Chapter 11, Exercise 74, and look ahead to Chapter 16, Exercise 3.)

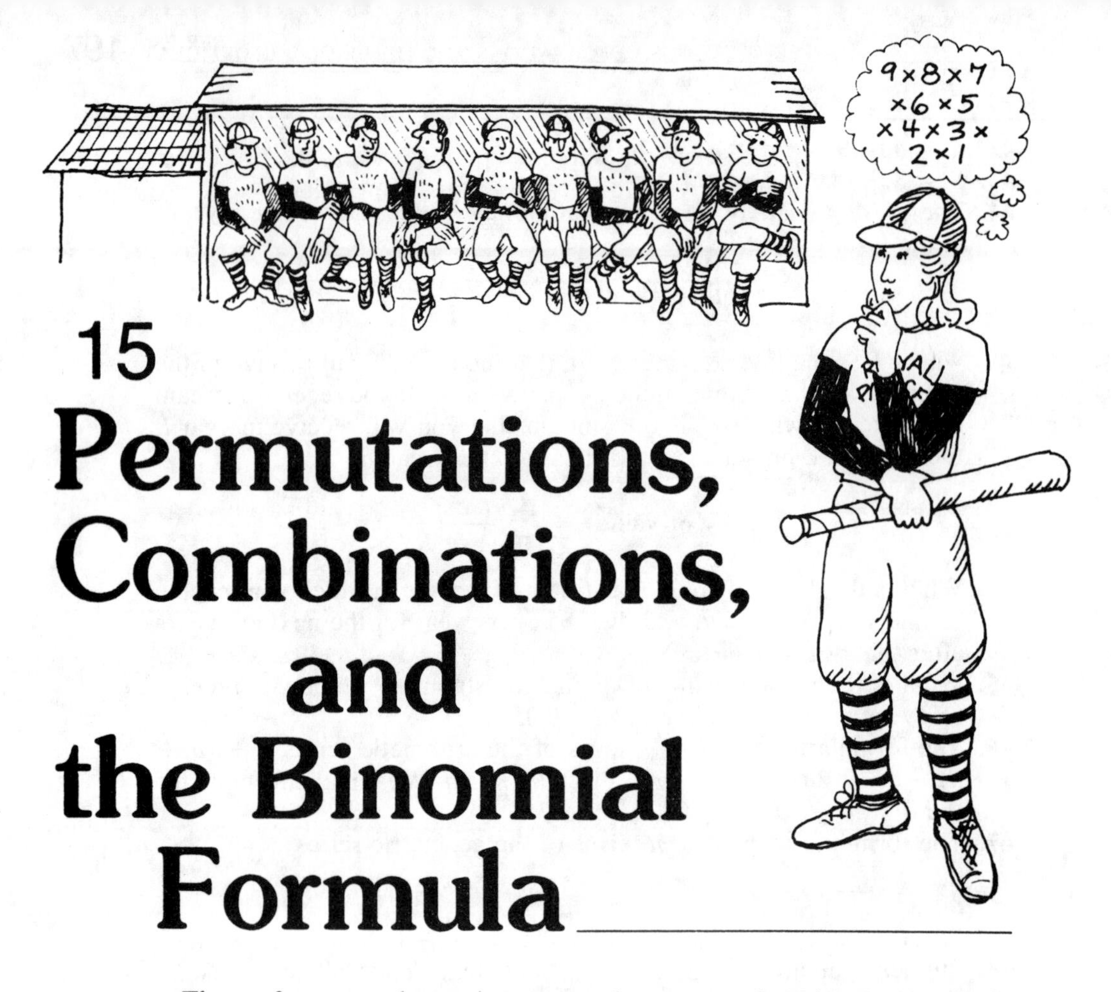

15 Permutations, Combinations, and the Binomial Formula

The professor was becoming extremely nervous. Opening Day was only one week away. "I don't know what our best batting order is," she fretted. "We have nine players, but I don't know in what order they should bat. I will tell the manager to try every single order and then we can determine which order will be best."

"How many possible batting orders are there?" Recordis asked. "Won't it take a long time to test every single possible order?"

"I'm sure we can derive a nice, simple formula," the professor said. "There clearly are nine possible choices for the batter who bats first, and then there must be nine possible choices for the batter who bats second, so there must be $9 \times 9 = 81$ possibilities for the first two batters."

"Wait a minute!" Recordis protested. "The first batter cannot bat again in the second position! Once you have chosen one player to be the first batter, then there are only 8 possibilities left for the second batter. So there are $9 \times 8 = 72$ ways of choosing the first two batters."

The professor blinked. "I hadn't thought of that," she said. "After the first two batters have been chosen, there must be 7 choices for the third batter, then 6 choices for the fourth batter, and so on." We figured out that after the first eight batters had been chosen, there was only one possible choice left for the ninth batter. Therefore, altogether there were

$$9 \times 8 \times 7 \times 6 \times 5 \times 4 \times 3 \times 2 \times 1$$

possible batting orders. Recordis carried out the multiplication, and came up with the result 362,880.

The professor stared in astonishment. "You are quite right—we will not play quite that many games this season. So there is no way that we will be able to test out every possible lineup." However, the professor was becoming extremely interested in the unusual expression

$$9 \times 8 \times 7 \times 6 \times 5 \times 4 \times 3 \times 2 \times 1$$

"Let's make up a shorter way to represent that expression," she said. "Let's write it like this:

$$9! = 9 \times 8 \times 7 \times 6 \times 5 \times 4 \times 3 \times 2 \times 1$$

We'll write the 9 to indicate that we want to multiply together all of the numbers from 1 up to 9. We'll put the exclamation mark there because this function can lead to some surprisingly large numbers."

Igor suggested that we use the name *factorial* to represent this function, so we made this definition:

FACTORIAL FUNCTION

The factorial of a whole number n (written $n!$ and read as "n-factorial") is the product of all of the whole numbers from 1 up to that number:

$$n! = n \times (n - 1) \times (n - 2) \times \cdots \times 4 \times 3 \times 2 \times 1$$

We calculated some examples of the factorial function:

$$1! = 1$$

$$2! = 2 \times 1 = 2$$

$$3! = 3 \times 2 \times 1 = 6$$

$$4! = 4 \times 3 \times 2 \times 1 = 24$$

$$5! = 120$$

$$6! = 720$$

$$69! = 1.711 \times 10^{98}$$

(The last calculation took an especially long time.)

"And we have already found a practical use for the factorial function: If you have n different objects, then there are $n!$ different ways of putting them in order," Recordis said.

"We should also define 0!," the professor said.

"That's easy," Recordis said. "$0! = 0$."

"No," the professor corrected. "We should have $0! = 1$. Suppose you have zero players on your team. Then there is only *one* possible way to put the team members in order, so we should say that $0! = 1$."

Recordis was still quite skeptical, but since he could not think of any reply to that logic he accepted the professor's definition that $0! = 1$.

For relaxation, the professor decided to help Pal play with his letter blocks. There were 26 blocks—1 for each letter. They began forming 4-letter words.

"I wonder how many different 4-letter words you can make with 26 letter blocks," the professor wondered. "There must be 26 possible choices for the first letter, then 25 choices for the second letter, then 24 choices for the third letter, and finally 23 choices for the last letter." She calculated that there would be

$$26 \times 25 \times 24 \times 23 = 358{,}800$$

possible words, so she quickly abandoned her original idea of making a list of all of the possibilities.

"I see a pattern," she realized. "There are 26 possible 1-letter words, $26 \times 25 = 650$ possible 2-letter words, $26 \times 25 \times 24 = 15{,}600$ possible 3-letter words, and so on." She looked for a simple formula for this result.

Once again the paradox appeared: in order to simplify the expression $26 \times 25 \times 24 \times 23$ we first had to make it more complicated by multiplying by 22!/22!:

$$26 \times 25 \times 24 \times 23 = \frac{26 \times 25 \times 24 \times 23 \times 22 \times 21 \times 20 \times \ldots \times 4 \times 3 \times 2 \times 1}{22 \times 21 \times 20 \times \ldots \times 4 \times 3 \times 2 \times 1}$$

$$\frac{26!}{22!}$$

Therefore, there are

$$\frac{26!}{22!} = 358{,}800$$

possible four letter words in which no letters repeat. In general, the number of j letter words with no repeats is given by the formula

$$\frac{26!}{(26 - j)!}$$

(Note: In general, if you select j objects from a group of n objects, then there are $n!/[n - j)!]$ different ways of putting these objects in order. This formula gives the number of *permutations* of n objects taken j at a time. In the formula for permutations, each different ordering of the objects is counted separately. For example, the word "bare" is counted separately from the word "bear," even though they have exactly the same letters.)

After the discoveries of the morning, Recordis and the professor decided to play a card game in which each player was dealt a 5-card hand from a 52-card deck. The professor was attempting to calculate mentally the probabilities associated with the various hands, but she found that she could not proceed unless she knew the number of possible 5-card hands that could be dealt from a 52-card deck.

"No problem," Recordis said. "There are 52 possibilities for the first card, then 51 possibilities left for the second card, then 50 possibilities, then 49, then 48, so the total number of hands is

$$52 \times 51 \times 50 \times 49 \times 48 = 311{,}875{,}200$$

"This is the same as the formula for the number of permutations with $n = 52$ and $j = 5$," Recordis noticed, "so we can write the number of different hands as $\frac{52!}{(52 - 5)!}$."

The professor decided to make a list of some of the possible hands (realizing that there were too many possibilities to list them all). Recordis dealt the professor five cards and she wrote them down in the order in which they were dealt:

king of hearts, king of diamonds, two of hearts,
two of spades, four of clubs

Then Recordis put those cards back in the deck and dealt five cards again:

king of diamonds, king of hearts, two of hearts,
two of spades, four of clubs

"You didn't shuffle the deck very well!" the professor screamed. "Those are exactly the same five cards you dealt me before! We shouldn't list this possibility separately."

"At least I didn't deal them to you in the same order as the first time," Recordis said defensively. "So you have to list this as a separate permutation, since the cards came in a different order."

The professor suddenly realized something was wrong. "Actually, our formula is correct," she decided, "but it answers the wrong question. We know that there are $52 \times 51 \times 50 \times 49 \times 48 = \frac{52!}{(52 - 5)!}$ different ways of dealing the five cards, provided that we count each possible ordering of those five cards separately. If we count the possibilities this way, then the two hands you dealt me would be counted separately. However, for many card games, we only care about the number of different ways of choosing the five cards in my hand. It doesn't matter in which order those five cards are actually dealt to the player."

"There are a lot of ways of putting these same five cards in different orders!" Recordis exclaimed. "In fact, we know that there are $5! = 120$ different ways of putting five objects in order."

"Precisely my point," the professor said. "The number 311,875,200 is 120 times too big, because it counts every group of five cards 120 times (one for each of the different orderings). Therefore, the number of five-card hands, when you don't want to count each different ordering separately, must be:

$$\frac{52 \times 51 \times 50 \times 49 \times 48}{5 \times 4 \times 3 \times 2 \times 1} = \frac{311{,}875{,}200}{120} = 2{,}598{,}960$$

"We can write that expression like this," the professor said.

$$\frac{52!}{47!\,5!}$$

And, in general, if we are selecting j objects from a group of n objects, then there will be

$$\frac{n!}{j!(n-j)!}$$

ways of doing that." The number of ways of selecting objects in this way is called the number of *combinations* of n objects taken j at a time. When you count the number of combinations, the order of the objects that you select does not matter. Notice that the number of combinations is therefore less than the number of permutations. We decided that it would be helpful to have a special notation to represent this formula, so we let the expression

$$\binom{n}{j}$$

be defined as follows:

$$\binom{n}{j} = \frac{n!}{(n-j)!j!}$$

The expression $\binom{n}{j}$ is sometimes read "n choose j," because it tells you the number of different ways of choosing j objects from a group of n objects. We put Builder to work making a large table containing some values for this formula.

At that moment Pal dropped most of his letter blocks behind the couch, and we didn't have time to retrieve them. He was left with only the first five letters. The professor decided that we should make a list of all of the permutations of five objects taken three at a time. She calculated that there were $\frac{5!}{(5-3)!} = \frac{5!}{2!} = \frac{(5 \times 4 \times 3 \times 2 \times 1)}{(2 \times 1)} = 5 \times 4 \times 3 = 60$ permutations, and $\frac{5!}{[3!(5-3)!]} = \frac{5!}{(3! \times 2!)} = \frac{(5 \times 4 \times 3 \times 2 \times 1)}{(3 \times 2 \times 1 \times 2 \times 1)} = \frac{(5 \times 4)}{(2 \times 1)}$ $= 10$ combinations. She listed all 60 permutations:

ABC	ACB	BAC	BCA	CAB	CBA
ABD	ADB	BAD	BDA	DAB	DBA
ABE	AEB	BAE	BEA	EAB	EBA
ACD	ADC	CAD	CDA	DAC	DCA
ACE	AEC	CAE	CEA	EAC	ECA
ADE	AED	DAE	DEA	EAD	EDA
BCD	BDC	CBD	CDB	DBC	DCB
BCE	BEC	CBE	CEB	EBC	ECB
BDE	BED	DBE	DEB	EBD	EDB
CDE	CED	DCE	DEC	ECD	EDC

"Notice that the 6 permutations in each row consist of the same three letters put in different orders." Then she listed the ten different combinations:

ABC ABD ABE ACD ACE ADE BCD BCE BDE CDE

PERMUTATIONS

If j objects are to be selected from a group of n objects, then there are $\frac{n!}{(n-j)!}$ ways of making the selection, if each possible ordering of the j objects is counted separately.

COMBINATIONS

If j objects are to be selected from a group of n objects, there are

$$\frac{n!}{(n-j)!j!}$$

ways of making the selection, if each possible ordering of the j objects is not counted separately. This formula is also symbolized as $\binom{n}{j}$.

We thought that we would have a chance for a brief break, but we were interrupted by the arrival of the gremlin!

"Fools!" he cried. "You still have not caught on to the idea. You still only develop formulas one step at a time, when an application is immediately obvious. You have yet to learn the necessity of searching for general formulas first, even if the applications do not become clear until later. So I am sure to win!"

"Name one formula that we can't discover," Recordis challenged him.

"I dare you to find a general formula for $(a + b)^n$ that will work for *any* whole number value of n!" the gremlin responded. Then there was a puff of wind and he whooshed out the window.

"If only we had a numerical value for the exponent n!" Recordis quaked. "But I don't have the faintest idea about how to proceed when we're not given the value for n."

"Let's try some concrete examples," the king said. We found that

$$(a + b)^0 = 1$$

$$(a + b)^1 = a + b$$

$$(a + b)^2 = a^2 + 2ab + b^2$$

$$(a + b)^3 = a^3 + 3a^2b + 3ab^2 + b^3$$

$$(a + b)^4 = a^4 + 4a^3b + 6a^2b^2 + 4ab^3 + b^4$$

$$(a + b)^5 = a^5 + 5a^4b + 10a^3b^2 + 10a^2b^3 + 5ab^4 + b^5$$

We were exhausted after calculating the last two examples. "We cannot go on like this," the king said. "We must soon find a general formula."

"Does anybody see any patterns?" the professor asked.

"I see an obvious pattern," Recordis said. "It is clear that the first term in each of the above examples is a^n. For example, the first term of $(a + b)^2$ is a^2, the first term of $(a + b)^3$ is a^3, and so on."

"And the last term is always b^n," the king noticed.

We continued to stare hard at the rows of algebraic expressions.

"Aha!" the professor said. "Look at the letter parts of the terms in the expression for $(a + b)^5$."

$$a^5 \quad a^4b \quad a^3b^2 \quad a^2b^3 \quad ab^4 \quad b^5$$

"We can write those like this."

$$a^5b^0 \quad a^4b^1 \quad a^3b^2 \quad a^2b^3 \quad a^1b^4 \quad a^0b^5$$

(We used the property that $x^0 = 1$ and $x^1 = x$ for any value of x.)

"Now look at what happens when you add together the exponent of a and the exponent of b for each term," the professor said.

$$5 + 0 = 5 \quad 4 + 1 = 5 \quad 2 + 3 = 5 \quad 1 + 4 = 5 \quad 0 + 5 = 5$$

"The sum is 5 in each case!" Recordis said. "I bet I can guess what the expression for $(a + b)^6$ will look like."

$$(a + b)^6 = a^6 + \_\_\_ a^5b + \_\_\_ a^4b^2 + \_\_\_ a^3b^3 + \_\_\_ a^2b^4 + \_\_\_ ab^5 + b^6$$

"I think you are right," the professor said. "We know what the letter part of each term will be. However, we don't know what coefficients we should put in the blanks."

"Let's arrange all of the coefficients that we do know in a big triangle and see if we can discover a pattern," the king suggested.

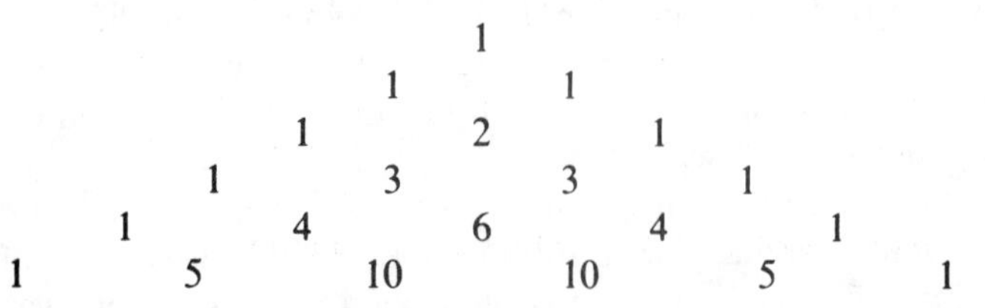

"I see a very strange pattern," Recordis said. "It looks as if any number in the triangle can be found by adding together the two numbers directly above it. (Except, of course, for the numbers around the edges, which are all 1's.) I bet I can guess that the numbers in the next row will be:

1 6 15 20 15 6 1

Therefore, I'm even willing to guess what the formula for $(a + b)^6$ looks like."

$$(a + b)^6 = a^6 + 6a^5b + 15a^4b^2 + 20a^3b^3 + 15a^2b^4 + 6ab^5 + b^6$$

(This interesting array of numbers is known as *Pascal's triangle,* named after Blaise Pascal, a 17th-century French mathematician.)

"That is a fascinating pattern," the professor agreed, "but we still need a *formula* that tells us what the coefficients will be." (She was still miffed because she had not seen the pattern first.)

At that moment Builder came by with the large etched table that contained all our results for the combinations formula

$$\binom{n}{j} = \frac{n!}{(n - j)!j!}$$

As we looked at the table, we noticed an amazing fact: The numbers were exactly the same as the numbers in Pascal's triangle! For example, the results for $n = 6$ were

$$\binom{6}{0} = \frac{6!}{6!\,0!} = 1 \qquad \binom{6}{1} = \frac{6!}{5!\,1!} = 6$$

$$\binom{6}{2} = \frac{6!}{4!\,2!} = \frac{6 \times 5}{2 \times 1} = 15 \qquad \binom{6}{3} = 20$$

$$\binom{6}{4} = 15 \qquad \binom{6}{5} = 6 \qquad \binom{6}{6} = 1$$

"So now we can write the formula for $(a + b)^n$," the king said. "We may as well call it the *binomial formula,* since it tells us how to find the powers of a binomial."

BINOMIAL FORMULA

$$(a+b)^n = \binom{n}{0}a^n + \binom{n}{1}a^{n-1}b + \binom{n}{2}a^{n-2}b^2$$

$$+ \binom{n}{3}a^{n-3}b^3 + \cdots + \binom{n}{n-1}ab^{n-1} + \binom{n}{n}b^n$$

Since $\binom{n}{0} = 1$, $\binom{n}{1} = n$, $\binom{n}{n-1} = n$, and $\binom{n}{n} = 1$ (see exercises 3 to 6), we can rewrite the formula as follows:

$$= a^n + na^{n-1}b + \binom{n}{2}a^{n-2}b^2 + \cdots nab^{n-1} + b^n$$

We can also write this with summation notation:

$$(a+b)^n = \sum_{k=0}^{n} \binom{n}{k} a^{n-k}b^k$$

For example, we calculated the coefficients for $n = 7$:

$$\binom{7}{0} = 1 \qquad \binom{7}{1} = 7 \qquad \binom{7}{2} = \frac{7 \times 6}{2 \times 1} = 21$$

$$\binom{7}{3} = \frac{7 \times 6 \times 5}{3 \times 2 \times 1} = 35$$

Recordis discovered an important work-saving result:

$$\binom{n}{n-j} = \binom{n}{j}$$

That result cut our calculations in half, since we could then tell that

$$\binom{7}{4} = \binom{7}{3} = 35 \qquad \binom{7}{5} = \binom{7}{2} = 21$$

$$\binom{7}{6} = \binom{7}{1} = 7 \qquad \binom{7}{7} = \binom{7}{0} = 1$$

Using the formula, we wrote the result for $(a + b)^7$:

$$(a + b)^7 = a^7 + 7a^6b + 21a^5b^2 + 35a^4b^3 + 35a^3b^4 + 21a^2b^5 + 7ab^6 + b^7$$

However, before we could celebrate, another note from the gremlin came sailing through the window: "But you have not *proved* this!" it said ominously.

Note to Chapter 15

The formulas for permutations and combinations are used extensively in the study of probability. For example, if you deal 5 cards from a 52 card deck, then the probability of obtaining exactly two hearts is given by the formula:

$$\frac{\binom{13}{2}\binom{39}{3}}{\binom{52}{5}}$$

See a book on probability or statistics for more details.

Exercises

1. Show that $n(n - 1)! = n!$

2. Show that $\binom{n}{j} = \binom{n}{n - j}$

Simplify each of these expressions:

3. $\binom{n}{0}$

4. $\binom{n}{1}$

5. $\binom{n}{n - 1}$

6. $\binom{n}{n}$

7. Show that $\binom{n}{j} + \binom{n}{j + 1} = \binom{n + 1}{j + 1}$

8. Here are the coefficients in the expansion of $(a + b)^6$. They are also the numbers in row 6 of Pascal's triangle.

1 6 15 20 15 6 1

Calculate the numbers in the next three rows of the triangle.

9. Write the binomial theorem using summation notation.
10. Calculate the sum of all the numbers in each row for the first five rows of Pascal's triangle. Can you find a simple expression for the sum of all of the numbers in a row of the triangle?
11. If you flip a coin 4 times, there are 2^4 possible outcomes. How many of these outcomes have 0 heads? How many have 1 head? 2 heads? 3 heads? 4 heads?
12. If you flip a coin 6 times, there are 2^6 possible outcomes. How many of these outcomes have 0 heads? 1 head? 2 heads? 3 heads? 4 heads? 5 heads? 6 heads?
13. If you flip a coin n times, how many outcomes will have j heads?
14. How many 13-card hands can you draw from a deck of 52 cards? How many 7-card hands are there? How many 4-card hands? How many 3-card hands?
15. How many different ways can you select a 5-member committee from a group of 20 people?

Use the binomial formula to write expressions for the following:

16. $(1 + x)^5$
17. $(1 + x)^n$
18. 11^4
19. 11^8
20. $(a + b)^5 - (a + b)^3$
21. $(a + 2)^6$
22. $(a + 3)^3$
23. $(a + 10)^7$
24. $(x - 1)^6$
25. $(x - 2)^6$
26. $(a - b)^n$
27. $(2 + 2)^5$
28. $(2x + 1)^5$
29. $(3a + 4b)^5$
30. $[(a + h)^n - a^n]/h$
31. What happens to the last expression as h becomes very close (but not quite equal) to zero?

❑ 32. Write a subroutine that calculates $n!$

❑ 33. Write a program that calculates $\binom{n}{j}$ by calling the subroutine from Exercise 32 three times.

❑ 34. Write a subroutine that calculates $\binom{n}{j}$ as efficiently as possible.

16 Proofs by Mathematical Induction

"How could we ever prove something as general and abstract as the binomial formula?" Recordis moaned. "I can easily show that the formula for $(a + b)^n$ works if $n = 5$, or $n = 6$, or $n = 10$, or any other value that we choose. However, I see no way in the world to show that is true for every value of n!"

"Maybe we should start with a much simpler proof," the king said. "Let's try to prove our formula for an arithmetic series:

$$a + (a + d) + (a + 2d) + \cdots + [a + (n - 1)d] = na + \tfrac{1}{2}n(n - 1)d$$

(See Chapter 14.)

If we can prove that this formula is true for every value of n, then we will at least have some practice with this type of proof."

"Let's start by proving that the formula works for some sample values of n," the professor said.

$$n = 1: \quad a = a$$

$$n = 2: \quad a + (a + d) = 2a + d$$

$$n = 3: \quad a + (a + d) + (a + 2d) = 3a + 3d$$

"These results do not help us prove that the formula is true for every single value of n!" Recordis said. "There is an infinite number of numbers, so it would take forever to prove that the formula is true. Not that I doubt the formula for a moment—after all, remember who discovered it—but I just don't know how to prove it."

"Let's see if we can prove that the formula works for $n = 4$ without going through the entire calculation," the king said. "We know that the formula is true for $n = 3$. If only there were some way we could use that fact to show that the formula works for $n = 4$."

$$\text{known fact: } a + (a + d) + (a + 2d) = 3a + \left(\frac{1}{2} \times 3 \times 2 \times d\right)$$

$$\text{fact we want to prove: } a + (a + d) + (a + 2d) + (a + 3d) = 4a + \frac{1}{2} \times 4 \times 3 \times d$$

"Drat it!" Recordis said. "I wanted to fit the entire expression all on one line." He suddenly had an idea. "We know that $a + (a + d) + (a + 2d) = 3a + \frac{1}{2} \times 3 \times 2 \times d$, using the formula for $n = 3$, so we can rewrite the fact we want to prove like this."

$$\left[3a + \frac{1}{2} \times 3 \times 2 \times d\right] + (a + 3d) = 4a + \frac{1}{2} \times 4 \times 3 \times d$$

"That equation clearly is true!" the professor said. "Both sides are equal to $4a + 6d$. However, I don't think you should take a short-cut like that—using the formula for $n = 3$ as if it had already been proved while we are still trying to prove the formula for $n = 4$."

Suddenly the king realized, "That's exactly what we *do* have to do!" he shouted. "We have to use the proof of the formula for smaller values of n to prove that the formula works for larger values of n."

"How?" the professor asked suspiciously.

"Let's suppose that we've proved two things," the king said.

(a) The formula works for $n = 1$.

(b) *If* the formula works for a particular number, *then* it must also work for the *next* consecutive number. In symbols: If the formula works for $n = j$, then it must work for $n = j + 1$.

"If both of these statements are true, then we can easily prove that the formula is true for $n = 2$," the king said. "We know that the formula works if $n = 1$. Then statement (b) says that it must also work for $n = 2$, because 2 is the next number after 1."

"And if it works for $n = 2$, then it must work for $n = 3$," Recordis said. "And if it works for $n = 3$, then it must work for $n = 4$. And so on and so on. In fact, if we have shown that both statement (a) and statement (b) are true, then we have effectively shown that the formula works for every single number in the whole world!"

"I hate to throw water on this celebration," the professor said. "but we have yet to prove that statement (b) is true. We have not yet shown that the mere fact that the formula works for a number j automatically means that it must work for $j + 1$."

"Let's assume that the formula does work for some arbitrary number j," the king said. We used S_j to represent the sum of the first n terms of the arithmetic series.

$$S_j = a + (a + d) + \cdots + [a + (j - 1)d] = ja + \frac{1}{2}j(j - 1)d$$

"Now, let's use that expression to find an expression for

$$S_{j+1} = a + (a + d) + \cdots + [a + (j - 1)d] + (a + jd)$$

We can rewrite that expression as

$$S_{j+1} = S_j + (a + jd)$$

by referring to the above expression for S_j."

We inserted the formula for S_j:

$$S_{j+1} = ja + \frac{1}{2}j(j - 1)d + (a + jd)$$

"We can simplify that, using Standard Operating Procedures," Recordis said.

$$S_{j+1} = (j + 1)a + \left[\frac{1}{2}j^2 - \frac{1}{2}j + j\right] d$$

$$S_{j+1} = (j + 1)a + \left[\frac{1}{2}j^2 + \frac{1}{2}j\right] d$$

$$S_{j+1} = (j + 1)a + \frac{1}{2}j(j + 1)d$$

"That's exactly what our formula predicted that the result would be!" the king said triumphantly. (See Chapter 14, Exercise 46.)

Igor suggested that we call this method of proof the method of *mathematical induction*.

THE METHOD OF MATHEMATICAL INDUCTION

This method can be used to prove that a particular formula is true for every whole number. The steps of the method are:

(1) Show that the method works for the number 1.

(2) Next, assume that the formula is true for an arbitrary number j.

(3) Show that *if* the formula is true for the number j, then it must also be true for the number $j + 1$.

(4) Once you have completed these steps, then the formula has been proved to be true for all whole numbers.

"Let's try another example to see if this method really works," the professor said skeptically. "Let's see if we can prove the formula for the sum of a geometric series."

$$a + ar + ar^2 + ar^3 + \cdots + ar^{n-1} = \frac{a(r^n - 1)}{r - 1}$$

First, we showed that the formula was true for $n = 1$:

$$a = \frac{a(r^1 - 1)}{r - 1}$$

Next, we assumed that the formula was true for $n = j$:

$$S_j = a + ar + ar^2 + \cdots + ar^{j-1} = \frac{a(r^j - 1)}{r - 1}$$

Now, we tried to find a formula for S_{j+1}:

$$S_{j+1} = a + ar + ar^2 + \cdots + ar^{j-1} + ar^j$$

$$S_{j+1} = S_j + ar^j$$

Using the formula for S_j:

$$S_{j+1} = \frac{a(r^j - 1)}{r - 1} + ar^j$$

"Now we had better be able to simplify that formula to show that

$$S_{j+1} = \frac{a\,(r^{j+1} - 1)}{r - 1}$$

It would be very embarrassing to try to prove the formula and then find out that we can't do it," Recordis said. We proceeded with the simplification:

$$S_{j+1} = \frac{a(r^j - 1)}{r - 1} + \frac{ar^j(r - 1)}{r - 1}$$

$$= \frac{a(r^j - 1) + ar^j(r - 1)}{r - 1}$$

$$= \frac{a\,(r^j - 1 + r^{j+1} - r^j)}{r - 1}$$

$$= \frac{a(r^{j+1} - 1)}{r - 1}$$

"Wonderful!" the professor said, by now convinced that mathematical induction worked. "However, now we have to do the biggie—can we prove the binomial formula?"

We had already shown that the formula worked for $n = 1$, $n = 2$, $n = 3$, $n = 4$, and $n = 5$. Next, we assumed that the formula worked for $n = j$:

$$(a + b)^j = a^j + ja^{j-1}b + \binom{j}{2}a^{j-2}b^2 + \cdots + \binom{j}{k}a^{j-k}b^k + \cdots + jab^{j-1} + b^j$$

(Remember that $\binom{j}{k}$ is a notation that means $\frac{j!}{k!(j-k)!}$. See Chapter 15.) Notice that we included a general formula for the term with $a^{j-k}b^k$. Recordis was still very uncomfortable with the use of the three dots to indicate that not all of the terms were included, but he reluctantly agreed that we had no other choice if we wanted a formula that would be totally general. Since we did not know the specific value of j in the general formula, there was no way that we could explicitly list all of the terms in the sum.

"Now we need to see if the formula is true for $j + 1$," Recordis said, beginning to see the general idea of the method of proofs by mathematical induction. "If the binomial formula really does work for $j + 1$, this is what it would look like:

$$(a + b)^{j+1} =$$

$$a^{j+1} + (j + 1)a^jb + \binom{j+1}{2}a^{j-1}b^2 + \cdots + \binom{j+1}{k}a^{j+1-k}b^k + \cdots + (j + 1)ab^j + b^{j+1}$$

"Now we need to see if we can use the formula for $(a + b)^j$ to prove the formula works for $(a + b)^{j+1}$," the professor said.

$$(a + b)^{j+1} = (a + b)(a + b)^j = a(a + b)^j + b(a + b)^j$$

We worked on simplifying this formula. We became lost in a sea of algebra:

$$(a + b)^{j+1}$$

$$= a\left[a^j + ja^{j-1}b + \binom{j}{2}a^{j-2}b^2 + \cdots + \binom{j}{k}a^{j-k}b^k + \cdots + jab^{j-1} + b^j\right]$$

$$+b\left[a^j + ja^{j-1}b + \binom{j}{2}a^{j-2}b^2 + \cdots + \binom{j}{k}a^{j-k}b^k + \cdots + jab^{j-1} + b^j\right]$$

Multiplying out the a and b in front of the brackets:

Row Number	Term from $a(a + b)^j$	Term from $b(a + b)^j$
0.	a^{j+1}	
1.	ja^jb	a^jb
2.	$\binom{j}{2}a^{j-1}b^2$	$ja^{j-1}b^2$
3.	$\binom{j}{3}a^{j-2}b^3$	$\binom{j}{2}a^{j-2}b^3$
4.	$\binom{j}{4}a^{j-3}b^4$	$\binom{j}{3}a^{j-3}b^4$
5.	$\binom{j}{5}a^{j-4}b^5$	$\binom{j}{4}a^{j-4}b^5$
	$\vdots$	$\vdots$
$j - 3.$	$\binom{j}{j-3}a^4b^{j-3}$	$\binom{j}{j-3-1}a^4b^{j-3}$
$j - 2.$	$\binom{j}{j-2}a^3b^{j-2}$	$\binom{j}{j-2-1}a^3b^{j-2}$
$j - 1.$	ja^2b^{j-1}	$\binom{j}{j-1-1}a^2b^{j-1}$
$j.$	ab^j	jab^j
$j + 1.$		b^{j+1}

Recordis realized that our only hope was to stay carefully organized, so he wrote the terms in the answer in two columns. The left hand column contained the terms found by multiplying a by our expression for $(a + b)^j$; the right hand column contained the terms found by multiplying b by $(a + b)^j$. He was careful to line up the terms so that terms with the same power of a and b were in the same row, and he decided to number the rows (starting with 0).

"Now we need to look carefully at the pattern of these results, and find a general formula for row k," the professor said. She suggested:

$$\left[\binom{j}{k}+\binom{j}{k-1}\right]a^{j-k+1}b^{k}$$

"The parts involving a and b are just perfect," Recordis said. "The sum of the exponents is always $(j+1)$, just as the formula says it should be. Now all we have to do is make sure the coeffficients are correct."

We could see that the coefficient of a^{j+1} was 1, just as it should be. The next coefficient was supposed to be $(j+1)!/j!1! = j+1$. We found that

$$\frac{j!}{(j-1)!1!}+1=j+1$$

which is just what we wanted. The professor worked out the general form for the coefficient for the term with $a^{j-k+1}b^{k}$, hoping the result would be $\binom{j+1}{k}$

$$\binom{j}{k}+\binom{j}{k-1}=\frac{j!}{(j-k)!k!}+\frac{j!}{(j-k+1)!(k-1)!}$$

$$=\frac{j!(j-k+1)}{(j-k+1)(j-k)!k!}+\frac{j!k}{(j-k+1)!k(k-1)!}$$

Note $(j-k+1)(j-k)! = (j-k+1)!$

$$=\frac{j!(j-k+1)}{(j-k+1)!k!}+\frac{j!k}{(j-k+1)!k!}$$

$$=\frac{j!(j-k+1+k)}{(j-k+1)!k!}$$

Note $(j+1)j! = (j+1)!$

$$=\frac{(j+1)!}{(j-k+1)!k!}$$

$$=\left.\binom{j+1}{k}\right]$$

"Just what we want!" Recordis said with relief.

"A picture-perfect example of a proof by mathematical induction," the professor said. "We showed that the binomial formula is true for $n = 1$ (and a few more numbers, for good measure). Then we showed that *if* it is true for $n = j$, then it must also be true for $n = j + 1$. And *voila*!—we have proved that the formula is true for every value of n."

"We have made significant progress in our understanding of mathematics," the king said. "In the old days, we simply observed the behavior of numbers. Now we can prove laws about their behavior. I sleep better at night knowing that once we have discovered a relationship, we can prove that it is *always* true."

Exercises

Prove these formulas by mathematical induction:

1. $1 + 2 + 3 + \cdots + n = (1/2)n(n + 1)$
2. $2 + 4 + 6 + \cdots + n = (1/2)n(n/2 + 1)$
3. $1^2 + 2^2 + 3^2 + \cdots + n^2 = (n/6)(n + 1)(2n + 1)$
4. $1^3 + 2^3 + 3^3 + \cdots + n^3 = (n^2/4)(n + 1)^2$
5. $1 + 3 + 5 + 7 + \cdots + (2n - 1) = n^2$
6. $1/2 + 1/(2 \times 3) + 1/(3 \times 4) + \cdots + 1/[n(n + 1)] = n/(n + 1)$
7. Prove that if n is odd, then n^2 is odd.
8. Prove that if n is even, then n^2 is even.

17 Exponential Functions and Logarithms

Recordis was becoming interested in geneaology and decided to make a large chart showing his family tree. "First, I have two parents," he said, "then four grandparents, then eight great-grandparents, and so on." (See Figure 17–1.) "I would like to carry this chart back for eleven generations," he said.

"You will need a very big chart in order to include all of the ancestors in the eleventh generation!" the professor exclaimed.

"How many ancestors are there?" Recordis wondered.

"We can easily make up a function that tells how many ancestors there are for each generation," the professor said. "There are two ancestors in the first generation, then four ancestors in the second generation, and so on. It seems that the number of ancestors is always going to be a power of 2." The professor suggested the function

$$y = f(x) = 2^x$$

where y = number of ancestors and x = generation number.

We quickly calculated that the eleventh generation would contain $2^{11} = 2{,}048$ ancestors. Recordis whistled. "This will be a much bigger project than I had anticipated!"

The professor was more interested in the function we had just invented. "Let's call it an *exponential function*", she suggested, "since the independent variable x appears as an exponent." Igor drew a graph. (See Figure 17–2.)

The graph was difficult to draw because the values of the exponential function quickly became very large. We decided to call the number 2 the *base* of the exponential function $f(x) = 2^x$. The professor realized that

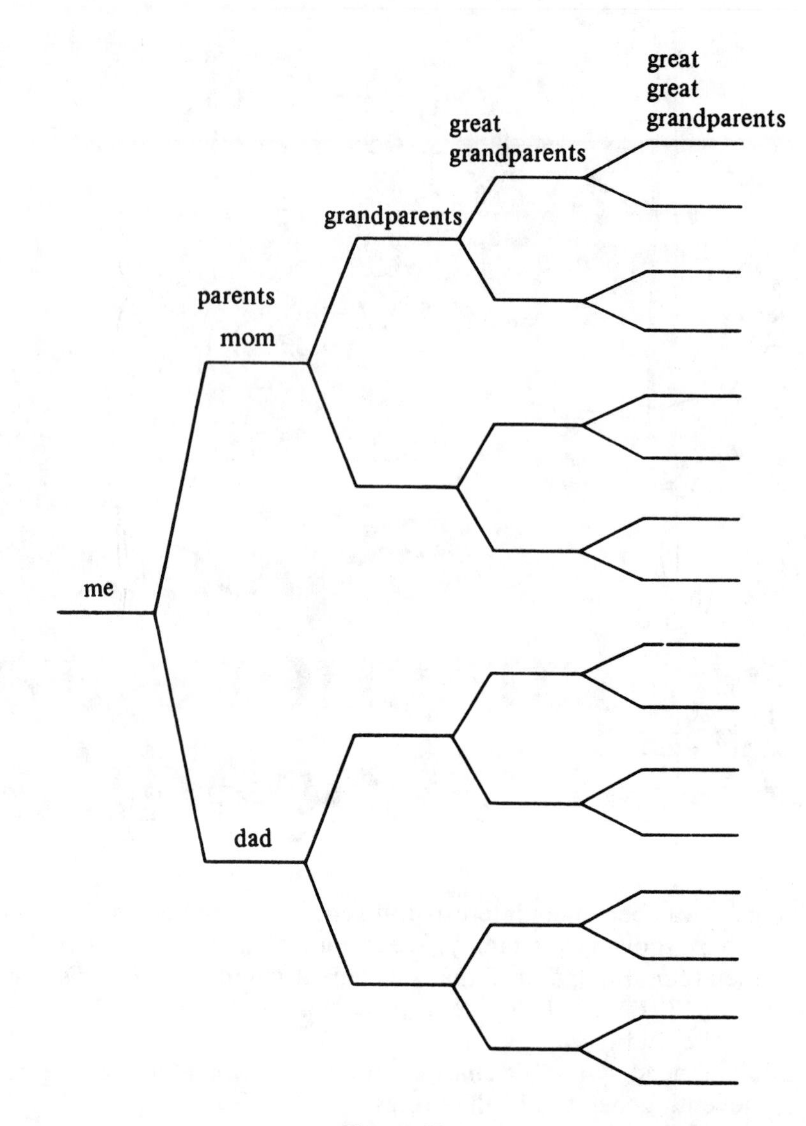

Figure 17–1

we could have exponential functions with many different bases—$f(x) = 3^x$, $f(x) = 10^x$, and so on.

At that instant the Royal Biologist came into the Main Conference Room carrying a large chart. "I have discovered a devastating new peril to the kingdom!" she reported. "I have been investigating the rate of growth of bacteria in Scenic Lake, and I have noticed a very disturbing trend. The rate of growth of the bacteria seems to be increasing constantly." She showed us the graph. (See Figure 17–3.) "If we don't do anything, we will soon be overrun by bacteria!"

"The rate of growth of the bacteria clearly follows an exponential function!" the professor said excitedly. We found that the function $f(x) = 10^x$ fit the growth of the bacteria perfectly.

However, Recordis had a problem with this graph. "We have drawn this graph as a smooth curve, but that implies that we are using some values of the exponent x that are fractions. For example, we have $10^{1/2}$. I know what 10^2 or 10^3 means, but what does $10^{1/2}$ mean?"

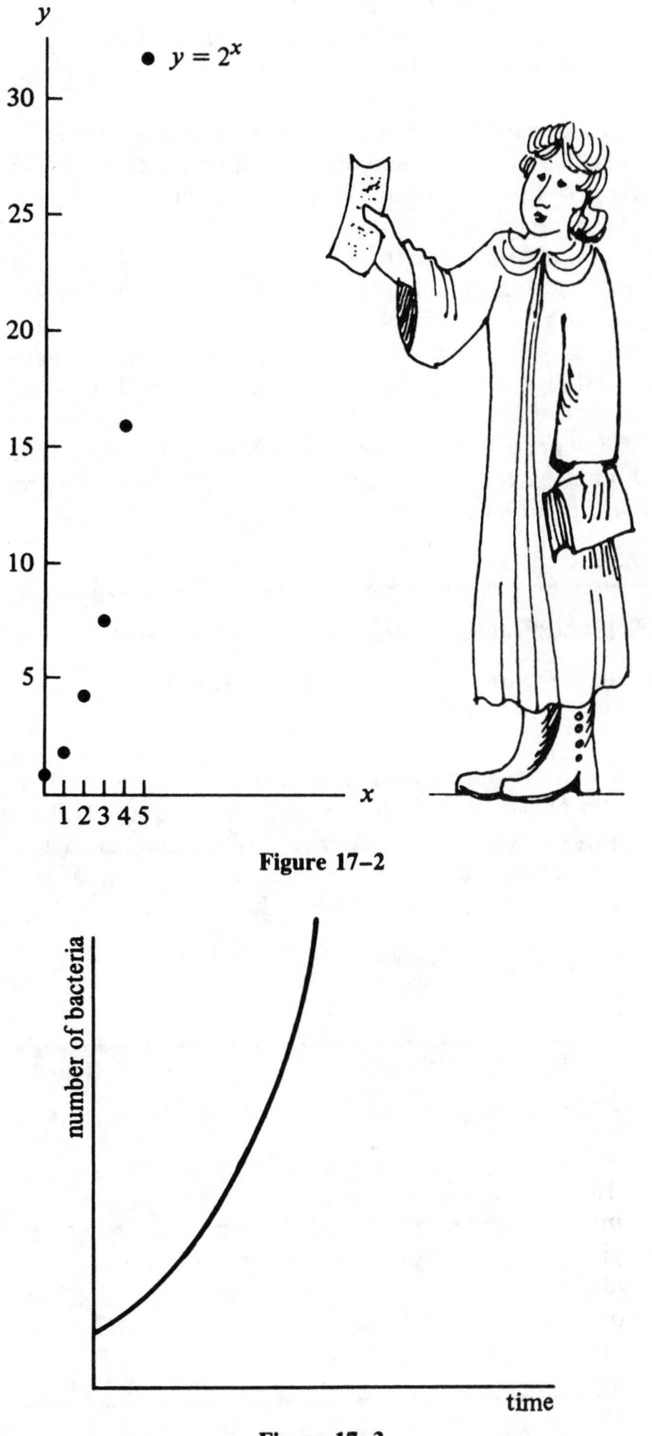

Figure 17–2

Figure 17–3

"Remember my brilliant idea for fractional exponents?" the professor said. "We decided that $10^{1/2}$ means $\sqrt{10}$, and, in general, $n^{a/b} = \sqrt[b]{n^a}$. (See Chapter 6.)

"I remember," Recordis agreed. "I concede that you can use any rational number as the value of the exponent. However, I just thought of something else. By drawing a smooth curve, you're implying that you're calculating the value for 10^x in some cases where x is an irrational number, such as $\sqrt{2}$. Now don't tell me that you know of a meaningful definition for $10^{\sqrt{2}}$?

The professor had to admit that she was stumped there. "We could calculate $10^{1.41}$ if we had to, and that's pretty close to $10^{\sqrt{2}}$," the professor said lamely. We decided not to worry too much about the exact meaning of irrational exponents.

"I see an interesting property of exponential functions," the king said. "For every single exponential function, $f(0) = 1$, since $b^0 = 1$ regardless of the value of b.

"We can also calculate what the function will be for negative values of x," the professor said, "using my idea that $b^{-a} = 1/b^a$. If x is less than zero, then 10^x is less than 1."

"One other property is clear," Recordis said. "The value of an exponential function can never be negative." We listed these properties for exponential functions:

EXPONENTIAL FUNCTIONS

An exponential function is a function of the form

$$y = f(x) = b^x$$

Here b is a constant number called the base of the function; b can be a positive number (except 1). The domain of the function is the set of all real numbers. The range of this function is the set of all positive numbers. (In other words, the value of the function can never be negative or 0.) Regardless of the value of b, it is always true that $f(0) = 1$. If $b > 1$, then

$$f(x) > 1 \quad \text{if } x > 0$$
$$f(x) < 1 \quad \text{if } x < 0$$

We also added some properties based on the laws of exponents we had derived in Chapter 5:

$$b^x b^y = b^{x+y}$$
$$b^x / b^y = b^{x-y}$$
$$(b^x)^n = b^{nx}$$

"Now we need to have a very accurate graph of the 10^x function, so we can determine the precise nature of the threat from the bacteria." Builder got out his best etching equipment while we calculated a table of values. (We first concentrated on the interval where x was between 0 and 1. These calculations were very tedious. It is much easier to do this type of calculation in our country when you can use a calculator or a computer.)

x	10^x	x	10^x	x	10^x
0.00	1.0000	0.35	2.2387	0.70	5.0119
0.05	1.1220	0.40	2.5119	0.75	5.6234
0.10	1.2589	0.45	2.8184	0.80	6.3096
0.15	1.4125	0.50	3.1623	0.85	7.0795
0.20	1.5849	0.55	3.5481	0.90	7.9433
0.25	1.7783	0.60	3.9811	0.95	8.9125
0.30	1.9953	0.65	4.4668	1.00	10.0000

"Understanding the nature of exponential functions helps make the urgency of our position more clear, but it still has not told us how to solve the problem of the bacteria growth. I will have to return to my lab and work on that."

The biologist left while we nervously waited in the Main Conference Room. "I wonder if bacteria eat you quickly, or nibble you in little bits?" Recordis wondered. However, he decided that it would be better to keep busy rather than spend all day thinking about the danger. So he explained the problem he was having with the Royal Ping Pong Tournament.

"The concept of the tournament is simple, but in practice it is complicated," Recordis said. "The winner of each match advances to the next round and the loser of the match is eliminated. For example, if you have four players, then you need two rounds of games. (See Figure 17–4) "It also turns out that we need three rounds if there are eight players. As it turns out, we have 128 players in the tournament. I need to know how many rounds will be required until everyone has been eliminated (except the winner.) Do you know how to calculate that?"

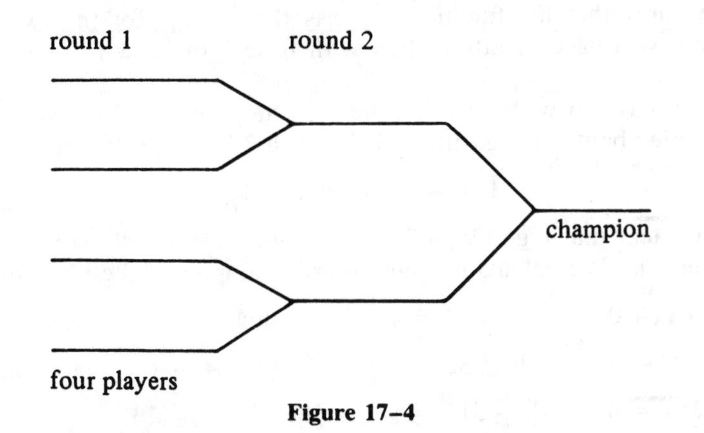

Figure 17–4

"You're lucky that the number of players (128) is a power of 2," the professor said, "since otherwise you would not be guaranteed that there would be an even number of players still in competition each round."

The king realized that there was a very simple relation between the number of rounds and the number of players:

1 round: 2 players
2 rounds: 4 players
3 rounds: 8 players
4 rounds: 16 players

"It seems clear that, if x is the number of rounds and y is the number of players, then $y = 2^x$," he said. "In fact, when we write the number of players as a function of the number of rounds, it turns out to be an exponential function with base 2."

"That's backwards, though!" Recordis fumed. "I don't need to know the number of players as a function of the number of rounds. I *know* that the number of players is 128, but I need to know the number of rounds. If you're going to do me any good, you're going to have to tell me the number of rounds as a function of the number of players."

"That's it!" the professor said. "We need to work the exponential function backwards. In other words, calculate an inverse function for the exponential function. Suppose $y = f(x) = 2^x$. Then there must be some function $g(y)$ such that $x = g(y)$."

"What will we call that function?" Recordis asked.

We were unable to reach a consensus. However, at that moment we were interrupted by a loud crashing sound. When we looked out the window we saw that the large pile of logs that Builder had constructed had come loose and crashed to the ground. "The dragon is trying to sabotage our lookout tower!" Recordis cried in dismay.

The king stared sadly out at the fallen logs. "We may as well call our function the *log* function," he said.

The professor thought the word "log" sounded too uncivilized, so she suggested that we use the word *logarithm* as the formal name for this function. "The word 'logarithm' has a nicer rhythm then the word 'log'," she said. However, we decided to use "log" as an abbreviation for logarithm. So we wrote:

$x = \log y$ to mean "x is the logarithm of y."

"Wait!" the king said. "There are many different exponential functions! That means we will need more than one logarithm function. How do we know that the function $\log x$ is the inverse for the exponential function with base 2, rather than with base 3 or base 15 or something else?"

"We'll have to write a little 2 next to the word 'log' so we know we are talking about the logarithm to the base 2," the professor suggested.

$$\text{If } y = 2^x, \text{ then } x = \log_2 y$$

"It is clear that $\log_2 128$ is 7, so the tournament will take 7 rounds," the king said. We calculated some more examples of the $\log_2 x$ function:

$\log_2 1 = 0$	$\log_2 2 = 1$	$\log_2 4 = 2$	$\log_2 8 = 3$
$\log_2 16 = 4$	$\log_2 32 = 5$	$\log_2 64 = 6$	$\log_2 128 = 7$
$\log_2 256 = 8$	$\log_2 512 = 9$	$\log_2 1{,}024 = 10$	
$\log_2(1/2) = -1$	$\log_2(1/4) = -2$	$\log_2(1/8) = -3$	

We also calculated some values for logarithm functions with different bases:

$\log_2 2 = 1$	$\log_3 3 = 1$	$\log_4 4 = 1$	$\log_5 5 = 1$	$\log_{10} 10 = 1$
$\log_2 4 = 2$	$\log_3 9 = 2$	$\log_4 16 = 2$	$\log_5 25 = 2$	$\log_{10} 100 = 2$
$\log_2 8 = 3$	$\log_3 27 = 3$	$\log_4 64 = 3$	$\log_5 125 = 3$	$\log_{10} 1{,}000 = 3$
$\log_2 16 = 4$	$\log_3 81 = 4$	$\log_4 256 = 4$	$\log_5 625 = 4$	$\log_{10} 10{,}000 = 4$
$\log_2 32 = 5$	$\log_3 243 = 5$	$\log_4 1{,}024 = 5$	$\log_5 3{,}125 = 5$	$\log_{10} 100{,}000 = 5$
$\log_2 64 = 6$	$\log_3 729 = 6$	$\log_4 4{,}096 = 6$	$\log_5 15{,}625 = 6$	$\log_{10} 1{,}000{,}000 = 6$
$\log_2 128 = 7$	$\log_3 2{,}187 = 7$	$\log_4 16{,}384 = 7$	$\log_5 78{,}125 = 7$	$\log_{10} 10{,}000{,}000 = 7$
$\log_2 256 = 8$	$\log_3 6{,}561 = 8$	$\log_4 65{,}536 = 8$	$\log_5 390{,}625 = 8$	$\log_{10} 100{,}000{,}000 = 8$
$\log_2 512 = 9$	$\log_3 19{,}683 = 9$	$\log_4 262{,}144 = 9$	$\log_5 1{,}953{,}125 = 9$	$\log_{10} 1{,}000{,}000{,}000 = 9$

(Compare this table with the table on page 52.)

"And I'm sure that we can find some properties for logarithm functions that correspond to the properties we found for exponential functions," the professor said. We discovered:

- $\log (xy) = \log x + \log y$

For example,

$$\log_2(8 \times 16) = \log_2(8) + \log_2(16) = 3 + 4 = 7$$

$$\log_2(128) = 7$$

- $\log (x/y) = \log x - \log y$

For example,

$$\log_2(256/8) = \log_2 256 - \log_2 8 = 8 - 3 = 5$$

- $\log (x^n) = n \log x$

For example,

$$\log_2(16^3) = 3 \log_2 16 = 3 \times 4 = 12$$

(These properties are true for logarithms with any base. They correspond to the properties given for exponential functions on page 218.)

"We are still in a very tight quandray," the king said. "We know how to calculate $\log x$ if the result is an integer or a convenient fraction. But what if we have to calculate a logarithm when the result is not simple? For example, what is $\log_{10} 2$?"

"We need to find a number y such that $10 = 2^y$," the professor said confidently, but then she suddenly realized that she knew of no such number.

"It's clear that $\log_{10} 2$ can't be an integer, and I bet it's not even a rational number," Recordis said. "We could perhaps calculate a decimal approximation for it, but I haven't the slightest idea how to do that."

At that moment Pal returned carrying Builder's graph of the exponential function $y = f(x) = 10^x$. (See Figure 17–5.)

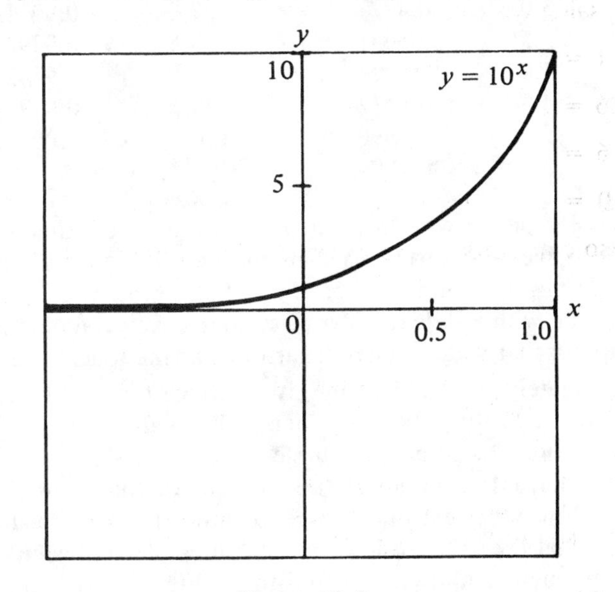

Figure 17–5

However, Pal slipped as he was carrying the heavy glass plate. The glass came tumbling to the ground. Pal made a desperate effort to save it, but it landed backwards from the way it had been. (See Figure 17–6.)

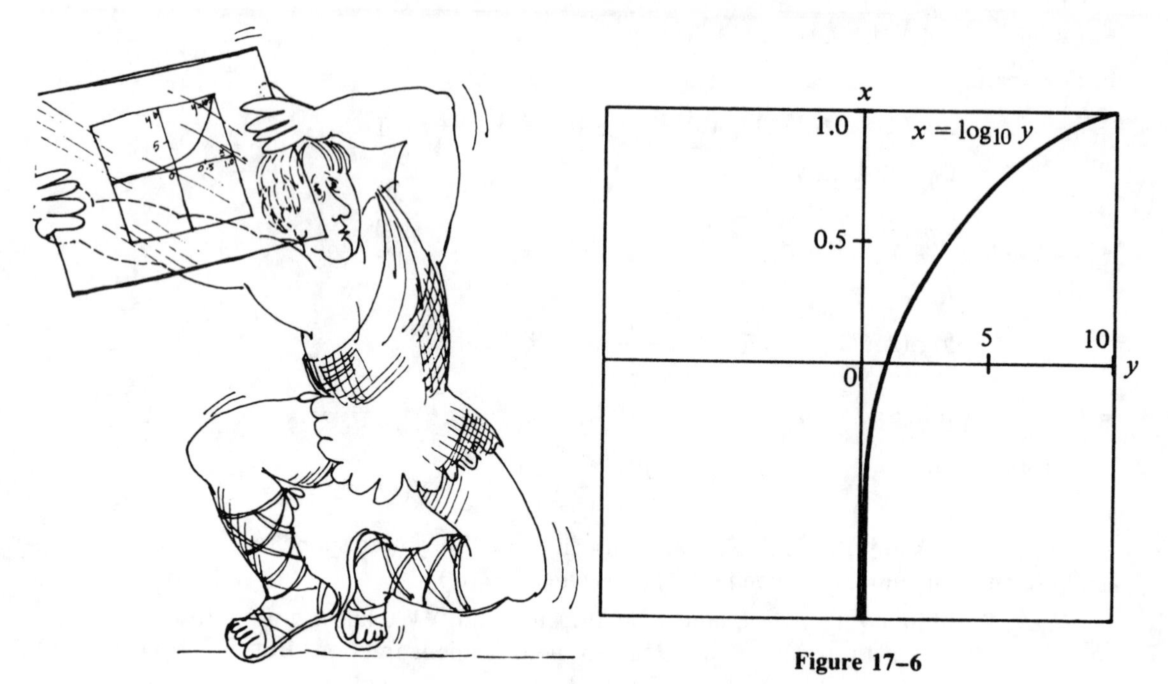

Figure 17–6

"That's a graph of the logarithm function!" the king said excitedly. We made a table of values that we read off the curve:

y	$\log y$	y	$\log y$
1.0	0.0000	6.0	0.7782
1.5	0.1761	6.5	0.8129
2.0	0.3010	7.0	0.8451
2.5	0.3979	7.5	0.8751
3.0	0.4771	8.0	0.9031
3.5	0.5441	8.5	0.9294
4.0	0.6021	9.0	0.9542
4.5	0.6532	9.5	0.9777
5.0	0.6990	10.0	1.0000
5.5	0.7404		

(We put Builder to work designing a more detailed table of logarithms. This table is included at the back of the book. In the old, pre-calculator days, the only way to find the logarithm of a number was to look it up in a table or find it with a slide rule. Now, of course, you can obtain a calculator or a computer that will calculate logarithms at the touch of a button.)

"Unfortunately, our table only gives values for $x = \log_{10} y$ if y is between 1 and 10," the king cautioned. "We will need to perform many more calculations to extend the table."

Recordis started to panic at the thought of the work that would be involved. However, just in the nick of time, he made a discovery that showed us that the table of $\log_{10} y$ for values of y between 1 and 10 was complete enough to allow us to find the value for any logarithm.

"Suppose we need to find $\log_{10} 20$," he said. "We can write that expression like this:

$$\log_{10} 20 = \log_{10}(2 \times 10)$$

Using the fundamental laws of logarithms:

$$\log_{10} 20 = \log_{10} 2 + \log_{10} 10$$

We know from the table that $\log_{10} 2 = 0.3010$, and $\log_{10} 10 = 1$. Therefore, $\log_{10} 20 = 1 + 0.3010 = 1.3010$."

"We can express any number as the product of a power of 10 and a number from 1 to 10," the professor realized. (See the discussion of scientific notation in Chapter 5.) "Since the logarithms of powers of 10 are trivial, we can find the log of any number."

For example:

$$\begin{aligned} \log_{10} 156 &= \log_{10}(1.56 \times 10^2) \\ &= \log_{10}(1.56) + \log_{10}(10^2) \\ &= 0.1931 + 2 \\ &= 2.1931 \end{aligned}$$

$$\begin{aligned} \log_{10}(535{,}000) &= \log_{10}(5.35 \times 10^5) \\ &= 5.7284 \end{aligned}$$

$$\begin{aligned} \log_{10}(0.034) &= \log_{10}(3.4 \times 10^{-2}) \\ &= 0.5315 - 2 \\ &= -1.4685 \end{aligned}$$

"This result demonstrates that logarithms to the base 10 will be by far the most convenient to use," Recordis said. Therefore, we decided to use the term *common logarithm* to represent logarithms to the base 10. To save us some writing, we decided that we could leave off the little 10 next to the word log. In other words, we decided $\log x$ means the same as $\log_{10} x$.

By now it was time for us to retire for the evening, since the next day would be the long-awaited opening day.

LOGARITHMS

If $x = b^y$, then $y = \log_b x$.
(y is the logarithm to the base b of x.) Logarithms to the base 10 are called common logarithms and are written as $y = \log x$. They are very convenient to work with because the powers of 10 are easy to calculate.

These properties are true for logarithms to any base:

$$\log(xy) = \log x + \log y$$

$$\log(x/y) = \log x - \log y$$

$$\log(x^n) = n \log x$$

Notes to Chapter 17

- In our country, common logarithms have played an important role as computational aids. The slide rule, a very useful computational device, is based on logarithms. However, now that electronic calculators are so common, this particular use of logarithms has become of less importance.
- Suppose you know what the value of $\log_a x$ is, but you want to know the value of $\log_b x$. (In other words, you want to know the logarithm of the same number but with a different base.) In this case you can use the change-of-base formula:

$$\log_b x = \frac{\log_a x}{\log_a b}$$

For example, $\log_2 64 = 6$ and $\log_2 4 = 2$, so $\log_4 64 = (\log_2 64)/(\log_2 4) = 6/2 = 3$. Or, since $\log_{10} 2 = 0.3010$, that means that

$$\log_2 x = \log_{10} x/0.3010.$$

See Figure 17–7 and Exercise 74.

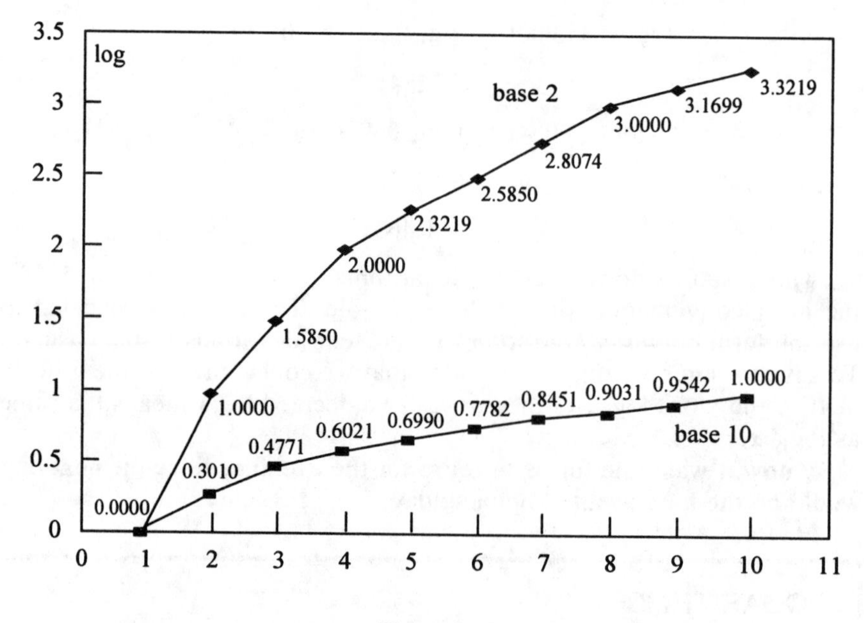

Figure 17–7

- In calculus the most useful logarithm function has its base equal to a special transcendental number called e, which is approximately 2.71828 . . . This type of logarithm is called the *natural logarithm*, and it is often written as $\ln x = \log_e x$. We can find that $\log e = 0.4343$, so the change-of-base formula tells us how to compare values for common logarithms and natural logarithms:

$$\ln x = \frac{\log x}{\log e} = \frac{\log x}{0.4343}$$

Exercises

Find the common logarithms of these numbers (see the table in the chapter):

1. log 100
2. log 100,000
3. log 10^{32}
4. log 15
5. log 25
6. log 5000
7. log 0.065
8. log 750
9. log(9.5×10^{10})
10. log 8,500
11. The integer part of a common logarithm is called the *characteristic*, and the part to the right of the decimal point is called the *mantissa*. Identify the characteristic and the mantissa for the logarithms in the previous set of exercises.

Rewrite these equations using logarithm notation:

12. $2^5 = 32$
13. $3^5 = 243$
14. $3^4 = 81$
15. $\sqrt{16} = 4$
16. $\sqrt{2} = 1.41$

Rewrite these equations using logarithm notation and then find the value for x. (Use the table of powers of whole numbers in Chapter 5.)

17. $2^x = 256$
18. $2^x = 1{,}024$
19. $10^x = 10{,}000$
20. $5^x = 3125$
21. $8^x = 4{,}096$
22. $3^x = 729$
23. $5^x = 15{,}625$
24. $3^x = 1/81$
25. $5^x = 0.04$
26. $10^x = 0.0001$
27. What is the total number of ancestors Recordis has had in the last 11 generations? (Count 2 in the first generation, 4 in the second generation, and so on.)
28. Why can't 1 be the base for a logarithm function? Why can't a negative number be the base for a logarithm function?

Simplify these expressions:

29. $\log \sqrt{x}$
30. $\log \sqrt{x + y}$
31. $\log(x^n)$
32. $\log[(x - 4)(x + 5)]$
33. $\log \left(\dfrac{a + b}{c + d}\right)$
34. $\log \left(\dfrac{x - 5}{x - 29}\right)$
35. $\log \left[\dfrac{(x + 15)(2x + 3)}{(x + 16)}\right]$
36. $\log \left[\dfrac{(x + 4)^3 (a + b)}{(x - 1)^2 (a^2 + b^2)}\right]$
37. $\log x^a y^b$
38. Calculate the natural logarithms (*i.e.*, the logarithms to the base $e = 2.71828 \ldots$) for the whole numbers from 1 to 10.

You will need a calculator capable of calculating logarithms for the remaining exercises.

The loudness of a sound is measured by the *decibel* system. The faintest sound the human ear can hear measures 0 decibels. If a sound is 10 times louder than that sound, then it measures 10 decibels. If a sound is 100 times louder than the faintest sound, then it measures 20 decibels. A sound 10,000 times louder measures 40 decibels. What will be the decibel measure of these sounds:

39. 1,000 times louder than the faintest sound
40. 10,000,000 times louder than the faintest sound
41. 2 times louder than the faintest sound
42. 3 times louder than the faintest sound
43. 5 times louder than the faintest sound
44. 500 times louder than the faintest sound

How many times louder than the faintest sound are these sounds:

45. 90 decibels
46. 100 decibels
47. 2 decibels
48. 5 decibels
49. 67 decibels
50. 65 decibels
51. 35 decibels

The brightness of stars is measured by the system of stellar magnitudes. The faintest stars that can be seen by the unaided human eye measure magnitude 6. A star 100 times brighter than the faintest star measures magnitude 1, and a star 10,000 times brighter measures magnitude -4. (Note that brighter stars have *smaller* magnitudes. A star fainter than magnitude 6 cannot be seen without a telescope.) Determine the magnitudes of these stars:

52. 100 times fainter then magnitude 6
53. 10,000 times fainter
54. 10 times brighter
55. 5 times brighter
56. 50 times brighter

Here are the brightnesses of some stars and other objects. Express each object's brightness as a multiple of the brightness of a 6th-magnitude star.

57. sun: -26
58. moon: -12
59. Venus: -4.4
60. Mars: -2.8
61. Jupiter: -2.5
62. Saturn: -0.4
63. Pluto: 14.9
64. Sirius (the brightest star): -1.4
65. Alpha Centauri (the nearest bright star): 0.01

66. Polaris (the north star): 1.99
67. Mizar (a star in the Big Dipper's handle): 2.26
68. the Andromeda galaxy: 3.5

The Richter scale measures the intensity of earthquakes. An earthquake of magnitude 6 is 10 times stronger than one of magnitude 5, and an earthquake of magnitude 7 is 100 times stronger than magnitude 5. How many times stronger than a magnitude 5 earthquake are these earthquakes:

69. magnitude 8
70. magnitude 9
71. magnitude 5.5
72. magnitude 7.3
73. magnitude 6.8
74. Prove the change-of-base formula.
75. Subtraction is the opposite operation to addition, and division is the opposite operation to multiplication. However, the operation of exponentiation has two opposites: taking roots and taking logarithms. Why does exponentiation need two opposite operations while addition and multiplication need only one opposite?

❑ **76.** Write a program that calculates the natural logarithm of x (with base e; see note on page 220) using this formula:

$$\ln(1 + x) = x - \frac{x^2}{2} + \frac{x^3}{3} - \frac{x^4}{4} + \frac{x^5}{5} - \cdots$$

18 Multiple Simultaneous Equations

Opening day dawned bright and clear. The professor woke early after a sleepless night of extreme nervousness. "I wish I could think of something else besides the game," she said.

A messenger came running from the stadium. "The concession stands have been struck by a mysterious bandit!" he cried.

We quickly raced to the stadium where we found the detective already searching for clues. The manager of the concession stand stood wringing his hands, surrounded by shredded boxes of hot dogs, pretzels, and ice cream bars.

"The thief has taken many boxes of hot dogs, pretzels, and ice cream bars!" he moaned.

"Exactly how many of each?" the detective asked.

"I don't know!" the manager said. "I know that altogether 10 boxes were taken. I know that the total weight of the stolen merchandise is 27 pounds, and I know that the total volume is 19 gallons. However, I don't know how many of those boxes were hot dog boxes, how many were pretzel boxes, and how many were ice cream bar boxes."

The detective made an instant deduction. "There is no doubt in my mind that this thief is the dragon that we are after. If we can figure out exactly how much of each item he has taken, then we will know how much mustard and water he will need to get."

"We can set up an equation system!" the professor said excitedly. "Do you know the weight and volume of the three types of boxes?"

The manager told us that the hot dog boxes weighed 3 pounds and had a volume of 1 gallon; the pretzel boxes weighed 2 pounds and had a volume of 3 gallons; and the ice cream bar boxes weighed 4 pounds and had a volume of 2 gallons.

"Then let's let x represent the number of hot dog boxes stolen, y represent the number of pretzel boxes stolen, and z represent the number of ice cream bar boxes stolen," the professor continued. "Then we know that these three equations must be true:

$$x + y + z = 10$$

(because there were 10 boxes taken)

$$x + 3y + 2z = 19$$

(because the total volume of the stolen boxes was 19 gallons)

$$3x + 2y + 4z = 27$$

(because the total weight of the stolen boxes was 27)

"Now all we have to do is find the values for x, y, and z that make those three equations true simultaneously!" the professor said.

"Which we could do if only we had the slightest idea of how to do it!" Recordis objected. "We've never seen an equation system with three equations and three unknowns like that before. How do we even know that there *is* a solution? Or, if there is a solution, how do we know that it will be a unique solution?"

"I'm sure that there will be a unique solution," the professor said. "Remember that one equation with one unknown usually has a unique solution; and a two-equation system with two unknowns usually has a unique solution. (See Chapter 10.) So I'm sure that a three-equation system with three unknowns must also have a solution."

Recordis started to draw a graph of the equation $x + y + z = 10$, since he knew that a graph of a two-variable equation helped us visualize the solutions to the equation. He drew the x axis and the y axis on his paper, but then he suddenly stopped. "I ran out of directions," he complained. "I have the x axis pointing east and the y axis pointing north, just as we have always done, but now I don't know where to draw the z axis." No matter how hard he tried, he could not find a way to draw the z axis on his paper.

"Then we must find a solution algebraically," the professor said. "And we must hurry, before the dragon has a chance to escape again."

$$x + y + z = 10$$

$$x + 3y + 2z = 19$$

$$3x + 2y + 4z = 27$$

"I suppose we could solve the system by repeated application of the substitution principle," Recordis said. "But that will be very tedious."

"There must be a way to simplify the system," the king said hopefully.

"We can do anything we want to the equations, just so long as we follow the Golden Rule of Equations," the professor said. She got a shrewd idea. "Let's multiply both sides of the middle equation by 3."

$$x + y + z = 10$$

$$3x + 9y + 6z = 57$$

$$3x + 2y + 4z = 27$$

We all agreed that the new equation system was equivalent to the original equation system.

"Now let's subtract the middle equation from the last equation to create a new last equation," the professor said.

old last equation:	$3x + 2y + 4z = 27$
subtract middle equation:	$-3x - 9y - 6z = -57$
new last equation:	$-7y - 2z = -30$

"Why can we do that?" Recordis asked. "The Golden Rule only lets us do exactly the same thing to both sides."

"But, according to the middle equation, the expression $3x + 9y + 6z$ is exactly the same as 57," the professor said. "Therefore, subtracting $3x + 9y + 6z$ from the left-hand side of the last equation is the same as subtracting 57 from the right-hand side." (This procedure was used in the elimination method to solve two-equation systems. See Chapter 10.)

"We got rid of the x in the last equation!" Recordis said gleefully. "That will be a big improvement."

The system now looked like this:

$$x + y + z = 10$$
$$3x + 9y + 6z = 57$$
$$-7y - 2z = -30$$

"Now if we could only get rid of some more variables!" Recordis said hopefully. We decided to try to get rid of the x in the middle equation. First, we divided the middle equation by three, and the system became

$$x + y + z = 10$$
$$x + 3y + 2z = 19$$
$$-7y - 2z = -30$$

Then we subtracted the first equation from the middle equation:

old middle equation:	$x + 3y + 2z = 19$
subtract first equation:	$-x - y - z = -10$
new middle equation:	$2y + z = 9$

Now the system looked like this:

$$x + y + z = 10$$
$$2y + z = 9$$
$$-7y - 2z = -30$$

Since all of the moves we had made were allowed by the Golden Rule of Equations, we were confident that this system was equivalent to the original system (in other words, the values of x, y, and z that formed the solution to this system would also be the solution to our original system).

"Now, let's use the last two equations to pin down the values for y and z!" Recordis said. "The last two equations form a regular two-equation, two-unknown system."

$$2y + z = 9$$
$$-7y - 2z = -30$$

We multiplied the top equation by 2 to form a new system:

$$4y + 2z = 18$$
$$-7y - 2z = -30$$

Then, we added the two equations together, and the result was

$$-3y = -12$$

That equation told us that $y = 4$. Once we knew the value of y, we could use the equation

$$2y + z = 9$$

to calculate that $z = 1$.

"So there were 4 pretzel boxes taken and 1 ice cream bar box," the detective said. "The only remaining unknown is the number of hot dog boxes taken."

The first equation of the system told us that

$$x + y + z = 10$$

After inserting the values $y = 4$ and $z = 1$ into this equation, we could see that $x = 5$. We checked to make sure that the solution $x = 5$, $y = 4$, and $z = 1$ was indeed a solution to the original three-equation system that we started with:

$$x + y + z = 10$$
$$x + 3y + 2z = 19$$
$$3x + 2y + 4z = 27$$

Fortunately, our solution was correct.

"Now we spring into action!" The detective said. He quickly started calculating. Then he led us through the stadium to the mustard dispenser. "According to my calculations, the dragon will need to show up for more mustard in precisely 45 minutes."

The long wait started. "I can't wait to get my hands on that criminal!" Recordis said excitedly.

In order to pass the time, the professor suggested that we investigate some more properties of three-equation systems with three unknowns. The king asked us to work on a problem that the Royal Economist had presented. "We use Y to stand for the total income in the country, C to stand for the total level of consumption spending, I to stand for the total level of investment spending, and G to stand for the level of government spending. Then we know from national income accounting that

$$Y = C + I + G$$

"We also happen to known that $I = 30$ and $G = 50$, so we can rewrite that equation:

$$Y = C + 80$$

"We also know how people behave. They will spend more money on consumption if they have a higher after-tax income. In fact, their spending is given by the formula

$$C = 10 + \frac{4}{5}(Y - T)$$

"In that equation, T is the total level of taxes, which can be found from the formula

$$T = 10 + \frac{1}{4}Y$$

"We can solve for Y, C, and T!" the professor said, "since we now have three equations with three unknowns."

$$Y = C + 80$$

$$C = 10 + \frac{4}{5}(Y - T)$$

$$T = 10 + \frac{1}{4}Y$$

"This system will be much easier than the other one," Recordis said, "because some of the equations don't contain all of the variables. In my opinion, the fewer variables an equation contains, the better."

Recordis suggested that we could solve the system directly by substitution. First, he suggested putting the expression for C from the second equation into the first equation:

$$Y = \left[10 + \frac{4}{5}(Y - T)\right] + 80$$

Then, he suggested putting the expression for T from the third equation into this equation:

$$Y = 10 + \frac{4}{5}\left[Y - \left(10 + \frac{1}{4}Y\right)\right] + 80$$

"Now we have just what we want!" Recordis said. "We have a single equation with a single unknown. We can solve that easily."

$$Y = 10 + \frac{4}{5}\left(Y - 10 - \frac{1}{4}Y\right) + 80$$

$$Y = 90 + \frac{4}{5}\left(\frac{3}{4}Y - 10\right)$$

$$Y = 90 + \left(\frac{4}{5}\right)\left(\frac{3}{4}\right)Y - \left(\frac{4}{5}\right)10$$

$$Y = 90 + \frac{3}{5}Y - 8$$

$$Y - \frac{3}{5}Y = 82$$

$$\frac{2}{5}Y = 82$$

$$Y = 205$$

"I like this kind of puzzle," Recordis said, "because once we find one part of the solution it's easy to find the rest of the solution."

$$T = 10 + \frac{1}{4}Y = 10 + \frac{1}{4}(205) = 61\frac{1}{4}$$

$$C = 10 + \frac{4}{5}(Y - T) = 10 + \frac{4}{5}\left(205 - 61\frac{1}{4}\right) = 125$$

The professor agreed that, in situations where some equations did not contain all of the variables, the use of the substitution principle would likely be the quickest way to find the solution. She suggested that we investigate another system in which some of the equations did not contain all of the variables:

$$x + 2y - \frac{1}{2}z = 12$$

$$10y - 6z = -32$$

$$12z = 144$$

"That system will be easy!" Recordis exclaimed. "Why, it's practically solved already! The bottom equation contains only one variable, so we can easily see that $z = 144/12 = 12$. Then, since the middle equation contains only y and z, we use the value for z to find the solution for y:

$$10y - 6z = -32$$

$$10y - 72 = -32$$

$$10y = 40$$

$$y = 4$$

And, knowing y and z, we can easily use the first equation to find x."

$$x + 2y - \frac{1}{2}z = 12$$

$$x + 8 - 6 = 12$$

$$x = 10$$

"I suggest we call a system in this form a *triangular system,* since it looks a bit like a triangle," the professor said. "And I think that when we are faced with a three-equation system, our goal should be to convert it into a triangular system."

SOLVING THREE-EQUATION, THREE-UNKNOWN SYSTEMS BY TRIANGULARIZATION

A three-equation system with three unknowns (x, y, and z) is triangular if z is the only unknown in the bottom equation and y and z are the only unknowns in the middle equation.

In general, a three-equation three-unknown system can be solved by converting it into an equivalent system that is triangular. According to the Golden Rule of Equations, these are the moves that can legally be made to transform the system:

(1) You can multiply both sides of one of the equations by a number.

(2) You can add one equation to another equation. (Of course, you can also subtract one equation from another equation.)

(3) It is also possible to perform a combination move: You can add to one equation a multiple of another equation.

Once the system has been triangularized, use the last equation to solve for z; then insert that value into the second equation to solve for y; then insert those two values into the first equation to solve for x.

"Maybe this method will help me design a new diet," Recordis said. "I have decided that from now on I will have milk, eggs, and apples for lunch. I would like to consume exactly 515 calories, 36.5 grams of protein, and 733 grams of calcium." He made a table showing how much of each nutrient each food contained:

	milk	egg	apple
calories	90	110	70
protein	9	7	0
calcium	250	50	8

"We can set up an equation system, using x to stand for the amount of milk, y to stand for the number of eggs, and z to stand for the number of apples."

calories:

$$90x + 110y + 70z = 515$$

protein:

$$9x + 7y = 36.5$$

calcium:

$$250x + 50y + 8z = 733$$

We went through the steps necessary to triangularize this system.

Move 1: To eliminate x from the last equation, subtract 250/90 times the top equation from the last equation:

$$90x + 110y + 70z = 515$$
$$9x + 7y = 36.5$$
$$-255\tfrac{5}{9}y - 186\tfrac{4}{9}z = -697\tfrac{5}{9}$$

Move 2: To eliminate x from the middle equation, subtract 1/10 times the top equation from the middle equation:

$$90x + 110y + 70z = 515$$
$$-4y - 7z = -15$$
$$-255\tfrac{5}{9}y - 186\tfrac{4}{9}z = -697\tfrac{5}{9}$$

Move 3: To eliminate y from the last equation, subtract $(255\tfrac{5}{9})/4$ times the middle equation from the last equation:

$$90x + 110y + 70z = 515$$
$$-4y - 7z = -15$$
$$260\tfrac{7}{9}z = 260\tfrac{7}{9}$$

"That's a triangular system if I ever saw one," Recordis said. "The last equation makes it plain that $z = 1$." We were also able to find that $y = 2$ and $x = 2.5$, so the optimal diet for Recordis included 2 1/2 cups of milk, 2 eggs, and one apple.

Recordis continued his futile attempt to think of a way to draw a graph of an equation with three variables. Suddenly we heard a loud shout from a room next to us. We went to investigate and found a man in the room with a television monitor shouting into a radio.

"No! You're too high!" the man shouted.

"I moved to the point $x = 30$, $y = 80$, just as you said," came a voice over the radio.

"Perhaps you can help," the man told us. "I am directing the television coverage of the game, and I am trying to give directions to our camera in the balloon. Mr. Builder was very generous in letting us use the balloon he had built, and then he explained how to give directions to the balloon by using x and y coordinates. So I put the origin at home plate, with the x axis along the line to first base and the y axis along the line to third base. I thought everything would work perfectly, but when I give the coordinates to the balloon it never goes quite where I want it—it is always too high or too low."

"I have an idea," the king suggested. "The problem is that you are giving the balloon only two coordinates. However, the balloon moves around in three-dimensional space, so in order to identify the location of a particular point, you need to list three coordinates: the x coordinate, the y coordinate, and a third coordinate that tells how high the balloon should be (we may as well call that the z coordinate). That way, you can distinguish between a point close to the ground, such as $x = 30$, $y = 80$, $z = 5$, and a point that is high in the sky, such as $x = 30$, $y = 80$, $z = 100$."

The director decided how high he wanted the balloon, and then he radioed the x, y, and z coordinates to the pilot. The pilot used the altimeter on the balloon to find the correct location, and then the picture on the television monitor was exactly what the director wanted. "Thank you very much," he told us. (See Figure 18–1.)

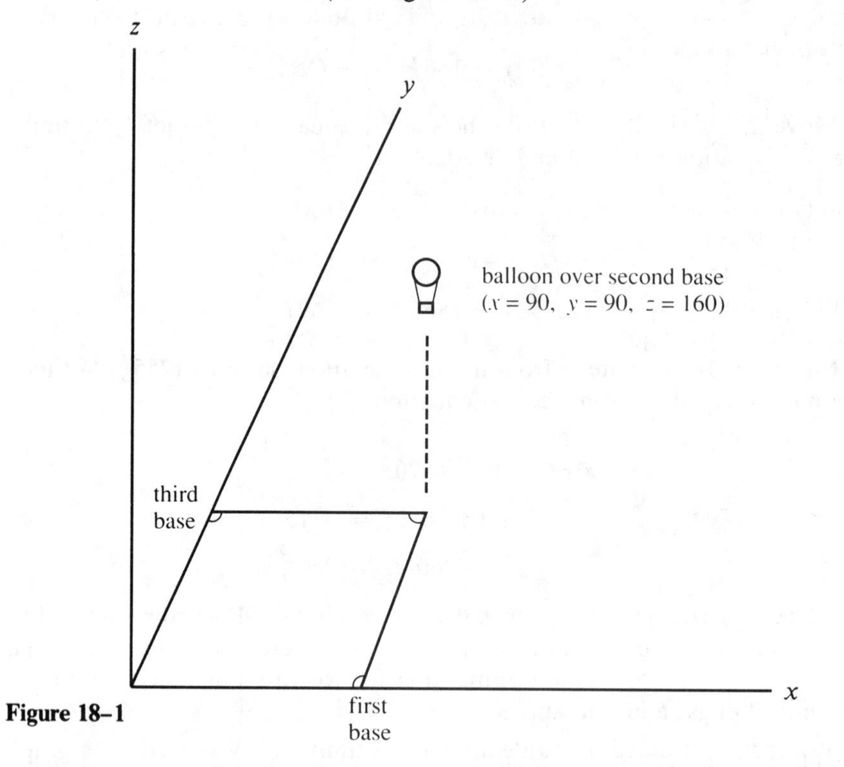

Figure 18–1

"The z coordinate of the balloon will always be positive or zero," the king realized. "However, in theory we could have negative values of z for points which are *below* the origin." (See Figure 18–2.)

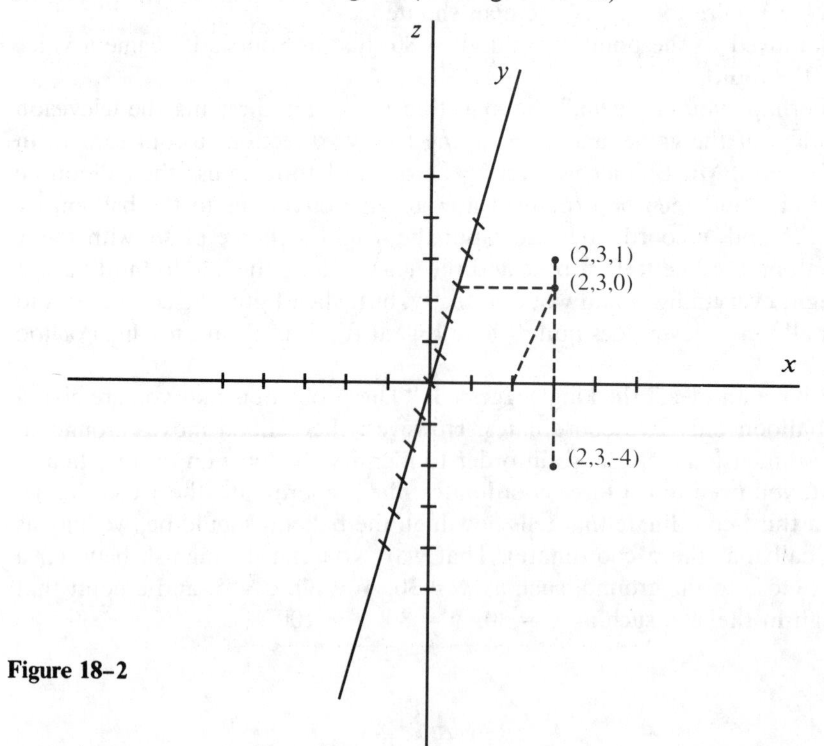

Figure 18–2

"Now we know how to draw a graph of equations with three variables!" the professor said excitedly. "We know that we can use x to represent the distance east, y to represent the distance north, and z to represent the distance up."

"Not so fast!" Recordis complained. "My paper only has two dimensions, so I can't very easily draw an upward-pointing z-axis on the paper."

Builder had an idea. He quickly hooked together three rods (one for the x-axis, one for the y-axis, and one for the z-axis). Then he installed an array of tiny light bulbs held in place by invisible wires to represent points in three-dimensional space. "This is my 3D-Space Model. All you have to do is tell me which points you want to light up, and then I turn on the switches that light up those points."

"Let's graph the equation $x + y + z = 10$," the professor suggested. "I predict that the graph of the equation will be a straight line, since the graph of the equation $x + y = 10$ is a straight line in two dimensions."

Recordis made a list of some points that satisfied the equation $x + y + z = 10$:

0, 0, 10
1, 0, 9 0, 1, 9
2, 0, 8 1, 1, 8 0, 2, 8
3, 0, 7 2, 1, 7 1, 2, 7 0, 3, 7
4, 0, 6 3, 1, 6 2, 2, 6 1, 3, 6 0, 4, 6
5, 0, 5 4, 1, 5 3, 2, 5 2, 3, 5 1, 4, 5 0, 5, 5
6, 0, 4 5, 1, 4 4, 2, 4 3, 3, 4 2, 4, 4 1, 5, 4 0, 6, 4
7, 0, 3 6, 1, 3 5, 2, 3 4, 3, 3 3, 4, 3 2, 5, 3 1, 6, 3 0, 7, 3
8, 0, 2 7, 1, 2 6, 2, 2 5, 3, 2 4, 4, 2 3, 5, 2 2, 6, 2 1, 7, 2 0, 8, 2

In each case the value of x is listed first, then y, and finally z. Recordis had only listed points where x, y, and z were all positive whole numbers. It was clear from these points that the graph was not a straight line. We gave this list of coordinates to Builder, and he switched on the light bulbs at those coordinates. When all these points were lit up, we noticed an astonishing fact: they formed a flat surface. We found that we could place a flat piece of plastic where it would touch all of the light bulbs that were lit up. (See Figure 18–3).

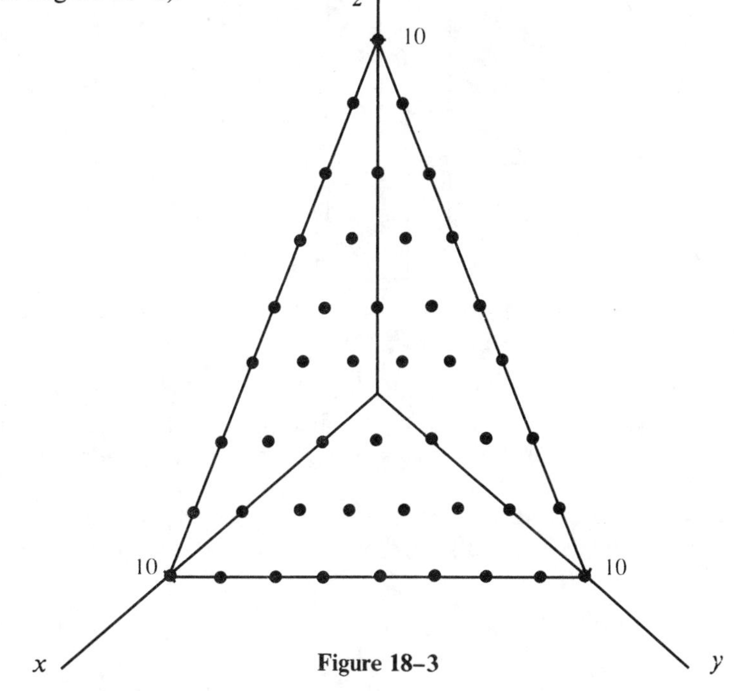

Figure 18–3

"The graph of the equation is a plane!" the professor said in awe.

"An airplane?" Recordis asked in puzzlement.

"No—a geometric plane," the professor reminded him. "Remember that a plane is an infinite flat surface, like a very large table top. We investigated planes when we studied geometry."

"I remember," Recordis said. "For example, Endless Plain in eastern Carmorra, where the land is flat as far as the eye can see, reminds me of geometric plains."

We investigated a few more three-variable equations. In each case we made a list of some points that solved the equation, and then we had Builder light up those points in his 3D-Space Model. We found that the graph of any equation of the form $ax + by + cx = d$ is a plane (where a, b, c, and d are known constants; and a, b, and c are not all equal to zero). The three equations we had originally looked at:

$$x + y + z = 10$$

$$x + 3y + 2z = 19$$

$$3x + 2y + 4z = 27$$

all fit this form, so we knew that the graph of each equation was a plane.

"We know that two lines intersect at one point," the professor said thoughtfully, "unless they happen to be parallel. We found that if we had two equations in two variables, we could draw graphs of the two lines representing the equations, and the point where the two lines crossed represented the solution to the system of equations. (See Chapter 10.) Now suppose we have three equations in three variables. Each equation will be represented by a plane in three-dimensional space. I bet that three planes will intersect at one point. For example, think of the floor, the north wall, and the east wall of a house as being portions of planes. They intersect at one point." (See Figure 18–4.) (However, we soon found that three planes did not inevitably intersect at a single point.)

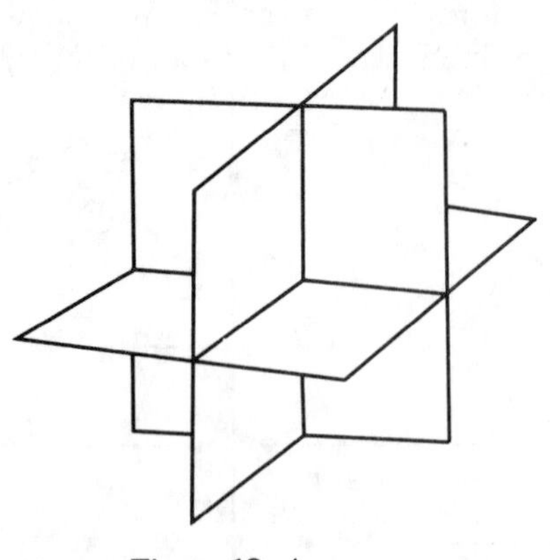

Figure 18–4

"We are very fortunate that Builder invented his 3D-Space Model,' Recordis said. "However, I can't carry it around with me like I carry my trusty notebook. I wish there was a way to draw three-dimensional graphs on paper."

"Actually, you can draw a perspective diagram that looks three dimensional, even though it is really two dimensional," the king realized. "After all, that is what any picture is: a two-dimensional representation of a three-dimensional object. I do realize, however, that it is not easy to draw that type of diagram." (Notice that the pages of this book are two-dimensional, so the diagrams included in this chapter are all perspective diagrams that try to create the illusion of three-dimensional space. The ability to understand three-dimensional graphs helps to clarify the nature of the solutions to a three-variable equation, although the difficulty of drawing the graphs means that this does not represent a very practical way of locating the solution.)

"I fear the situation will be even worse when we investigate equations with four variables," the professor said sadly. "It is impossible to draw a graph that looks four-dimensional. However, I have some other three-equation systems that I think we should investigate."

The professor suggested another system:

$$16x + 33y + 78z = -10$$
$$2x + 4y - z = 6$$
$$2x + 4y - z = 7$$

Move 1: To eliminate x from the last equation, subtract the middle equation from the last equation.

However, when we tried that, we were left with $0 = 1$ as the last equation.

"That's impossible!" Recordis said.

"This system must not have a solution," the professor said. When we stared closely at the system, we could see clearly what was wrong.

"The last two equations are inconsistent!" Recordis said. "There is no way that $2x + 4y - z$ can equal both 6 and 7 simultaneously!"

The professor noted, "Therefore, it is possible for a three equation, three-unknown system to have no solutions at all. Geometrically, I believe that this means the planes representing the solutions to two of the equations are parallel. If two of the equations are inconsistent, then there will be no solution for the system, no matter what the third equation says." (See Figure 18–5.)

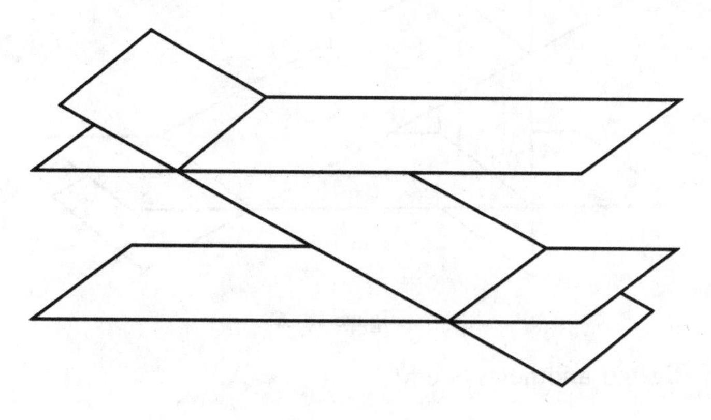

Figure 18–5

The professor suggested another system:

$$\begin{aligned} 2x + y + 3z &= 2 \\ -x + 7y + z &= 5 \\ 4x + 17y + 11z &= 1 \end{aligned}$$

These equations were not obviously inconsistent, but when we tried to triangularize the system we ended up with a problem.

Move 1: Add 4 times the middle equation to the last equation:

$$\begin{aligned} 2x + y + 3z &= 2 \\ -x + 7y + z &= 5 \\ 45y + 15z &= 21 \end{aligned}$$

Move 2: Add 1/2 times the top equation to the middle equation:

$$\begin{aligned} 2x + y + 3z &= 2 \\ 7.5y + 2.5z &= 6 \\ 45y + 15z &= 21 \end{aligned}$$

Move 3: Subtract 45/7.5 times the middle equation from the last equation:

$$\begin{aligned} 2x + y + 3z &= 2 \\ 7.5y + 2.5z &= 6 \\ 0 &= -15 \end{aligned}$$

"What's happening here," the professor said, "is that, even though no pairs of equations are inconsistent, the three equations taken together are inconsistent." She guessed that geometrically this meant that the planes representing the equations intersected so as to form three parallel lines. (See Figure 18–6.)

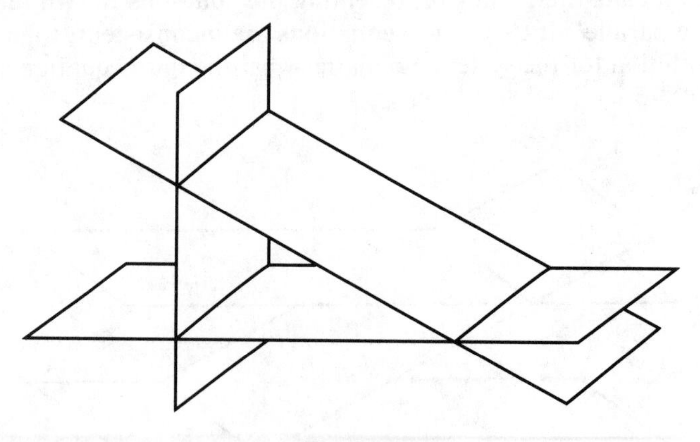

Figure 18–6

We investigated another system:

$$\begin{aligned} -3x + 4y - 6z &= 5 \\ -6x + 8y - 12z &= 10 \\ -9x + 12y - 18z &= 15 \end{aligned}$$

When we tried to triangularize this system, we ended up with $0 = 0$ as the last equation. "That is perfectly true," Recordis said, "but it doesn't help us find the solutions to the system."

"I bet this means that this sytem has an infinite number of solutions," the professor said.

The king noticed that the three equations were all equivalent, since the second equation was twice the first equation and the third equation was three times the first one. "Therefore, any point that is a solution to one of the equations will be a solution to all of them. There will be an infinite number of solutions all lying along one plane." We made a list of two possible solutions:

$$x = 0, \quad y = 0, \quad z = -5/6$$

$$x = 1, \quad y = 1, \quad z = -2/3$$

The next system the professor suggested was

$$\begin{aligned} x + 2y + 3z &= 10 \\ 3x + 6y + 9z &= 30 \\ x - y + 6z &= 1 \end{aligned}$$

"Hold it!" Recordis said. "The top two equations there are equivalent equations—which means that we really only have two distinct equations:

$$\begin{aligned} x + 2y + 3z &= 10 \\ x - y + 6z &= 1 \end{aligned}$$

We'll have a real problem finding a solution now that we have more unknowns then we do equations."

"The problem won't be finding a solution," the professor said. "The problem will be finding a unique solution. I think that there will be an infinite number of solutions. Geometrically, this situation can be represented as the intersection of two planes, so I bet that the solutions will consist of all of the points that lie along a line." (See Figure 18–7.)

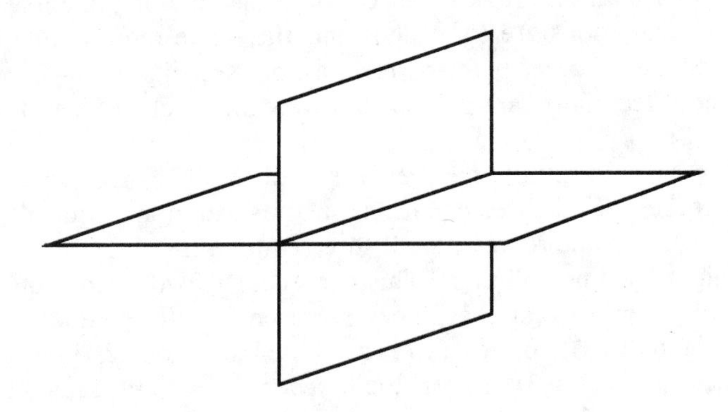

Figure 18–7

"How do we find which points those are?" the king asked. Recordis suggested that we multiply the bottom equation by 2:

$$\text{top equation:} \qquad x + 2y + 3z = 10$$
$$\text{bottom equation:} \qquad 2x - 2y + 12z = 2$$

Then we added these two equations together:

$$3x + 15z = 12$$

"Therefore, any (x, y, z) triple that satisfies the first two equations must also satisfy the last equation." We made a list of three of the solutions to that equation:

$$x = 4, z = 0; \qquad x = -1, z = 1; \qquad x = 0, z = \frac{4}{5}$$

Once we had found the values of x and z, we could find the value of y from the other two equations:

$$(4, 3, 0); \qquad (-1, 4, 1); \qquad \left(0, \frac{19}{5}, \frac{4}{5}\right)$$

"I hope you play better than this in the game!" Recordis told the professor. "You have now struck out four straight times. None of the last four systems you have suggested has had a unique solution." Recordis entered in his record book:

An equation in three variables of this form:

$$ax + by + cz = d$$

is called a linear equation (a, b, c, and d are known constants; a, b, and c are not all zero).
It can be represented in three-dimensional space by a plane.

When you have a system of three simultaneous linear equations in three variables, there are four possibilities for the solution:

- The system can have a unique solution (x, y, z). This means that the three planes intersect at one point (See Figure 18–4).
- The system can have an infinite number of solutions all along one line. This can occur if two equations are equivalent and the plane representing those equations intersects the other plane to form a line (see Figure 18–7) or this can occur if the three planes are distinct but they intersect to form one line (see Figure 18–8).
- The system can have an infinite number of solutions all lying on one plane. This situation occurs if all three equations in the system are equivalent, in which case the same plane represents all three equations.
- The system might have no solutions. This can occur if two of the equations contradict each other, so that the planes representing those equations are parallel (see Figure 18–5); or it can occur if the three planes intersect to form three lines (see Figure 18–6); or it can occur if the three equations are contradictory, so that all three planes are parallel (see Figure 18–9).

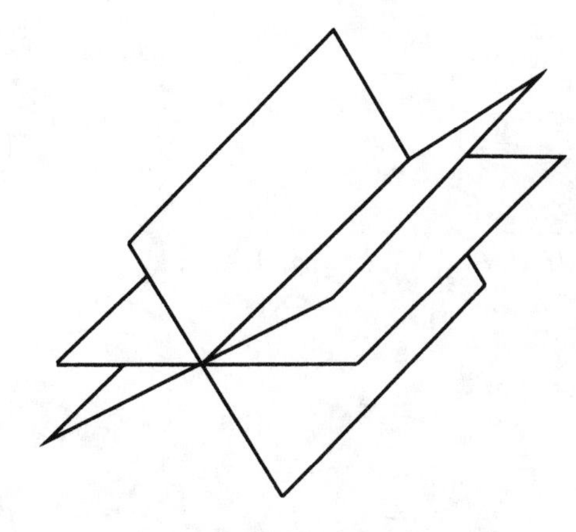

Figure 18–8

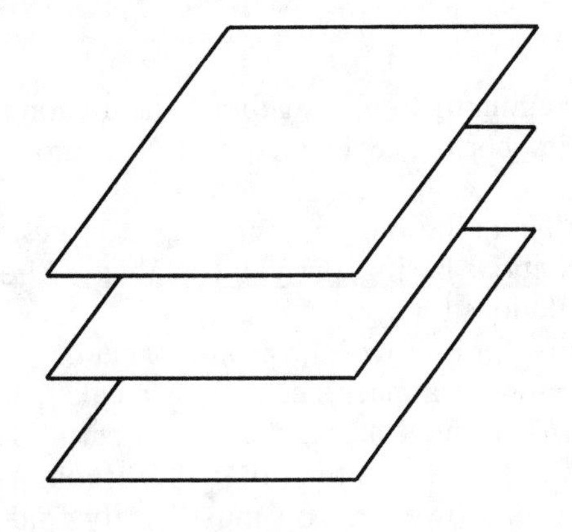

Figure 18–9

"Only two more minutes until the dragon will be here!" the detective warned us.

We all silently hid behind the mustard dispenser waiting for the dragon to appear. Then, suddenly, there he was, sneaking stealthily along the corridor of the stadium. The moment he reached out to grab the mustard, we all jumped out and grabbed him. This time the surprised dragon had no chance to escape.

"We have you now, you thief!" the king told him.

The dragon began to cry. "I didn't mean to scare you," he sobbed. "I was only having fun. And I thought you would learn something about algebra along the way."

"What an amazingly effective new form of terrorism!" Recordis said. "Forcing people to learn algebra like that! You don't have any idea what a fright you gave us!" he scolded.

"But aren't you beginning to like algebra?" the dragon pleaded.

"Algebra is a very boring, tedious subject," Recordis said. "How could anybody like it?"

"Only parts of algebra are boring," the dragon protested. "Part of that is only because arithmetic is boring, but you can always hire an arithmetic servant to do the calculations for you."

"Recordis does that for us now," the professor said.

"I was thinking more of a machine, such as a calculator or a computer," the dragon said. "All right, I admit that some parts of algebra itself are very tedious. For example, I admit that the binomial formula is very tedious to work with. However, you must surely find baseball practice to be tedious at times, but you still think it is worthwhile, don't you?" he asked the professor.

"Sometimes practice is tedious," the professor said. "But most of the time it isn't."

"Most of our algebraic derivations have not been that long," the king said, "with only a few exceptions."

"But there is an inherent beauty in watching a baseball in flight!" the professor said.

"And you can describe the flight of the baseball with algebra!" the dragon said. "Don't you find beauty in drawing a graph of a parabola? Or an ellipse? Or a fourth-degree polynomial? Don't you find beauty in coming up with a nice, simple formula for the orbits or the planets?"

Recordis was being swept along in the dragon's poetic discourse. "Still, you must admit that there is no beauty in square root signs," he said.

"True," the dragon admitted. "But then, there are also many practical advantages to learning algebra. Even if you never appreciate the beauty, you should learn algebra just for what it can do for you."

"We have solved many practical problems with algebra," the king agreed. "We never would have been able to do so without algebra."

"And there are many more things that you will be able to do in the future," the dragon continued earnestly. He held out pictures of rockets and televisions and computers and bridges and airplanes. "Eventually you will be able to build all of these things, but you need to know algebra first."

"But what of the books you stole?" Recordis demanded.

"I just took some old-fashioned books that gave the impression that mathematics was nothing but a lot of memorization," the dragon responded.

"But who sent you to our country in the first place?" the king demanded.

"The person who originally sent me wanted to terrorize you so he could take over the kingdom. But I didn't have the heart to take part in that scheme. Perhaps you know this chap—a vile creature with a black cape."

"The gremlin!" we all gasped in astonishment.

"Yes, the gremlin!" a familiar voice called out. We ran onto the field and found the gremlin standing at home plate.

"So that fool of a dragon has failed me!" he said. "This time, though, I shall win for sure! I admit that you understand the principles of algebra. However, I can still wear you out with pure tedium." He pointed to a large bacteria cloud rising out of Scenic Lake and descending on the stadium.

"So you're responsible for the bacteria in Scenic Lake!" the king gasped.

"You have your choice!" the gremlin hissed. "Give up to me—or give up to the bacteria!"

"Stop!" The Royal Biologist came running onto the field. "I have completed the calculations for the chemical that will destroy the bacteria!" she said defiantly. "It was a simple matter of solving an equation system to get the formula right."

She sprayed a chemical at the bacteria and the cloud vanished. But the gremlin only laughed. "So you have succeeded—once," he said. "But can you perform the same calculations many times a day? I have many different kinds of bacteria. To defeat them all, you will have to solve many different equation systems. Unless you think you can solve multiple simultaneous equation systems very rapidly, you can give up now!" he cackled.

Exercises

Solve these equation systems:

1. $a + b = 15$
$a + c = 16$
$b + c = 11$

2. $6a - 8b + 10c = 6$
$7a - 7b + 10c = 25$
$4a - 3b + 6c = 28$

3. $16a + 46b + 24c = 4{,}548$
$32a + 15b + 22c = 3{,}386$
$52a + 56b + 16c = 6{,}752$

4. $12a + 11b + 10c = 146$
$10a + 8b - 9c = 65$
$-16a - 12b + 14c = -102$

5. $12x - 9y + 4z = 16$
$2x + 8y - 6z = 12$
$12x + 12y + 11z = 20$

6. $4x - 6z = 8$
$2y + 2z = 16$
$x - z = 5$

7. $a + b + c = 100$

$b = 6 + 2a + 4c$

$c = 12 + 8a$

8. Solve this equation system:

$Y = C + I + G$ (Y = national income)

$C = 11 + (4/5)(Y - T)$ (C = consumption)

$I = 5 + (1/5)Y$ (I = investment)

$T = 20 + (1/4)Y$ (T = total taxes)

$G = 30$ (G = government spending)

9. Solve this system. (Y, C, I, and T are unknown, the other letters are all known.)

$Y = C + I + G$

$C = C_0 + b(Y - T)$

$I = I_0 + dY$

$T = T_0 + tY$

Find a second-degree polynomial that passes through each of these groups of three points:

10. (1, 7); (2, 8); (3, 13)
11. (1, 0); (5, 40); (−3, −8)
12. (3, 33); (4, 54.5); (−2, 15.5)
13. (6, 65); (7, 90); (10, 189)
14. (1, 1); (2, 23); (3, 55)

The following table lists the atomic number and the approximate atomic weight for several chemical elements:

Element	Symbol	Atomic number	Approximate weight
Hydrogen	H	1	1
Carbon	C	6	12
Nitrogen	N	7	14
Oxygen	O	8	16
Sodium	Na	11	23
Sulfur	S	16	32
Chlorine	Cl	17	35.5
Potassium	K	19	39
Chromium	Cr	24	52

A molecule consists of several atoms of different elements joined together. In each problem below, you are told what elements make up a particular molecule, the total atomic weight of each molecule, and the sum of the atomic numbers of all of the atoms in that molecule. Determine how many of each kind of atom are included in the molecule.

Elements	Total atomic number	Total weight
15. Na, Cl	28	58.5
16. H, O	10	18
17. N, H	10	17
18. C, H	10	16
19. S, O	32	64

In each exercise below, you are told the number of atoms in the molecule, in addition to the total atomic number and atomic weight.

	Elements	Number of atoms	Total atomic number	Total weight
20.	K, Cl, O	5	60	122.5
21.	K, Cr, O	11	142	294
22.	K, O, H	3	28	56
23.	K, S, O	7	86	174
24.	C, H, O	9	26	46
25.	C, H, O	14	50	92
26.	C, H, O	10	32	58

Solve these equation systems. Note that these systems involve equations that are not linear. You can solve them either by substitution or by adding the equations together in such a fashion that you can cancel out one of the variables.

27. $x^2 + y^2 = 25$
$x = y$

28. $x^2 + y^2 = 16$
$x - 2y = 4$

29. $y = -10x^2 + 5$
$4y + 5x = 10$

Identify the center and radius of the circle that passes through each triple of points:

30. $(-1, 24)$, $(14, 19)$, $(6, 23)$
31. $(12, 2)$, $(17, 27)$, $(24, 10)$
32. $(2, 21)$, $(3, 4)$, $(4, 3)$
33. $(6, 11)$, $(12, -3)$, $(2, 16)$

19 Matrices and Determinants

"Don't panic!" the professor said. "We know how to solve systems of equations using the triangularization process."

"But that is a lot of work!" Recordis moaned.

We hurried to meet the threat before the game was scheduled to begin. A new batch of bacteria appeared and the biologist quickly performed some measurements. "We need to solve this system." she told us.

$$\begin{aligned} x + y - z &= 6 \\ 2x - y + 3z &= 11 \\ 4x + 2y - 3z &= 14 \end{aligned}$$

"Remember the basic move we can make in the triangularization process," the professor said. "We can add to one of the equations a multiple of one of the other equations."

Recordis prepared to perform the necessary operations. "If only we didn't have to write all those letters and plus signs and equal signs!" he sighed. He stopped and thought a moment. "I have an idea!" he exclaimed. "Let's just write the numbers, like this."

$$\begin{matrix} 1 & 1 & -1 & 6 \\ 2 & -1 & 3 & 11 \\ 4 & 2 & -3 & 14 \end{matrix}$$

"That's just a meaningless jumble of numbers!" the professor said.

"No, it's not," Recordis said. "If you use your imagination, you can pretend to see the x's, y's, z's, plus signs, and equal signs."

$$\begin{aligned} 1x + 1y - 1z &= 6 \\ 2x - 1y + 3z &= 11 \\ 4x + 2y - 3z &= 14 \end{aligned}$$

"This is a great idea!" the king agreed with Recordis. "Each row represents an equation. The first column contains the coefficients of x; the second column contains the coefficients of y; the third column contains the coefficients of z; and the last column contains the terms from the right-hand side of the equations."

"We can perform our regular triangularization operations on this array of numbers," Recordis said. "We can add a multiple of one row to another row."

We decided to call a rectangular array of numbers like this one a *matrix* and to put big parentheses around it. We tried to solve the equation system represented by this matrix:

$$\begin{pmatrix} 1 & 1 & -1 & 6 \\ 2 & -1 & 3 & 11 \\ 4 & 2 & -3 & 14 \end{pmatrix}$$

First, we added -2 multiplied by the middle row to the last row, leaving the new last row with a zero in the column representing the coefficient of x:

$$\begin{pmatrix} 1 & 1 & -1 & 6 \\ 2 & -1 & 3 & 11 \\ 0 & 4 & -9 & -8 \end{pmatrix}$$

Next, we added -2 multiplied by the top row to the middle row, leaving the new middle row with a zero in the first column:

$$\begin{pmatrix} 1 & 1 & -1 & 6 \\ 0 & -3 & 5 & -1 \\ 0 & 4 & -9 & -8 \end{pmatrix}$$

Now, to complete the triangularization process, we merely had to add $4/3$ multiplied by the middle row to the bottom row:

$$\begin{pmatrix} 1 & 1 & -1 & 6 \\ 0 & -3 & 5 & -1 \\ 0 & 0 & -2\frac{1}{3} & -9\frac{1}{3} \end{pmatrix}$$

"This matrix tells us that we have now converted our original equation system into this equivalent equation system." Recordis said.

$$\begin{aligned} x + y - z &= 6 \\ -3y + 5z &= -1 \\ -2\tfrac{1}{3}z &= -9\tfrac{1}{3} \end{aligned}$$

The professor conceded that the matrix contained exactly the same information as was contained by the equation system. The only difference was that the matrix required much less writing. Now that we had triangularized the system, we proceeded to solve it. We found that $z = 4$, $y = 7$, and $x = 3$.

The biologist quickly mixed the proper chemicals and sprayed them at the bacteria. However, by this time she had discovered several other types of bacteria, and she gave us several more equation systems that we would need to solve.

Recordis wrote down the general solution procedure:

PROCEDURE TO SOLVE 3 LINEAR EQUATIONS IN 3 UNKNOWNS

- Write down a matrix with 3 rows and 4 columns to represent the equation. Each row represents one equation. The first column contains the coefficients of x; the second column contains the coefficients of y; the third column contains the coefficients of z; and the last column contains the terms from the right-hand side of the system.
- Next, triangularize the matrix. The basic move is to muliply one row by a number and then add that row to another row. Once the triangularization has been completed, solve for z first; then y; and then x.

Here are the other systems the biologist gave us:

system:

$$\begin{aligned} 3x - y + 4z &= 0 \\ x + 2y + 3z &= 15 \\ -x + y + 5z &= 12 \end{aligned}$$

matrix:

$$\begin{pmatrix} 3 & -1 & 4 & 0 \\ 1 & 2 & 3 & 15 \\ -1 & 1 & 5 & 12 \end{pmatrix}$$

triangularized form:

$$\begin{pmatrix} 3 & -1 & 4 & 0 \\ 0 & 2.333 & 1.667 & 15 \\ 0 & 0 & 5.857 & 7.714 \end{pmatrix}$$

solution:

$$x = 0.073; \quad y = 5.488; \quad z = 1.317$$

system:

$$\begin{aligned} x - y - z &= 5 \\ x + y + z &= 4 \\ -x - y + z &= 3 \end{aligned}$$

matrix:

$$\begin{pmatrix} 1 & -1 & -1 & 5 \\ 1 & 1 & 1 & 4 \\ -1 & -1 & 1 & 3 \end{pmatrix}$$

triangularized form:

$$\begin{pmatrix} 1 & -1 & -1 & 5 \\ 0 & 2 & 2 & -1 \\ 0 & 0 & 2 & 7 \end{pmatrix}$$

solution:

$$x = 4.5; \qquad y = -4; \qquad z = 3.5$$

system:

$$20x + 20y - z = 30$$

$$\frac{1}{2}x + 2y + z = 5$$

$$14x - 2y - 3z = 0$$

matrix:

$$\begin{pmatrix} 20 & 20 & -1 & 30 \\ \frac{1}{2} & 2 & 1 & 5 \\ 14 & -2 & -3 & 0 \end{pmatrix}$$

triangularized form:

$$\begin{pmatrix} 20 & 20 & -1 & 30 \\ 0 & 1.5 & 1.025 & 4.25 \\ 0 & 0 & 8.633 & 24.333 \end{pmatrix}$$

solution:

$$x = 0.7335; \qquad y = 0.9073; \qquad z = 2.8185$$

We were granted a temporary respite while the gremlin worked on developing some more types of bacteria. The professor decided to investigate some more properties of matrices.

"We should classify matrices according to the number of rows and columns they contain," she said.

Examples:

2-by-2 matrix (2 rows and 2 columns)

$$\begin{pmatrix} 2 & 10 \\ -1 & 5 \end{pmatrix}$$

2-by-3 matrix (2 rows and 3 columns)

$$\begin{pmatrix} 10 & 0 & 16 \\ 5 & 2 & 11 \end{pmatrix}$$

3-by-3 matrix (3 rows and 3 columns)

$$\begin{pmatrix} 1 & 0 & 0 \\ -1 & 1 & -1 \\ 0 & -1 & 1 \end{pmatrix}$$

4-by-3 matrix (4 rows and 3 columns)

$$\begin{pmatrix} 1 & 2 & 5 \\ 12 & 0 & 6 \\ -4 & -7 & -3 \\ 0 & 1 & 5 \end{pmatrix}$$

We decided to use boldface letters, such as **A** or **B**, to represent matrices (in the same way as regular letters represent numbers). For example, we may have

$$\mathbf{A} = \begin{pmatrix} 5 & 3 & 2 \\ 3 & 6 & 1 \\ 0 & 5 & 0 \end{pmatrix} \qquad \mathbf{B} = \begin{pmatrix} 1 & 12 & 5 \\ 4 & 5 & 6 \end{pmatrix}$$

(In these examples, **A** is a 3-by-3 matrix and **B** is a 2-by-3 matrix. In identifying the size of a matrix, we decided that we would always write the number of rows first. If the number of rows is the same as the number of columns, then the matrix is called a *square* matrix.)

"We'll call each number in a matrix an *element* of the matrix," the professor decided. "And we can easily add two matrices together by adding together their corresponding elements—provided that each matrix has the same number of rows and columns. For example, suppose

$$\mathbf{A} = \begin{pmatrix} 10 & 15 \\ 12 & 17 \end{pmatrix} \quad \text{and} \quad \mathbf{B} = \begin{pmatrix} 5 & 3 \\ 4 & 2 \end{pmatrix}$$

Then

$$\mathbf{A} + \mathbf{B} = \begin{pmatrix} 10 + 5 & 15 + 3 \\ 12 + 4 & 17 + 2 \end{pmatrix} = \begin{pmatrix} 15 & 18 \\ 16 & 19 \end{pmatrix}$$

"And a matrix consisting of all zero's will have a very special property," the professor said. "Let's call such a matrix a *zero matrix* and represent it by a boldface zero: **0**. Then, if **A** is any other matrix that is the same size as **0**, **0** + **A** = **A**."

$$\begin{pmatrix} 0 & 0 \\ 0 & 0 \end{pmatrix} + \begin{pmatrix} 4 & 5 \\ 6 & 7 \end{pmatrix} = \begin{pmatrix} 4 & 5 \\ 6 & 7 \end{pmatrix}$$

Next, the professor wanted to investigate how to multiply two matrices. She wanted to define multiplying matrices to mean simply multiplying all of the corresponding elements if the two matrices had the same size. However, this time Recordis successfully objected. "You've had your way long enough, so now I want to make up a definition. I want to define matrix multiplication in such a way as to make it as easy as possible to write down equation systems. Suppose we have this equation system:

$$a_1x + b_1y + c_1z = d_1$$

$$a_2x + b_2y + c_2z = d_2$$

$$a_3x + b_3y + c_3z = d_3$$

"Let's write the a's, b's, and c's as part of a 3-by-3 matrix called **A**:

$$\mathbf{A} = \begin{pmatrix} a_1 & b_1 & c_1 \\ a_2 & b_2 & c_2 \\ a_3 & b_3 & c_3 \end{pmatrix}$$

and we can write the d's as a 3-by-1 matrix:

$$\mathbf{d} = \begin{pmatrix} d_1 \\ d_2 \\ d_3 \end{pmatrix}$$

"Let's make x, y, and z the elements of a 3-by-1 matrix **x**:

$$\mathbf{x} = \begin{pmatrix} x \\ y \\ z \end{pmatrix}$$

"It would help a lot if we could write the equation system using matrix notation in a very simple form, like this:

$$\mathbf{Ax} = \mathbf{d}$$

"What does **Ax** mean?" the king asked.

"**Ax** means that the matrix **A** is to be multiplied by the matrix **x**," Recordis said. "Since we use implied multiplication when we want to multiply together two variables, we may as well use it when we want to multiply together two matrices."

"But how can **A** and **x** be multiplied?" the professor demanded. "They are not even the same size. **A** is a 3-by-3 matrix, and **x** is a 3-by-1 matrix."

"They can be multiplied as long as we define matrix multiplication to mean what I want it to mean," Recordis said. "In order to make it easy to write the equation system, I want **Ax** to mean this:

$$\begin{pmatrix} a_1 & b_1 & c_1 \\ a_2 & b_2 & c_2 \\ a_3 & b_3 & c_3 \end{pmatrix} \begin{pmatrix} x \\ y \\ z \end{pmatrix} = \begin{pmatrix} a_1x + b_1y + c_1z \\ a_2x + b_2y + c_2z \\ a_3x + b_3y + c_3z \end{pmatrix}$$

"What a bizarre definition!" the professor exclaimed.

"It may look strange now, but I guarantee you this definition will help in the long run," Recordis said.

MATRIX MULTIPLICATION

Consider two matrices **A** and **B**. Suppose that **A** has m rows and n columns and **B** has n rows and p columns. Then you can find the matrix product **AB**. (Note that the product **AB** only exists if the number of *columns* of **A** is the same as the number of *rows* of **B**.)

Here is the procedure to find the matrix product:

- Start with the first *row* of **A** and the first *column* of **B**. (Note that both of these will contain n elements.)
- Multiply the corresponding elements of this row and this column.
- Add all the products together. The result will be the element in the first row and first column of the product matrix **AB**.
- Repeat this process to find each element of **AB**, using the ith row of **A** and the jth column of **B** to get the element in the ith row and jth column of **AB**. (The resulting matrix will have m rows and p columns.)

We worked an example of multiplying a 3-by-3 matrix **A** times a 3-by-1 matrix **B**:

$$\mathbf{A} = \begin{pmatrix} 3 & 7 & 11 \\ 9 & 4 & 8 \\ 5 & 6 & 10 \end{pmatrix} \quad \mathbf{B} = \begin{pmatrix} 20 \\ 40 \\ 30 \end{pmatrix}$$

We started by taking the first row of **A** and the first (in this case, only) column of **B**:

$$3 \times 20 + 7 \times 40 + 11 \times 30 = 670$$

The result of this calculation was the element in row 1, column 1 of the product matrix **AB**. To find the element in row 2, column 1 of the product matrix, we needed to take the second row of **A** and the first column of **B**:

$$9 \times 20 + 4 \times 40 + 8 \times 30 = 580$$

Then we needed to take the third row of **A** and the first column of **B**:

$$5 \times 20 + 6 \times 40 + 10 \times 30 = 640$$

Therefore we could write the complete product matrix:

$$\mathbf{AB} = \begin{pmatrix} 3 & 7 & 11 \\ 9 & 4 & 8 \\ 5 & 6 & 10 \end{pmatrix} \begin{pmatrix} 20 \\ 40 \\ 30 \end{pmatrix} = \begin{pmatrix} 670 \\ 580 \\ 640 \end{pmatrix}$$

"The product of a 3-by-3 matrix and a 3-by-1 matrix is a 3-by-1 matrix, just as I said," Recordis said.

We worked an example of multiplying a 4×3 matrix by a 3×2 matrix:

$$\mathbf{A} = \begin{pmatrix} 11 & 12 & 13 \\ 21 & 22 & 23 \\ 31 & 32 & 33 \\ 41 & 42 & 43 \end{pmatrix} \qquad \mathbf{B} = \begin{pmatrix} 1 & 2 \\ 100 & 200 \\ 10{,}000 & 20{,}000 \end{pmatrix}$$

To help us keep things straight, we wrote the dimensions of each matrix below the name of the matrix:

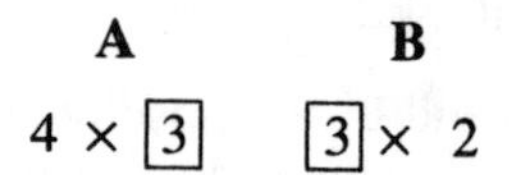

"The number of columns in **A**, 3, matches the number of rows of **B**, so all is well and good," Recordis said. "The resulting matrix will have 4 rows and 2 columns." We worked out the steps:

first row, first column:

$$\begin{array}{ccc} \boxed{11 \quad 12 \quad 13} \\ 21 \quad 22 \quad 23 \\ 31 \quad 32 \quad 33 \\ 41 \quad 42 \quad 43 \end{array} \qquad \begin{array}{|r|r} \hline 1 & 2 \\ 100 & 200 \\ 10{,}000 & 20{,}000 \\ \hline \end{array}$$

$$11 \times 1 + 12 \times 100 + 13 \times 10{,}000 = 131{,}211$$

second row, first column:

$$\begin{array}{ccc} 11 \quad 12 \quad 13 \\ \boxed{21 \quad 22 \quad 23} \\ 31 \quad 32 \quad 33 \\ 41 \quad 42 \quad 43 \end{array} \qquad \begin{array}{|r|r} \hline 1 & 2 \\ 100 & 200 \\ 10{,}000 & 20{,}000 \\ \hline \end{array}$$

$$21 \times 1 + 22 \times 100 + 23 \times 10{,}000 = 232{,}221$$

third row, first column:

$$\begin{array}{ccc} 11 & 12 & 13 \\ 21 & 22 & 23 \\ \multicolumn{3}{c}{\boxed{31 \quad 32 \quad 33}} \\ 41 & 42 & 43 \end{array} \qquad \boxed{\begin{array}{r} 1 \\ 100 \\ 10{,}000 \end{array}} \quad \begin{array}{r} 2 \\ 200 \\ 20{,}000 \end{array}$$

$$31 \times 1 + 32 \times 100 + 33 \times 10{,}000 = 333{,}231$$

fourth row, first column:

$$\begin{array}{ccc} 11 & 12 & 13 \\ 21 & 22 & 23 \\ 31 & 32 & 33 \\ \multicolumn{3}{c}{\boxed{41 \quad 42 \quad 43}} \end{array} \qquad \boxed{\begin{array}{r} 1 \\ 100 \\ 10{,}000 \end{array}} \quad \begin{array}{r} 2 \\ 200 \\ 20{,}000 \end{array}$$

$$41 \times 1 + 42 \times 100 + 43 \times 10{,}000 = 434{,}241$$

first row, second column:

$$\begin{array}{ccc} \multicolumn{3}{c}{\boxed{11 \quad 12 \quad 13}} \\ 21 & 22 & 23 \\ 31 & 32 & 33 \\ 41 & 42 & 43 \end{array} \qquad \begin{array}{r} 1 \\ 100 \\ 10{,}000 \end{array} \quad \boxed{\begin{array}{r} 2 \\ 200 \\ 20{,}000 \end{array}}$$

$$11 \times 2 + 12 \times 200 + 13 \times 20{,}000 = 262{,}422$$

second row, second column:

$$\begin{array}{ccc} 11 & 12 & 13 \\ \multicolumn{3}{c}{\boxed{21 \quad 22 \quad 23}} \\ 31 & 32 & 33 \\ 41 & 42 & 43 \end{array} \qquad \begin{array}{r} 1 \\ 100 \\ 10{,}000 \end{array} \quad \boxed{\begin{array}{r} 2 \\ 200 \\ 20{,}000 \end{array}}$$

$$21 \times 2 + 22 \times 200 + 23 \times 20{,}000 = 464,\ 442$$

third row, second column:

$$\begin{array}{ccc} 11 & 12 & 13 \\ 21 & 22 & 23 \\ \multicolumn{3}{c}{\boxed{31 \quad 32 \quad 33}} \\ 41 & 42 & 43 \end{array} \qquad \begin{array}{r} 1 \\ 100 \\ 10{,}000 \end{array} \quad \boxed{\begin{array}{r} 2 \\ 200 \\ 20{,}000 \end{array}}$$

$$31 \times 2 + 32 \times 200 + 33 \times 20{,}000 = 666{,}462$$

fourth row, second column:

$$\begin{array}{ccc} 11 & 12 & 13 \\ 21 & 22 & 23 \\ 31 & 32 & 33 \\ \multicolumn{3}{c}{\boxed{41 \quad 42 \quad 43}} \end{array} \qquad \begin{array}{r} 1 \\ 100 \\ 10{,}000 \end{array} \quad \boxed{\begin{array}{r} 2 \\ 200 \\ 20{,}000 \end{array}}$$

$$41 \times 2 + 42 \times 200 + 43 \times 20{,}000 = 868{,}482$$

Therefore, the result was:

$$\mathbf{AB} = \begin{pmatrix} 131{,}211 & 262{,}422 \\ 232{,}221 & 464{,}442 \\ 333{,}231 & 666{,}462 \\ 434{,}241 & 868{,}482 \end{pmatrix}$$

The professor sneakily suggested another example:

$$\mathbf{A} = \begin{pmatrix} 16 & 25 & 54 \\ 32 & 9 & 18 \end{pmatrix} \qquad \mathbf{B} = \begin{pmatrix} 19 & 7 \\ 2 & 84 \\ 45 & 31 \\ 74 & 14 \end{pmatrix}$$

$$\begin{array}{cc} \mathbf{A} & \mathbf{B} \\ 2 \times \boxed{3} & \boxed{4} \times 2 \end{array}$$

"Wait a minute!" Recordis exclaimed, noticing the trick the professor was trying to play on him. "The number of columns of **A**, 3, does not match the number of rows of **B**, 4, so these two matrices cannot be multiplied."

We worked some more examples:

$$\begin{pmatrix} 3 & 4 \\ 5 & 6 \end{pmatrix}\begin{pmatrix} 3 \\ 1 \end{pmatrix} = \begin{pmatrix} 3 \times 3 + 4 \times 1 \\ 5 \times 3 + 6 \times 1 \end{pmatrix} = \begin{pmatrix} 13 \\ 21 \end{pmatrix}$$

$$\begin{pmatrix} 3 & 0 \\ 4 & 1 \end{pmatrix}\begin{pmatrix} 6 & 1 \\ 8 & 2 \end{pmatrix} = \begin{pmatrix} 18 & 3 \\ 32 & 6 \end{pmatrix}$$

$$\begin{pmatrix} a & b \\ c & d \end{pmatrix}\begin{pmatrix} e \\ f \end{pmatrix} = \begin{pmatrix} ae + bf \\ ce + df \end{pmatrix}$$

$$\begin{pmatrix} a & b \\ c & d \end{pmatrix}\begin{pmatrix} e & f \\ g & h \end{pmatrix} = \begin{pmatrix} ae + bg & af + bh \\ ce + dg & cf + dh \end{pmatrix}$$

$$\begin{pmatrix} 1 & 0 \\ 0 & 1 \end{pmatrix}\begin{pmatrix} a & b \\ c & d \end{pmatrix} = \begin{pmatrix} a & b \\ c & d \end{pmatrix}$$

The professor was still highly skeptical that this definition of matrix multiplication was a good idea, but she admitted that with some practice she probably would become used to it.

"However, note that if the number of columns in **A** is different from the number of rows in **B**, then the product matrix **AB** won't exist," she said. "And that leads to one distressing fact—it means that matrix multiplication doesn't obey the order-doesn't-make-a-difference property. (We called that the *commutative property*.) The product matrix **AB** won't necessarily be the same as **BA**. For example, if **A** is a 3-by-2 matrix and **B** is a 2-by-2 matrix, then **AB** exists and is a 3-by-2 matrix, but **BA** doesn't even exist." We also tried an example where **BA** did exist but was different from **AB**:

$$\mathbf{A} = \begin{pmatrix} 6 & 12 \\ 4 & 2 \end{pmatrix} \qquad \mathbf{B} = \begin{pmatrix} 11 & 6 \\ 1 & 7 \end{pmatrix}$$

$$\mathbf{AB} = \begin{pmatrix} 78 & 120 \\ 46 & 38 \end{pmatrix} \qquad \mathbf{BA} = \begin{pmatrix} 90 & 144 \\ 34 & 26 \end{pmatrix}$$

"The matrix

$$\begin{pmatrix} 1 & 0 \\ 0 & 1 \end{pmatrix}$$

seems interesting," the king said. "Whenever you multiply that matrix by another matrix, the other matrix remains unchanged."

$$\begin{pmatrix} 1 & 0 \\ 0 & 1 \end{pmatrix}\begin{pmatrix} a & b \\ c & d \end{pmatrix} = \begin{pmatrix} 1 \times a + 0 \times c & 1 \times b + 0 \times d \\ 0 \times a + 1 \times c & 0 \times b + 1 \times d \end{pmatrix} = \begin{pmatrix} a & b \\ c & d \end{pmatrix}$$

"It acts like the number 1 in arithmetic," the professor said. "We will call it the *identity matrix*." We decided to use the letter **I** to stand for the identity matrix.

"However, there are different identity matrices," the king said. "There is a 2-by-2 identity matrix, a 3-by-3 identity matrix, and so on. (An identity matrix is always a *square* matrix—one with the same number of rows and columns.)"

3-by-3 identity matrix:

$$\begin{pmatrix} 1 & 0 & 0 \\ 0 & 1 & 0 \\ 0 & 0 & 1 \end{pmatrix}$$

4-by-4 identity matrix:

$$\begin{pmatrix} 1 & 0 & 0 & 0 \\ 0 & 1 & 0 & 0 \\ 0 & 0 & 1 & 0 \\ 0 & 0 & 0 & 1 \end{pmatrix}$$

We wrote down this property:

IDENTITY MATRIX PROPERTY

If I is an identity matrix, and A is any other matrix with the same number of rows as I, then IA = A. Also, if B has the same number of columns as I, then BI = B.

"Since there is a matrix corresponding to the number 1 in ordinary arithmetic, I wonder if we can find a matrix that acts as the *inverse* of a matrix," the professor mused. "In other words, suppose **A** is a matrix. Do you think we can find another matrix (call it $\mathbf{A}^{-1}$) such that $\mathbf{A} \cdot \mathbf{A}^{-1} = \mathbf{I}$?"

"Why did you call it $\mathbf{A}^{-1}$?" Recordis asked.

"Suppose a is a regular number. Then we know that $a \cdot a^{-1} = 1$, so it seems natural that if **A** is a matrix, then $\mathbf{A} \cdot \mathbf{A}^{-1} = \mathbf{I}$."

(Mathematicians define inverses only for square matrices, and not all of them have inverses. When the inverse exists, $\mathbf{A}^{-1} \cdot \mathbf{A}$, as well as $\mathbf{A} \cdot \mathbf{A}^{-1}$, is equal to **I**.)

Calculating matrix inverses proved to be a tricky problem. We found that the inverse of the identity matrix was also the identity matrix:

$$\mathbf{I}^{-1} = \mathbf{I}$$

The professor decided to look for a general formula for the inverse of a 2-by-2 matrix. She let a, b, c, and d represent the known values of the elements of the original matrix, and w, x, y, and z represent the unknown values of the inverse matrix. Then she wrote the equation connecting the two matrices:

$$\begin{pmatrix} a & b \\ c & d \end{pmatrix}\begin{pmatrix} w & x \\ y & z \end{pmatrix} = \begin{pmatrix} 1 & 0 \\ 0 & 1 \end{pmatrix}$$

She used matrix multiplication to write this system of equations:

$$aw + by = 1 \qquad ax + bz = 0$$

$$cw + dy = 0 \qquad cx + dz = 1$$

The lower left equation shows that $y = -cw/d$. Inserting this value into the upper left equation gives an equation for w:

$$aw - bcw/d = 1$$

$$w(a - bc/d) = 1$$

$$w(ad - bc)/d = 1$$

$$w = d/(ad - bc)$$

We were able to use the same type of procedure to find formulas for x, y, and z:

$$x = -b/(ad - bc)$$

$$y = -c/(ad - bc)$$

$$z = a/(ad - bc)$$

"The expression $(ad - bc)$ appears in each formula," the professor said, filled with curiosity. (At that time we did not yet suspect that $(ad - bc)$ would turn out to be the crucial quantity that determines the nature of the matrix.) She made a definition: $T = ad - bc$ and then she wrote an equation for the inverse matrix:

$$\begin{pmatrix} a & b \\ c & d \end{pmatrix}^{-1} = \begin{pmatrix} d/T & -b/T \\ -c/T & a/T \end{pmatrix}$$

We worked an example of the formula:

$$\begin{pmatrix} 2 & 1 \\ -4 & -1 \end{pmatrix}^{-1} = \begin{pmatrix} -\frac{1}{2} & -\frac{1}{2} \\ 2 & 1 \end{pmatrix}$$

We verified the result by multiplying these two matrices together to make sure that the product matrix was indeed the identity matrix:

$$\begin{pmatrix} 2 & 1 \\ -4 & -1 \end{pmatrix}\begin{pmatrix} -\frac{1}{2} & -\frac{1}{2} \\ 2 & 1 \end{pmatrix} = \begin{pmatrix} 2 \times (-\frac{1}{2}) + 1 \times 2 & 2 \times (-\frac{1}{2}) + 1 \times 1 \\ -4 \times (-\frac{1}{2}) - 1 \times 2 & -4 \times (-\frac{1}{2}) - 1 \times 1 \end{pmatrix}$$

$$= \begin{pmatrix} 1 & 0 \\ 0 & 1 \end{pmatrix}$$

(It can be very complicated to calculate the inverse of a matrix that is larger than 2-by-2. The note at the end of the chapter explains how this can be done, although in our country this type of work is best left to a computer.)

"If we calculate the inverse matrix $\mathbf{A}^{-1}$, then we have practically solved the equation system $\mathbf{Ax} = \mathbf{d}$," Recordis realized. "Let's multiply both sides of that matrix equation by $\mathbf{A}^{-1}$."

$$\mathbf{A}^{-1}\mathbf{Ax} = \mathbf{A}^{-1}\mathbf{d}$$

Since $\mathbf{A}^{-1}\mathbf{A} = \mathbf{I}$,

$$\mathbf{Ix} = \mathbf{A}^{-1}\mathbf{d}$$

Since $\mathbf{Ix} = \mathbf{x}$,

$$\mathbf{x} = \mathbf{A}^{-1}\mathbf{d}$$

"Therefore, if we know $\mathbf{A}^{-1}$, we can just calculate the matrix product $\mathbf{A}^{-1}\mathbf{d}$, and then we will have the solution for $\mathbf{x}$! (Although how exactly we will find $\mathbf{A}^{-1}$ is beyond me.)"

Recordis made up an equation system that we could use as an example:

$$2x + y = 26$$

$$-4x - y = -40$$

We wrote this system using matrix notation:

$$\begin{pmatrix} 2 & 1 \\ -4 & -1 \end{pmatrix}\begin{pmatrix} x \\ y \end{pmatrix} = \begin{pmatrix} 26 \\ -40 \end{pmatrix}$$

We multiplied both sides of the equation by the inverse of the matrix of coefficients (which we had just calculated a couple minutes earlier):

$$\begin{pmatrix} -\frac{1}{2} & -\frac{1}{2} \\ 2 & 1 \end{pmatrix}\begin{pmatrix} 2 & 1 \\ -4 & -1 \end{pmatrix}\begin{pmatrix} x \\ y \end{pmatrix} = \begin{pmatrix} -\frac{1}{2} & -\frac{1}{2} \\ 2 & 1 \end{pmatrix}\begin{pmatrix} 26 \\ -40 \end{pmatrix}$$

$$\begin{pmatrix} x \\ y \end{pmatrix} = \begin{pmatrix} -\frac{1}{2} \times 26 - \frac{1}{2} \times (-40) \\ 2 \times 26 + 1 \times (-40) \end{pmatrix} = \begin{pmatrix} 7 \\ 12 \end{pmatrix}$$

We verified that $x = 7$, $y = 12$ was a solution to the equation system.

While we were waiting for the next bacterial onslaught, the professor decided to look for a general formula for the solution to the system:

$$\begin{pmatrix} a_1 & b_1 & c_1 \\ a_2 & b_2 & c_2 \\ a_3 & b_3 & c_3 \end{pmatrix}\begin{pmatrix} x \\ y \\ z \end{pmatrix} = \begin{pmatrix} d_1 \\ d_2 \\ d_3 \end{pmatrix}$$

I won't describe the arduous algebraic labors that went into this calculation. When we were finally finished, the result was a horribly long formula (although it actually wasn't quite as long as we were afraid it would be):

$$x = \frac{d_1b_2c_3 + d_2b_3c_1 + d_3b_1c_2 - d_3b_2c_1 - d_2b_1c_3 - d_1b_3c_2}{a_1b_2c_3 + a_2b_3c_1 + a_3b_1c_2 - a_3b_2c_1 - a_2b_1c_3 - a_1b_3c_2}$$

$$y = \frac{a_1d_2c_3 + a_2d_3c_1 + a_3d_1c_2 - a_3d_2c_1 - a_2d_1c_3 - a_1d_3c_2}{a_1b_2c_3 + a_2b_3c_1 + a_3b_1c_2 - a_3b_2c_1 - a_2b_1c_3 - a_1b_3c_2}$$

$$z = \frac{a_1b_2d_3 + a_2b_3d_1 + a_3b_1d_2 - a_3b_2d_1 - a_2b_1d_3 - a_1b_3d_2}{a_1b_2c_3 + a_2b_3c_1 + a_3b_1c_2 - a_3b_2c_1 - a_2b_1c_3 - a_1b_3c_2}$$

The professor noticed that the denominators were all the same. "This long, peculiar quantity,

$$a_1b_2c_3 + a_2b_3c_1 + a_3b_1c_2 - a_3b_2c_1 - a_2b_1c_3 - a_1b_3c_2$$

seems to be an interesting function of all of the elements in this matrix:

$$\begin{pmatrix} a_1 & b_1 & c_1 \\ a_2 & b_2 & c_2 \\ a_3 & b_3 & c_3 \end{pmatrix}$$

I wonder if we should give a special name to that quantity?"

Igor suggested that we call this quantity the *determinant* of the matrix. We made this definition:

Start with the 3-by-3 matrix

$$\begin{pmatrix} a_1 & b_1 & c_1 \\ a_2 & b_2 & c_2 \\ a_3 & b_3 & c_3 \end{pmatrix}$$

The determinant of the matrix is calculated from the formula

$$a_1b_2c_3 + a_2b_3c_1 + a_3b_1c_2$$
$$-a_3b_2c_1 - a_2b_1c_3 - a_1b_3c_2$$

Note that the determinant is a number, not a matrix. We decided to symbolize a determinant by writing all the elements inside straight lines:

$$\det \mathbf{A} = \begin{vmatrix} a_1 & b_1 & c_1 \\ a_2 & b_2 & c_2 \\ a_3 & b_3 & c_3 \end{vmatrix}$$

The professor explained how we could use determinants to calculate the solution to the multiple equation system

$$\begin{pmatrix} a_1 & b_1 & c_1 \\ a_2 & b_2 & c_2 \\ a_3 & b_3 & c_3 \end{pmatrix} \begin{pmatrix} x \\ y \\ z \end{pmatrix} = \begin{pmatrix} d_1 \\ d_2 \\ d_3 \end{pmatrix}$$

SOLVING EQUATIONS BY DETERMINANTS

- Calculate the determinant of the matrix of coefficients. That determinant will be the denominator for each solution.
- To find the solution for x, replace the coefficients of x in the coefficient matrix with the column of constants from the right-hand side. Then, calculate the determinant of that matrix. That determinant will be the numerator of the solutions for x. Follow the same procedure to find the solutions for y and z:

$$x = \frac{\begin{vmatrix} d_1 & b_1 & c_1 \\ d_2 & b_2 & c_2 \\ d_3 & b_3 & c_3 \end{vmatrix}}{\begin{vmatrix} a_1 & b_1 & c_1 \\ a_2 & b_2 & c_2 \\ a_3 & b_3 & c_3 \end{vmatrix}} \qquad y = \frac{\begin{vmatrix} a_1 & d_1 & c_1 \\ a_2 & d_2 & c_2 \\ a_3 & d_3 & c_3 \end{vmatrix}}{\begin{vmatrix} a_1 & b_1 & c_1 \\ a_2 & b_2 & c_2 \\ a_3 & b_3 & c_3 \end{vmatrix}} \qquad z = \frac{\begin{vmatrix} a_1 & b_1 & d_1 \\ a_2 & b_2 & d_2 \\ a_3 & b_3 & d_3 \end{vmatrix}}{\begin{vmatrix} a_1 & b_1 & c_1 \\ a_2 & b_2 & c_2 \\ a_3 & b_3 & c_3 \end{vmatrix}}$$

(This solution method is called *Cramer's rule*.)

The king noticed something even more astonishing. "We can use the exact same rule for 2-by-2 systems!" he exclaimed. "Let's define the determinant of a 2-by-2 matrix like this:

$$\begin{vmatrix} a_1 & b_1 \\ a_2 & b_2 \end{vmatrix} = a_1b_2 - a_2b_1$$

Then, our rule tells us that the solution to the system

$$\begin{pmatrix} a_1 & b_1 \\ a_2 & b_2 \end{pmatrix} \begin{pmatrix} x \\ y \end{pmatrix} = \begin{pmatrix} c_1 \\ c_2 \end{pmatrix}$$

is

$$x = \frac{\begin{vmatrix} c_1 & b_1 \\ c_2 & b_2 \end{vmatrix}}{\begin{vmatrix} a_1 & b_1 \\ a_2 & b_2 \end{vmatrix}} = \frac{b_2c_1 - b_1c_2}{a_1b_2 - a_2b_1}$$

$$y = \frac{\begin{vmatrix} a_1 & c_1 \\ a_2 & c_2 \end{vmatrix}}{\begin{vmatrix} a_1 & b_1 \\ a_2 & b_2 \end{vmatrix}} = \frac{a_1c_2 - a_2c_1}{a_1b_2 - a_2b_1}$$

That is the same formula we found earlier." (See Chapter 10.)

We found a few more interesting properties of determinants:

$\det(\mathbf{A}^{-1}) = 1/\det(\mathbf{A})$

$\det(\mathbf{AB}) = \det(\mathbf{A})\det(\mathbf{B})$ (See Exercise 49.)

If $\det(\mathbf{A}) = 0$, then the matrix **A** has no inverse.

The value of the determinant is unchanged if you add a multiple of one row to another row. (See Exercise 47.)

"I'm really beginning to like algebra," the professor said. "These discoveries are all so interesting."

"The gremlin is making one final attack!" the biologist called out. A huge new cloud appeared. "This batch is much more complicated!" She quickly performed some measurements, and then she informed us that we would have to solve this system:

$$\begin{aligned}
2x_1 + x_2 + 3x_3 + 4x_4 + 6x_5 &= 17 \\
6x_1 + 7x_2 + x_3 + 2x_4 + 7x_5 &= 19 \\
x_1 + 11x_2 + 9x_3 + x_4 + 9x_5 &= 13 \\
8x_1 + 2x_2 + 7x_3 + 7x_4 + 2x_5 &= 5 \\
2x_1 + 3x_2 + x_3 + 6x_4 + 7x_5 &= 31
\end{aligned}$$

Recordis almost fainted. "That system has five equations and five unknowns!" We set up the matrix and performed the normal triangularization process. First, we eliminated the coefficients of x in every row except the first, by adding to each row a multiple of the first row. Then the matrix looked like this:

$$\begin{pmatrix}
2 & 1 & 3 & 4 & 6 & 17 \\
0 & 4 & -8 & -10 & -11 & -32 \\
0 & 10.5 & 7.5 & -1 & 6 & 4.5 \\
0 & -2 & -5 & -9 & -22 & -63 \\
0 & 2 & -2 & 2 & 1 & 14
\end{pmatrix}$$

Next, we went to work on the second column. By adding multiples of the second row, we were able to convert the second column into a form that was mostly zeros:

$$\begin{pmatrix} 2 & 1 & 3 & 4 & 6 & 17 \\ 0 & 4 & -8 & -10 & -11 & -32 \\ 0 & 0 & 28.5 & 25.25 & 34.875 & 88.5 \\ 0 & 0 & -9 & -14 & -27.5 & -79 \\ 0 & 0 & 2 & 7 & 6.5 & 30 \end{pmatrix}$$

Next, we worked on the third and fourth columns:

$$\begin{pmatrix} 2 & 1 & 3 & 4 & 6 & 17 \\ 0 & 4 & -8 & -10 & -11 & -32 \\ 0 & 0 & 28.5 & 25.25 & 34.875 & 88.5 \\ 0 & 0 & 0 & -6.0263 & -16.4868 & -51.0526 \\ 0 & 0 & 0 & 0 & -10.2503 & -20.5007 \end{pmatrix}$$

Now that the matrix was triangularized, we could proceed to find the solution:

$$x_5 = 2;\ x_4 = 3;\ x_3 = -2;\ x_2 = 1;\ x_1 = -1$$

We were exhausted from our efforts, but the biologist quickly mixed the necessary chemicals and sprayed them at the bacteria. The cloud vanished. The gremlin had tried his hardest, but he did not have any more types of bacteria. No new cloud appeared. The gremlin disappeared, and the dragon was taken into custody. "The kingdom is safe!" Recordis breathed a sigh of relief.

We all settled down to watch the game. The professor pitched brilliantly, and the Royal Palace team won easily.

Notes to Chapter 19

- Here is one method to find the inverse of an n-by-n matrix. Form a new matrix with $2n$ rows and n columns consisting of the original matrix written above an identity matrix. Then, slowly convert the top half of this matrix into an identity matrix, performing the standard operation on the columns; Add to one column a multiple of another column. Once the top half of the matrix has been turned into an identity matrix. the bottom half of the matrix will be the inverse of the original matrix. In practice this calculation is difficult to carry out, and the computations are best left to a computer. The answer to Exercise 52 gives a computer program that uses this method to calculate the inverse of a matrix.
- Here is a method for finding the determinant of a matrix. The *minor* of a matrix element a_{ij} is the matrix consisting of all of the elements in the original matrix except for the elements in row i and column j. For example, the minor of the element 1 in the 3-by-3 matrix

$$\begin{pmatrix} 1 & 2 & 3 \\ 4 & 5 & 6 \\ 7 & 8 & 9 \end{pmatrix}$$

is the matrix

$$\begin{pmatrix} 5 & 6 \\ 8 & 9 \end{pmatrix}$$

To calculate the determinant of a matrix, start with the first element in the first row. Multiply that element by the determinant of its minor. Then subtract the product of the second element and the determinant of its minor. Then, add the product of the third element and its minor. Keep going like that. For example, a 3-by-3 determinant can be found from the expression:

$$\begin{vmatrix} a_1 & b_1 & c_1 \\ a_2 & b_2 & c_2 \\ a_3 & b_3 & c_3 \end{vmatrix} = a_1 \begin{vmatrix} b_2 & c_2 \\ b_3 & c_3 \end{vmatrix} - b_1 \begin{vmatrix} a_2 & c_2 \\ a_3 & c_3 \end{vmatrix} + c_1 \begin{vmatrix} a_2 & b_2 \\ a_3 & b_3 \end{vmatrix}$$

The determinant of a 4×4 matrix can be found from this formula:

$$\begin{vmatrix} a_1 & b_1 & c_1 & d_1 \\ a_2 & b_2 & c_2 & d_2 \\ a_3 & b_3 & c_3 & d_3 \\ a_4 & b_4 & c_4 & d_4 \end{vmatrix}$$

$$= a_1 \begin{vmatrix} b_2 & c_2 & d_2 \\ b_3 & c_3 & d_3 \\ b_4 & c_4 & d_4 \end{vmatrix} - b_1 \begin{vmatrix} a_2 & c_2 & d_2 \\ a_3 & c_3 & d_3 \\ a_4 & c_4 & d_4 \end{vmatrix} + c_1 \begin{vmatrix} a_2 & b_2 & d_2 \\ a_3 & b_3 & d_3 \\ a_4 & b_4 & d_4 \end{vmatrix} - d_1 \begin{vmatrix} a_2 & b_2 & c_2 \\ a_3 & b_3 & c_3 \\ a_4 & b_4 & c_4 \end{vmatrix}$$

You need to calculate four 3×3 determinants in order to find the 4×4 determinant. A similar approach works for matrices of any size but you can see that so much work is involved that it is best to use a computer for the calculations. However, you don't have to expand along the first row. If there is a row or column with a lot of zeros, use that one. Be careful to alternate the signs. The sign of the upper left hand element is always positive, and the sign of any other element is found by alternating according to this pattern:

$$\begin{pmatrix} + & - & + & - \\ - & + & - & + \\ + & - & + & - \\ - & + & - & + \end{pmatrix}$$

In general, to find an $n \times n$ determinant by expanding along row i:

$$\sum_{j=1}^{n} a_{ij}\overline{a_{ij}}(-1)^{i+j}$$

where a_{ij} is the number in row i, column j, $\overline{a_{ij}}$ is the determinant of the matrix formed of all numbers that are left after crossing out all elements in row i and column j, and $(-1)^{i+j}$ is included to make sure that the terms properly alternate in sign. A similar formula works if you want to expand along a column instead of a row. For example, when a matrix is mostly zeros along the last column, find the determinant by expanding along that column:

$$\begin{vmatrix} a_1 & b_1 & c_1 & 0 \\ a_2 & b_2 & c_2 & 0 \\ a_3 & b_3 & c_3 & 0 \\ a_4 & b_4 & c_4 & d_4 \end{vmatrix}$$

$$= -0 \begin{vmatrix} a_2 & b_2 & c_2 \\ a_3 & b_3 & c_3 \\ a_4 & b_4 & c_4 \end{vmatrix} + 0 \begin{vmatrix} a_1 & b_1 & c_1 \\ a_3 & b_3 & c_2 \\ a_4 & b_4 & c_4 \end{vmatrix} - 0 \begin{vmatrix} a_1 & b_1 & c_1 \\ a_2 & b_2 & c_2 \\ a_4 & b_4 & d_4 \end{vmatrix} + d_4 \begin{vmatrix} a_1 & b_1 & c_1 \\ a_2 & b_2 & c_2 \\ a_3 & b_3 & c_3 \end{vmatrix}$$

Exercises

Calculate these matrix products (if the two matrices can be multiplied):

1. $\begin{pmatrix} 1 & 2 \\ 3 & 4 \end{pmatrix} \begin{pmatrix} 5 & 6 \\ 7 & 8 \end{pmatrix}$

2. $\begin{pmatrix} 6 & 8 \\ 3 & 6 \end{pmatrix} \begin{pmatrix} 5 & 7 \\ 5 & 5 \end{pmatrix}$

3. $\begin{pmatrix} 4 & 12 \\ 4 & 7 \end{pmatrix} \begin{pmatrix} 9 & 5 \\ 8 & 6 \end{pmatrix}$

4. $\begin{pmatrix} 7 & 0 & 0 \\ 8 & 4 & 0 \\ 7 & 11 & 8 \end{pmatrix} \begin{pmatrix} 10 & 6 & 11 \\ 0 & 7 & 6 \\ 0 & 0 & 11 \end{pmatrix}$

5. $\begin{pmatrix} 1 & 0 & 1 \\ 0 & 1 & 0 \\ 1 & 0 & 1 \end{pmatrix} \begin{pmatrix} 0 & 1 & 0 \\ 1 & 0 & 1 \\ 0 & 1 & 0 \end{pmatrix}$

6. $\begin{pmatrix} 3 & 0 & 0 \\ 0 & 3 & 0 \\ 0 & 0 & 3 \end{pmatrix} \begin{pmatrix} 10 & 11 & 12 \\ 12 & 13 & 11 \\ 12 & 10 & 10 \end{pmatrix}$

7. $\begin{pmatrix} 3 \\ 8 \end{pmatrix} \begin{pmatrix} 5 & 4 \\ 1 & 6 \end{pmatrix}$

8. $\begin{pmatrix} 5 & 3 \\ 2 & 1 \end{pmatrix} \begin{pmatrix} 2 \\ 3 \end{pmatrix}$

9. $\begin{pmatrix} 7 & 1 \\ 5 & 9 \end{pmatrix} \begin{pmatrix} 3 & 5 & 2 \\ 6 & 0 & 1 \end{pmatrix}$

10. $\begin{pmatrix} 4 & 2 \\ 6 & 3 \\ 1 & 5 \end{pmatrix} \begin{pmatrix} 6 \\ 4 \\ 2 \end{pmatrix}$

11. $\begin{pmatrix} 11 \\ 2 \\ 5 \end{pmatrix} \begin{pmatrix} 6 & 3 \\ 1 & 2 \\ 1 & 5 \end{pmatrix}$

12. $\begin{pmatrix} 3 & 1 \\ 6 & 4 \\ 11 & 1 \end{pmatrix} \begin{pmatrix} 2 \\ 5 \end{pmatrix}$

13. $\begin{pmatrix} 0 & 0 & 0 \\ 1 & 0 & 0 \\ 0 & 1 & 0 \end{pmatrix} \begin{pmatrix} 1 & 2 & 3 \\ 11 & 12 & 13 \\ 21 & 22 & 23 \end{pmatrix}$

14. $\begin{pmatrix} 0 & 0 & 0 \\ 1 & 0 & 0 \\ 0 & 1 & 0 \end{pmatrix} \begin{pmatrix} 0 & 0 & 0 \\ 1 & 2 & 3 \\ 11 & 12 & 13 \end{pmatrix}$

15. $\begin{pmatrix} 0 & 0 & 0 \\ 1 & 0 & 0 \\ 0 & 1 & 0 \end{pmatrix} \begin{pmatrix} 0 & 0 & 0 \\ 0 & 0 & 0 \\ 1 & 2 & 3 \end{pmatrix}$

16. $\begin{pmatrix} 0 & 0 & 0 \\ 1 & 0 & 0 \\ 0 & 1 & 0 \end{pmatrix} \begin{pmatrix} a & b & c \\ d & e & f \\ g & h & i \end{pmatrix}$

17. $\begin{pmatrix} 0 & 0 & 0 \\ 1 & 0 & 0 \\ 0 & 1 & 0 \end{pmatrix} \begin{pmatrix} 0 & 0 & 0 \\ 1 & 0 & 0 \\ 0 & 1 & 0 \end{pmatrix}$

18. $\begin{pmatrix} a_1 & b_1 \\ a_2 & b_2 \end{pmatrix} \begin{pmatrix} c_1 & d_1 \\ c_2 & d_2 \end{pmatrix}$

19. $\begin{pmatrix} a & b \\ c & d \end{pmatrix} \begin{pmatrix} 1/a & 1/b \\ 1/c & 1/d \end{pmatrix}$

20. Express the product of these two matrices using summation notation:

$$\begin{pmatrix} a_{11} & a_{12} & a_{13} & \cdots & a_{1n} \\ a_{21} & a_{22} & a_{23} & \cdots & a_{2n} \\ \cdot & & & & \cdot \\ \cdot & & & & \cdot \\ a_{m1} & a_{m2} & a_{m3} & \cdots & a_{mn} \end{pmatrix} \begin{pmatrix} b_{11} & b_{12} & b_{13} & \cdots & b_{1p} \\ b_{21} & b_{22} & b_{23} & \cdots & b_{2p} \\ \cdot & & & & \cdot \\ \cdot & & & & \cdot \\ b_{n1} & b_{n2} & b_{n3} & \cdots & b_{np} \end{pmatrix}$$

Use the matrix triangularization process to solve these equation systems:

21. $6x + 3y + 8z = 98$
$5x + 11y - 9z = 104$
$7x - 12y - 9z = -114$

22. $8x - 7y + 5z = 15$
$-6x + 9y - 10z = -10$
$8x - 6y + 10z = 22$

23. $2x + 4y + 7z = x - 5y + 12z$
$7x + 16 - 3y = 5z + 11y - x$
$11x - 14 + 7z = x + y + z$

24. $12y - z = 24x + 16 - 3z + 5x$
$3x^2 - 4 + 12z = x(3x - 1) + 4y$
$y(3 + 4x) - 6z = 4xy - 11z + 6y - 4 + x$

25. $21x + 12y + 64z = 54$
$34x + 63y + 25z = 131$
$45x + 6y + 7z = 103$

26. $(1/2)x - (3/4)y + (6/7)z = 21$
$(7/12)x + (1/3)y + (1/14)z = 11\ 1/2$
$(2/3)x - (1/8)y + (2/7)z = 8\ 1/2$

27. $8x + 7y + 10z = 179$
$3x + 2y + 8z = 87$
$7x + 10y + 8z = 187$

28. $9x - 3y - 4z = 5$
$7x + 9y - 11z = 5$
$4x - 6y + 12z = 2$

Calculate these determinants:

29. $\begin{vmatrix} 4 & 6 \\ 2 & 3 \end{vmatrix}$

30. $\begin{vmatrix} 6 & 2 \\ 12 & 1 \end{vmatrix}$

31. $\begin{vmatrix} 2 & 31 \\ 11 & 3 \end{vmatrix}$

32. $\begin{vmatrix} 9 & 8 \\ 0 & 5 \end{vmatrix}$

33. $\begin{vmatrix} 6 & 11 \\ 12 & 22 \end{vmatrix}$

34. $\begin{vmatrix} 2 & 2 \\ 5 & 5 \end{vmatrix}$

35. $\begin{vmatrix} 1 & 2 \\ 3 & 4 \end{vmatrix}$

36. $\begin{vmatrix} 2 & 1 \\ 4 & 3 \end{vmatrix}$

37. $\begin{vmatrix} 1 & 2 \\ 10 & 20 \end{vmatrix}$

38. $\begin{vmatrix} 1 & 5 \\ 13 & 26 \end{vmatrix}$

Calculate these determinants. In each case, pick one row or column and expand along it, using the method described in the Note at the end of the chapter. (Note: the trick is picking the right row or column. In each case there will be one row or column that is the easiest to work with.)

39. $\begin{vmatrix} 3 & 0 & 0 \\ 6 & 4 & 5 \\ -1 & 2 & 4 \end{vmatrix}$

40. $\begin{vmatrix} 4 & 1 & 0 \\ 6 & 3 & 2 \\ -1 & -3 & 5 \end{vmatrix}$

41. $\begin{vmatrix} 10 & 0 & 0 & 0 \\ 3 & 6 & 5 & 1 \\ 2 & -1 & 4 & 0 \\ 6 & 5 & 1 & 3 \end{vmatrix}$

42. $\begin{vmatrix} 0 & 0 & 6 & 5 \\ 9 & 8 & 0 & 1 \\ 4 & 3 & 1 & 2 \\ 7 & 1 & 11 & 8 \end{vmatrix}$

43. $\begin{vmatrix} 6 & 1 & 4 & 0 \\ 2 & 3 & 1 & 0 \\ 6 & 8 & 2 & 0 \\ 7 & 5 & 1 & 1 \end{vmatrix}$

44. $\begin{vmatrix} 12 & 165 & 14 & 0 \\ 18 & 197 & 16 & 0 \\ 34 & 256 & 17 & 0 \\ 11 & 190 & 11 & 0 \end{vmatrix}$

Prove that these properties are true for 2-by-2 determinants. (Note: These properties are true for determinants of all sizes, but it is harder to prove that. Also, all of these properties will still be true if you substitute the word "row" in the place of the word "column.")

45. If you interchange two columns of the determinant, the value of the determinant changes sign.
46. If two columns of a determinant are the same, the value of the determinant is zero.
47. If you add to one column a multiple of another column, then the value of the determinant stays the same.
48. If you multiply one column of a determinant by a number, then the value of the determinant is multiplied by that same number.
49. If **A** and **B** are two matrices, then det (**AB**) = det (**A**) det (**B**).
50. Prove **IA** = **A** in the 3-by-3 case.
☐ **51.** Write a program that uses the matrix triangularization process to solve an n-by-n system of equations.
☐ **52.** Write a program that calculates the inverse of an n-by-n matrix.
☐ **53.** Write a program that calculates the determinant of a 4-by-4 matrix. (Note: the program should call a subroutine which calculates the determinant of a 3-by-3 matrix; that subroutine should call another subroutine that calculates the determinant of a 2-by-2 matrix.)
☐ **54.** Write a program that solves a simultaneous equation system using Cramer's rule. The answer section contains a program written in Pascal which uses recursion (which means that the function to calculate the determinant calls itself in the course of evaluating 3 by 3 or larger determinants).

20 Imaginary Numbers

We were so exhausted after saving the kingdom that we took a vacation at the Carmorra Beachfront hotel. Mrs. O'Reilly was glad to welcome us back.

While we were sunning ourselves along the beach one day the gremlin slowly walked up to us. He looked haggard and exhausted. As we would have imagined, trying to take over the kingdom was hard work. He spoke more softly then usual. "I concede now that you know most of elementary algebra," he conceded. "But I have one more challenge for you. I doubt that you have developed the necessary imagination to solve this one simple equation."

$$x^2 + x + 1 = 0$$

"That's a regular old quadratic equation!" Recordis said happily. "We can solve those as easily as . . ."

The gremlin laughed softly. "Nevertheless, I dare you to find a solution for this equation. I doubt that you are yet willing to admit the existence of a type of number that you cannot see." The gremlin slowly walked away.

"What does he mean—a type of number we cannot see?" Recordis asked. "We can represent any real number that we want on a number line, so there is no such thing as a number we cannot draw a picture of.

Anyway, to solve this equation we just need to use the quadratic formula."

$$ax^2 + bx + c = 0 \qquad\qquad x^2 + x + 1 = 0$$

$$x = \frac{-b \pm \sqrt{b^2 - 4ac}}{2a} \qquad\qquad x = \frac{-1 \pm \sqrt{1^2 - 4 \times 1 \times 1}}{2 \times 1}$$

$$x = \frac{-1 \pm \sqrt{1 - 4}}{2}$$

$$x = -\frac{1}{2} \pm \frac{1}{2}\sqrt{-3}$$

"No!" Recordis screamed in horror. "That equation has no solution! There is no such number as $\sqrt{-3}$. We have already shown that!" Recordis drew a quick diagram to show that the parabola represented by the equation $y = x^2 + x + 1$ did not cross the x axis, showing that the equation $x^2 + x + 1 = 0$ had no solutions. (See Figure 20–1.)

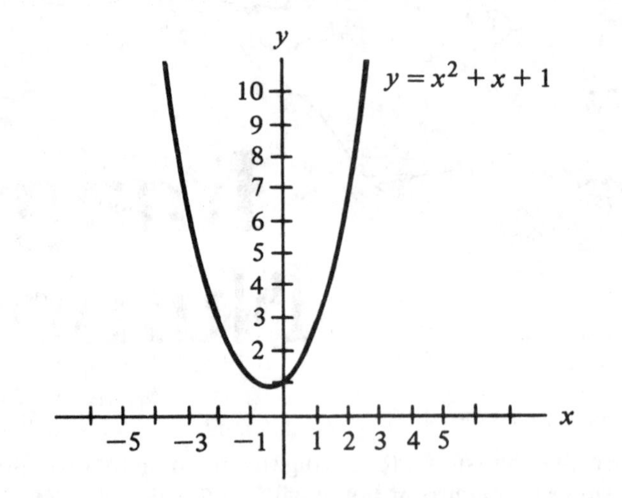

Figure 20–1

"That's what the gremlin wants us to say," the king observed. "The gremlin wants us to give up and say that this type of equation has no solution."

"A type of number that we cannot see," the professor mused. "Maybe we need to develop a new kind of number."

"No!" Recordis protested feverishly. "There is no such thing as a number that is not a real number!"

"But remember that 'number' is just an undefined term," the professor said. "We never defined the word number—we just said that intuitively we would know what a number is. We can create a new type of number and see how it behaves, if we want to."

We tried to simplify the expression $\sqrt{-3}$. We wrote:

$$\sqrt{-3} = \sqrt{3 \times (-1)}$$

Using one of the properties of square roots,

$$\sqrt{ab} = \sqrt{a} \cdot \sqrt{b}$$

we could write:

$$\sqrt{-3} = \sqrt{3} \cdot \sqrt{-1}$$

"We know what $\sqrt{3}$ is," the professor said. "So the main problem is finding out what $\sqrt{-1}$ is."

"We will have to make up a number—using our imagination," the king said. "Let's think of a symbol to represent the number that is equal to the square root of -1, and then see how it behaves."

Recordis started trembling violently. "I will have nothing to do with any imaginary numbers!" Then he screamed and fainted.

The professor rushed to his side to tend him. "He will soon be recovered," she said. "However, we had better hurry and complete our investigation of the properties of imaginary numbers before he wakes up. The worst thing that could happen to him now would be for him to wake up and find us still talking about imaginary numbers. The trauma would be too much for him."

We decided to use the letter i to represent the square root of negative 1:

$$i = \sqrt{-1}$$

"The number i is the fundamental imaginary number," the professor decided. "We can build up other imaginary numbers as multiples of i. For example, suppose we need to find $\sqrt{-64}$. That is $\sqrt{64 \cdot (-1)} = \sqrt{64} \cdot \sqrt{-1} = 8i$. $8i$ is an imaginary number, just as i and $3i$ are imaginary numbers."

"So it is clear what it means to multiply a real number by an imaginary number," the king said. "The result will be another imaginary number." We decided that an imaginary number could be written in the form bi, where b is a real number. "Now we can express the square root of any negative number as an imaginary number:

$$\sqrt{-b^2} = bi$$

$$\sqrt{-n} = \sqrt{n}\, i$$

We discovered that adding two imaginary numbers together was simple:

$$i + i = 2i$$

$$i + i + i = 3i$$

$$5i + 6i = 11i$$

$$b_1 i + b_2 i = (b_1 + b_2)i$$

However, we found that there was no simple way to add a real number to an imaginary number:

$$2 + 3i$$

As far as we could tell, that expression was as simple as it could get.

"What kind of number is $2 + 3i$?" the king asked. "It's clearly not a real number, but it's also clearly not an imaginary number."

"We'll have to create a new kind of number," the professor said. "When you add an imaginary number to a real number, the result will be a new, complicated kind of number."

We decided to use the term *complex number* for a number that was the sum of a real number and an imaginary number. The general form for a complex number is

$$a + bi$$

where a and b are both real numbers. (a is called the *real part* of the number and b is called the *imaginary part*.) Here are some examples of complex numbers:

$$2 + 3i, \quad 10 + 16i, \quad 43 - 3i, \quad 2 - 3i$$

"I can see that the set of complex numbers includes both the set of real numbers and the set of imaginary numbers," the king said. "Suppose we have a complex number $a + bi$ with $b = 0$. Then the number is just plain a—a real number. On the other hand, if $a = 0$, then the number is bi—an imaginary number." (We decided to use the term *pure imaginary number* for a number of the form bi.)

"How can we tell if one complex number is bigger or smaller than another?" the king asked. "For example, we know that 2 is bigger than 1. However, is i bigger than 1, or is it smaller than 1?"

We realized that, although we could put all the real numbers in order, it was impossible to put all the complex numbers in order. Therefore, the symbols "greater than" (>) and "less than" (<) have no meaning when applied to complex numbers. "We have learned something new," the professor said. "We have discovered one important way in which complex numbers are not the same as real numbers."

"Now, let's see what happens when we add two complex numbers," the professor said.

$$(2 + 3i) + (4 + 5i) = 6 + 8i$$

In general, we decided that the sum of the two complex numbers $(a + bi)$ and $(c + di)$ was the complex number $[(a + c) + (b + d)i]$. In other words, the sum of two complex numbers is another complex number. To find the sum, add the real parts together and then add the imaginary parts together.

The professor wanted to try multiplying two complex numbers:

$$(2 + 3i)(4 + 5i)$$

"It looks as if we are multiplying two binomials," the king said, "so we should be able to use the FOIL method."

$$8 + 10i + 12i + 15i^2$$

"We can simplify that expression, since $i^2 = -1$," the professor said.

$$8 + 22i - 15 = -7 + 22i$$

We found the general form for the product of two complex numbers:

$$(a + bi)(c + di) = ac + adi + bci + dbi^2$$
$$= (ac - db) + (ad + bc)i$$

We tried another example:

$$(a + bi)(a - bi) = a^2 - abi + abi - b^2i^2$$
$$= a^2 - b^2i^2$$

Since $i^2 = -1$:

$$(a + bi)(a - bi) = a^2 + b^2$$

"That product is interesting—the result is a real number," the king said. We decided to call the number $a - bi$ the *complex conjugate* of the number $a + bi$. In other words, to find the conjugate of a complex number, you simply reverse the sign of the imaginary part. Conjugates have the useful property that the product of any complex number and its conjugate is a non-negative real number.

"Let's see what happens when we raise imaginary numbers to powers," the professor said. We calculated some powers of i:

$i^0 = 1$ $\qquad$ $i^5 = i$
$i^1 = i$ $\qquad$ $i^6 = -1$
$i^2 = -1$ $\qquad$ $i^7 = -i$
$i^3 = -1 \times i = -i$ $\qquad$ $i^8 = 1$
$i^4 = -i \times i = 1$ $\qquad$ $i^9 = i$

"Amazing!" the professor said. "Those powers keep repeating the same pattern!"

Recordis was starting to moan feebly. "Hurry!" the professor said.

"The gremlin said that we would have to deal with numbers that we cannot picture," the king said. "However, I think we can draw a diagram to represent complex numbers. Let's consider the complex number 2 + 3i. Why can't we draw it on a diagram as the ordered pair (2, 3)?" The king drew a diagram. He labelled the vertical axis the "imaginary axis" and the horizontal axis the "real axis." Then he put a dot 2 units to the right and 3 units up. (See Figure 20–2.)

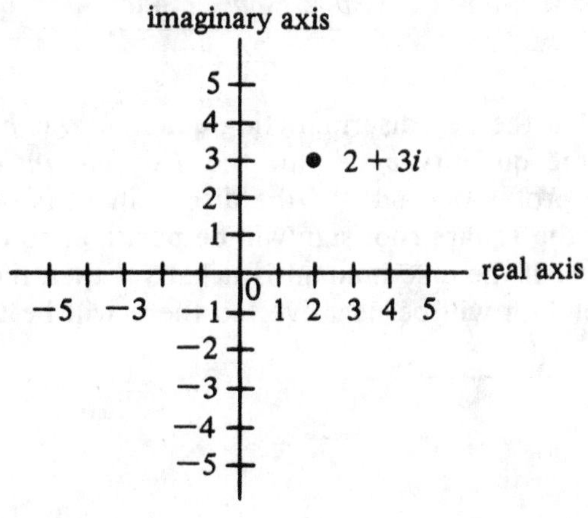

Figure 20–2

"We can represent any complex number that we want on this diagram," the king said. He drew some more examples.

"The diagram makes it clear why you cannot put the complex numbers in order," the professor said. "And now we can define the absolute value of a complex number. Since the absolute value of a real number is its distance from zero, that means that the absolute value of a complex number should also be its distance from zero. We can calculate that distance using the Pythagorean theorem." We calculated some examples (again using vertical lines | | to represent absolute values).

$$|3 + 4i| = 5$$
$$|6| = 6$$
$$|10i| = 10$$
$$|5 + 12i| = 13$$
$$|1 + i| = \sqrt{2}$$

We found that, in general, the absolute value of the complex number $a + bi$ is $\sqrt{a^2 + b^2}$. (See Figure 20–3.)

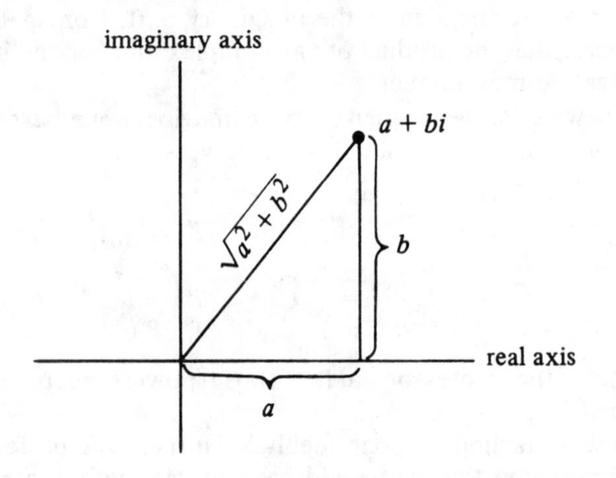

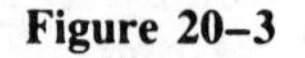
Figure 20–3

Recordis started to stir some more. "We only have one more thing to do," the professor said. "We need to investigate the roots of the quadratic equation $ax^2 + bx + c = 0$." From the quadratic formula,

$$x = \frac{-b \pm \sqrt{b^2 - 4ac}}{2a}$$

we could see that the key discriminating quantity was $b^2 - 4ac$.

"Let's call the quantity $b^2 - 4ac$ the *discriminant* of the quadratic equation," the professor said. "If the discriminant is positive, then the quantity under the square root sign will be positive, so the solutions will be real numbers. If the discriminant is negative, then the quantity under the square root sign will be negative, so there will be two complex solutions."

We listed the two complex solutions:

$$\frac{-b}{2a} + \frac{i\sqrt{|b^2 - 4ac|}}{2a}, \qquad \frac{-b}{2a} - \frac{i\sqrt{|b^2 - 4ac|}}{2a}$$

The gremlin reappeared. "Have you given up yet?"

"We did solve the equation!" the king told him. "We had to invent a new kind of number. We called it an imaginary number. We discovered how to add, subtract, and multiply imaginary numbers, and even how to take powers of imaginary numbers."

The king showed the gremlin the paper containing our work. The gremlin screamed and then started sobbing. "Curses! You have learned far more about algebra then I ever thought you could!" he cried. However, his face suddenly brightened. "Algebra is only the beginning!" he said. "There are many other subjects that are much harder. It is only a matter of time before you discover another subject, and then I shall win and rule Carmorra!" He vanished in a gust of wind.

We felt a deep sense of relief. We hid our work on imaginary numbers just before Recordis woke up. Eventually Recordis came to love algebra, and it even became his favorite subject. He especially liked drawing graphs of polynomials. However, he never did learn to like square-root signs, and he always fainted at the mere mention of the words "imaginary number."

The dragon was convicted, but since he did not seem to be malicious he was granted clemency—provided that he agreed to perform community service. He willingly agreed to serve as the new hot-water heater at the Carmorra Beachfront Hotel.

On the last day of our vacation we held a big party at the hotel. We were joined by the detective, Mrs. O'Reilly, the biologist, and everyone else who had helped us with our discoveries.

This brings to an end my part of the story. I spent several years in Carmorra after being shipwrecked there. Algebra was indeed only the beginning of our adventures. Trigonometry and calculus are two subjects that we discovered later. Both of these require a background in algebra.

I hope that this account has been fun for the reader, and I hope it will be beneficial for anyone interested in learning about the mysteries of algebra.

SUMMARY OF COMPLEX NUMBERS

The letter i is used to represent the square root of -1.
This square root is not a real number, since the square of any real number (positive or negative) is always positive.
The square root of a negative number $-n$ can be written as $i\sqrt{n}$ (where n is a positive real number).

A number of the form bi, where b is any real number (except 0), is called a pure imaginary number.

A number of the form $a + bi$, where a and b are both real numbers, is called a complex number.

Complex numbers can be added and multiplied as if they were ordinary binomials, provided you remember to replace i^2 with -1:

$$(a + bi) + (c + di) = (a + c) + (b + d)i$$
$$(a + bi)(c + di) = ac + adi + bci + bdi^2 = (ac - bd) + (ad + bc)i$$

The absolute value of the complex number $a + bi$ is $\sqrt{a^2 + b^2}$

Note to Chapter 20

If you are allowed to use complex numbers, you can factor any *n*th-degree polynomial as the product of *n* factors. For example:

$$x^2 + 1 = (x - i)(x + i)$$
$$x^4 - 1 = (x - 1)(x + 1)(x - i)(x + i)$$
$$x^3 - 3x^2 + 2x - 6 = (x - 3)(x + \sqrt{2}\,i)(x - \sqrt{2}\,i)$$

However, note that the factors are not necessarily all distinct. For example:

$$x^2 + 2x + 1 = (x + 1)(x + 1) = (x + 1)^2$$

The equation $x^2 + 2x + 1 = 0$ is said to have the *double root* $x = -1$. If double roots are counted as two roots, triple roots as three roots, and so on, it is easy to see that every *n*th-degree polynomial equation has *n* (real or complex) roots.

Exercises

Draw graphs representing these complex numbers:

1. $2 - 5i$
2. $-6i$
3. $(4 + 3i) + (-2 - i)$
4. $2 - 2i$
5. $2(6 - 4i)$
6. $12 + 16i$

Calculate the result of these operations:

7. $(54 - 6i) + (12 + 13i)$
8. $(196 - 0.34i) - (165 + 0.74i)$
9. $(363 - 57i) + (56 + 278i)$

10. $(7 + 15i) - (12 + 3i)$
11. $(5 + 4i)(9 + 16i)$
12. $(6 + 3i)(22 + 11i)$
13. $(4 + 1.95i)(4 - 1.95i)$
14. $(4 + 11i)(6 - 2i) + (3 + 5i)(10 + 7i)$
15. $(6 + 5i)[(78 + 7i) + (3 - 21i)]$

Calculate these powers:

16. i^8
17. i^9
18. i^{10}
19. i^{44}
20. i^{45}
21. i^{46}
22. i^{47}
23. i^{127}
24. i^{128}
25. i^{500}
26. $\sqrt{i}$
27. $1/i$
28. $(1 + i)^2$
29. $(2 + 3i)^2$
30. $(6 + 7i)^2$
31. $(3 - i)^2$
32. $(\sqrt{3} - 5i)^2$
33. $(a + bi)^2$
34. $(a + bi)^3$
35. $(a + bi)^4$
36. $(a + bi)^5$
37. $(a + bi)^n$

Find the solutions to these equations:

38. $x^2 + x + 4 = 0$
39. $x^2 - x + 8 = 0$
40. $x^2 + 4x + 5 = 0$
41. $x^2 + 3x + 4 = 0$
42. $x^2 + 8x + 25 = 0$
43. $(5/4)x^2 + 24x + 125 = 0$
44. $(1/4)x^2 + 5x + 169 = 0$
45. $6x^2 + x + 16 = 0$
46. $9x^2 + 7x + 6 = 0$

The conjugate of a complex number z is sometimes written with a star: z^*. Calculate the conjugates of these numbers. Then calculate the product zz^*.

47. $1 + i$
48. $16i$
49. 32
50. $10 - 5i$
51. $16 + 11i$
52. $33 + 2i$
53. $i - 10$
54. Express zz^* as a function of the absolute value of the complex number z.
55. Suppose $z^* = z$. What kind of number is z?
56. Suppose $z^* = -z$. What kind of number is z?

Suppose we have a fraction that has a complex number in the denominator, such as $5/(3 + 4i)$. In order to express this number in the normal form for a complex number, we need to get rid of the complex number in the denominator. To do that, we can multiply the top and bottom of the fraction by the conjugate of the denominator (in this case $3 - 4i$):

$$\frac{5}{3 + 4i} = \frac{5(3 - 4i)}{(3 + 4i)(3 - 4i)} = \frac{15 - 20i}{9 + 16} = \frac{3}{5} - \frac{4}{5}i$$

Simplify each of these fractions by removing the complex number from the denominator:

57. $1/(1 + i)$
58. $5/(6 + 4i)$
59. $(3 + 6i)/(4 + 2i)$
60. $(6 + 3i)/(6 - 4i)$
61. Find a general formula for $(a + bi)/(c + di)$.

Solve these equations for z. (z can be any complex number):

62. $(2 + 4i)z = 6 + 12i$
63. $(3 + 4i)z = 18 - i$
64. $(i + 1)z = 2$
65. $(i + 1)z = 9 + 3i$
66. $(9 - 3i)z = 84$
67. $(3 + 7i)z = 25 + 39i$

Complex numbers are used extensively in the analysis of electronic circuits. The *impedance* (Z) of a circuit is a complex number that measures the resistance of a circuit to alternating-current (AC) electricity. We'll consider three types of circuit elements: *resistors* with resistance R; *capacitors* with capacitance C; and *inductors* with inductance L. Suppose that the frequency of the current is f. The *angular frequency* ω is equal to $\omega = 2\pi f$.

The impedance of a circuit with just a resistor is $Z_R = R$; the impedance of a circuit with just a capacitor is $Z_C = 1/(i\omega C)$; and the impedance of a circuit with just an inductor is $Z_L = i\omega L$. (Note that the absolute value of the impedance becomes larger as the frequency becomes larger in an inductor circuit, but the impedance becomes smaller as the frequency becomes larger in a capacitor circuit.) The impedance is independent of the frequency in the resistor circuit.

Calculate the absolute value of the impedance for each of these situations:

68. A resistor and inductor in series: $Z = Z_R + Z_L$
69. A resistor and capacitor in series: $Z = Z_R + Z_C$
70. An inductor and capacitor in series: $Z = Z_C + Z_L$
71. A resistor and inductor in parallel with a capacitor:

$$Z = \frac{Z_C(Z_R + Z_L)}{Z_R + Z_L + Z_C}$$

The answer is

$$|Z| = \left[\frac{R^2 + (\omega L)^2}{[1 - \omega^2(LC)]^2 + (\omega RC)^2}\right]^{1/2}$$

If RC is much smaller than LC, then the impedance is at a minimum when $\omega = 1/\sqrt{LC}$. This important phenomenon is called *resonance*. Tuning a radio to the station that you want to listen to is an an example of changing C to find the correct resonant frequency.

Answers to Exercises

Chapter 1

1. $12 \times (6 \times 2) = 12 \times 12 = 144$
   $(12 \times 6) \times 2 = 72 \times 2 = 144$
2. $11 \times (5 \times 16) = 11 \times 80 = 880$
   $(11 \times 5) \times 16 = 55 \times 16 = 880$
3. $33 \times (1 \times 5) = 33 \times 5 = 165$
   $(33 \times 1) \times 5 = 33 \times 5 = 165$
4. $2 \times (2 \times 2) = 2 \times 4 = 8$
   $(2 \times 2) \times 2 = 4 \times 2 = 8$
5. $54 + (6 + 20) = 54 + 26 = 80$
   $(54 + 6) + 20 = 60 + 20 = 80$
6. $23 + (11 + 18) = 23 + 29 = 52$
   $(23 + 11) + 18 = 34 + 18 = 52$
7. $11 + (12 + 8) = 11 + 20 = 31$
   $(11 + 12) + 8 = 23 + 8 = 31$
8. $6 + (19 + 4) = 6 + 23 = 29$
   $(6 + 19) + 4 = 25 + 4 = 29$
9. $3 \times (4 + 10) = 3 \times 14 = 42$
   $(3 \times 4) + 3 \times 10 = 12 + 30 = 42$
10. $10 \times (1 + 1) = 10 \times 2 = 20$
    $10 \times 1 + 10 \times 1 = 10 + 10 = 20$
11. $6 \times (12 - 7) = 6 \times 5 = 30$
    $6 \times 12 - 6 \times 7 = 72 - 42 = 30$
12. $1 \times (6 + 8) = 1 \times 14 = 14$
    $1 \times 6 + 1 \times 8 = 6 + 8 = 14$

13. $2 \times (3 + 7) = 2 \times 10 = 20$
$2 \times 3 + 2 \times 7 = 6 + 14 = 20$
14. $3 \times (4 - 2) = 3 \times 2 = 6$
$3 \times 4 - 3 \times 2 = 12 - 6 = 6$
15. $3 \times (10 - 6) = 3 \times 4 = 12$
$3 \times 10 - 3 \times 6 = 30 - 18 = 12$
16. $3 \times 11 + 3 \times 6 = 33 + 18 = 51$
$3 + (11 + 6) = 3 \times 17 = 51$
17. $10 \times 6 + 10 \times 8 = 60 + 80 = 140$
$10 \times (6 + 8) = 10 \times 14 = 140$
18. $5 \times 6 + 5 \times 5 = 30 + 25 = 55$
$5 \times (6 + 5) = 5 \times 11 = 55$
19. $4 \times 11 + 4 \times 12 = 44 + 48 = 92$
$4 \times (11 + 12) = 4 \times 23 = 92$
20. Let p = pay. Then $p = 5 \times h$.
21. $p = 200 + 7\ 1/2 \times (h - 40)$
22. Let m = miles. Then $m = 55 \times h$.
23. $m = v \times h$
24. Let t = time. Then $t = d/55$
25. $t = d/v$
26. $a = b \times h$, where a = area, b and h are the lengths of the two sides.
27. m = miles, k = kilometers; $m = k/1.609$
28. c = centimeters, i = inches; $i = 0.3937 \times c$
29. k = kilograms, p = pounds; $k = p/2.205$
30. F = degrees Fahrenheit, C = degrees Celsius;
$C = (5/9) \times (F - 32)$
31. a = batting average, h = number of hits, n = number of times at bat; $a = h/n$
32. Use the commutative law and the distributive law.
33. Let $2 \times m$ and $2 \times n$ be two even numbers. Then
$(2 \times m) \times (2 \times n) = 2 \times (2 \times m \times n)$ which is even.
34. $(2 \times m) \times (2 \times n + 1) = 2 \times [m \times (2 \times n + 1)]$
The product of one odd number and one even number is even.
35.
$$(2 \times m + 1) \times (2 \times n + 1)$$
Use the distributive law.
$$(2 \times m + 1) \times 2 \times n + (2 \times m + 1)$$
This is the sum of an even number and an odd number, so it is odd. When you multiply two odd numbers, the result is an odd number.
36. Let the two numbers be $n \times k$ and $n \times h$, where h and k are any two whole numbers. Then

$$(n \times k) + (n \times h) = n \times (h + k)$$

The sum of two numbers divisible by n is divisible by n. For example, $20 + 16 = 36$ is divisible by 4.
37. A number not divisible by n can be written in the form $n \times k + m$, where k can be any whole number and m can be any whole number greater than zero but less than n. Add one number divisible by n and one number not divisible by n:

$$(n \times k + m) + (n \times h) = n \times (k + h) + m$$

which is not divisible by n. For example, $16 + 21 = 37$, which is not divisible by 4.

38. If you add two numbers not divisible by n, then you cannot say in general whether the result will be divisible by n. For example, $17 + 19 = 36$ is divisible by 4, but $21 + 26 = 47$ is not divisible by 4.
39. $10 + 12 \times 4 = 10 + 48 = 58$
40. $10 + (12 \times 4) = 10 + 48 = 58$
41. $(10 + 12) \times 4 = 22 \times 4 = 88$
42. $10 \times 12 + 4 = 120 + 4 = 124$
43. $10 \times (12 + 4) = 10 \times 16 = 160$
44. $(10 \times 12) + 4 = 120 + 4 = 124$
45. $3 + 4 \times 5 + 6 \times 7 + 8 \times 9 + 10 = 3 + 20 + 42 + 72 + 10 = 147$

Chapter 2

1. $x = 2$
2. $x = 4$
3. $x = 2$
4. $x = 10$
5. $x = 8$
6. $x = 3$
7. $x = 8$
8. $x = 3$
9. $x = 10$
10. $x = 20$
11. $x = 10$
12. $x = (16 - 12)/a = 4/a$
    Note: We will usually write division problems as fractions.
13. $x = (12 - b)/a$
14. $x = (c - b)/a$
15. $x = c/(a - b)$
16. $x = (c + d)/a$
17. $x = h/b$
18. $x = (d + b)/(a - c)$
19. 225 miles
20. 100 hours
21. 2 hours
22. $100/(v_1 + v_2)$
23. Hard job 4 hours, easy job 0 hours; hard 3, easy 2; hard 2, easy 4; hard 1, easy 6; hard 0, easy 8
24. 5 CDs, 0 pizzas; 4 CDs, 2 pizzas; 3 CDs, 4 pizzas; 2 CDs, 6 pizzas; 1 CD, 8 pizzas; 0 CDs, 10 pizzas.
25. 20 roosters, 12 horses
26. Let x be the number of trees on Maple street. Then $x + 2x = 18$, $3x = 18$, $x = 6$. There will be 12 trees on Elm street.
27. Chapter 1: 35 pages Chapter 2: 30 pages
28. Let x be the smaller number. Then $x + (x + 1) = 63$, $x = 31$. The two numbers are 31 and 32.
29. $x + (x + 1) + (x + 2) = 75$. The numbers are 24, 25, and 26.
30. Let x be the number of pages typed by J.R. Then T.J. will type $2x$ pages. Set up the equation: $x + 2x = 600$, $x = 200$, $2x = 400$.
31. 20 miles per hour.

Chapter 3

1. −5
2. 0
3. 4
4. −21
5. −4
6. −16
7. −3
8. −4
9. −32
10. −30
11. 30
12. −12
13. −30
14. 39
15. −108
16. 0
17. 103
18. 7
19. 0
20. 16
21. 17
22. 1365.4
23. 3
24. 3
25. 6
26. 120
27. 152
28. 17
29. 7
30. 7
31. 20
32. 20
33. 20
34. 20
35. 20
36. 12
37. 7

38. If $a = 2$, $b = 3$ (for example), then $|ab| = |6| = 6$, $|a||b| = |2||3| = 2 \times 3 = 6$. If $a = 2$, $b = -3$, then $|ab| = |2 \times (-3)| = |-6| = 6$, $|a||b| = |2||-3| = 2 \times 3 = 6$. If $a = -2$, $b = -3$, then $|ab| = |(-2) \times (-3)| = |6| = 6$, $|a||b| = |-2||-3| = 2 \times 3 = 6$.

39. It is true if a and b have the same sign.

40. $x = -10$

41. $x = 3$

42. $x = 5$

43. $x = -2$

44. $x = 0$

45. $x = -1$

46. $x = -4$

47. $x = -80$

48. $x = -4$

49. $x = -5$

50. If $x > 0$, this expression is equal to x. If $x < 0$, then this expression is equal to 0.

Chapter 4

1. 6/2 = 3

2. 3/6 = 1/2

3. 1/6

4. 3/12 = 1/4

5. 16/63

6. 75/77

7. 15/81 = 5/27

8. 4/14 = 2/7

9. 5/9

10. 41/224

11. 13/20

12. 4/15 + 7/48 = 64/240 + 35/240 = 99/240 = 33/80

13. 1/6

14. 3/49

15. 5/18

16. 1/2

17. 1/4

18. 1/2

19. 3/8

20. 1/13

21. 1/9

22. 1/2

23. 2/3

24. 3/4

25. 1/9

26. 5/32

27. If the numerator and denominator of a fraction are both even, you can divide both by 2. Hence the fraction could not be in simplest form.

28. $\frac{ad}{bd} = \frac{bc}{bd}$, $ad = bc$

29. Multiply by (−1)/1.

30. Multiply by 1/(−1).

31. Multiply by (−1)/(−1).

32. $(ac + b)/c$

33. 0.0625

34. 4.08

35. 4.56

36. 1.9881

37. 0.999

38. 128.44

39. 16.480

40. 0.875

41. 0.555 . . .

42. 0.625

43. 0.461538 461538 461538 . . .

44. $x = -2/3$

45. $x = 4$

46. $x = -24$

47. $x = -7$

48. $x = 5/3$

49. $x = 7/4$

50. $x = 11/3$

51. $x = 10$

52. $(x - 1)/x$

53. $(x + 1)/(x - 1)$

54. $xx/(x - 1) = x^2/(x - 1)$

55. $x + 1$

56. $(x - 1)(x + 1)$

57. $1/(x - 5)$

58. x

59. $(a + b)/ab$

60. $ab/(a + b)$

61. $(x + 3)/(x + 2)$

62. $(3a + 6)/25$

63. $9/20\ x$

64. $(4x + 5y)/20$

65. $(1 + 4abc)/abc$

66. $9/(a + b)$

67. $(bc + ac + ab)/abc$

68. $bx = a$

69. The larger size has a smaller per-unit price.

70. $Y = 360$

Chapter 5

1. 2^n
2. $52^5 = 380{,}204{,}032$
3. $26^4 = 456{,}976$
4. $30^3 = 27{,}000$
5. $(x^2 + y^2)/(xy)$
6. If a is even, it can be written as $a = 2b$, where b is an integer. Then $a^2 = 4b^2 = 2(2b^2)$, which is even.
7. $x > 1$ or $x < 0$
8. $0 < x < 1$
9. $x = 1$ or $x = 0$
10. $x > 1$ or $x < -1$
11. $-1 < x < 1$
12. x^2 if $x > 0$, $-x^2$ if $x < 0$.
13. 3^3
14. 2^4
15. $1/5$
16. $1/(1.16)^2$
17. $(3.45)^5$
18. y^3/x^3
19. $3/4\ r$
20. $10ab^2$
21. $2/cb^2$
22. r
23. x
24. $1/b^3$
25. c^2/a^2
26. $x^2/(x + 1)$
27. 15 years, since $1.050^{15} = 2.08$
28. 8 years, since $1.1^8 = 2.14$
29. 4 years, since $1.2^4 = 2.07$
30. Yes.
31. If $x < 0$ and n is an integer, then x^n is positive if n is even and x^n is negative if n is odd. If n is not an integer, then x^n will often not even be a real number.
32. 500 joules
33. 1.098×10^{12} cubic kilometers
34. 5.87×10^{12} miles
35. 25 times weaker
36. 16 times as much

Chapter 6

1. $2\sqrt{3}$
2. 10
3. 3/4
4. $2\sqrt{2}$
5. $6\sqrt{2}$
6. $4\sqrt{2}$
7. $2\sqrt{11}$
8. a^2x
9. $2x^3$
10. $2\sqrt{3}y$
11. $2a$
12. $$\frac{1}{(\sqrt{a}+\sqrt{b})} = \frac{1}{(\sqrt{a}+\sqrt{b})} \times \frac{(\sqrt{a}-\sqrt{b})}{(\sqrt{a}-\sqrt{b})} = \frac{(\sqrt{a}-\sqrt{b})}{a-b}$$
$$\frac{1}{(\sqrt{a}-\sqrt{b})} = \frac{1}{(\sqrt{a}-\sqrt{b})} \times \frac{(\sqrt{a}+\sqrt{b})}{(\sqrt{a}+\sqrt{b})} = \frac{(\sqrt{a}+\sqrt{b})}{a-b}$$
13. $\sqrt{5}/5$
14. $(4 - \sqrt{5})/11$
15. $2\sqrt{5}$
16. $(9 - \sqrt{6})/15$
17. $-3 - 2\sqrt{2}$
18. $\sqrt{5} + \sqrt{10} - \sqrt{2} - 1$
19. $(\sqrt{8} - \sqrt{2})/6$
20. $5(\sqrt{54} - \sqrt{6})/48$
21. Guess 1: 1.5
    Guess 2: 1.75
    Guess 3: 1.732143
    Guess 4: 1.732051
    The last value is extremely close to the true value.
22. 2
23. 3
24. 10
25. 1
26. 1/4
27. $\sqrt{2}$
28. $\sqrt{3}$
29. $\sqrt{2}$
30. 9
31. $\sqrt{abc^2}$
32. a^3/b^6
33. 60 m.p.h.: 0.99999995
    600 m.p.s.: 0.9999948
    1/10: 0.994987
    1/2: 0.8660
    9/10: 0.4359
34. $x^{a/b} = x^{(1/b)a} = (x^{1/b})^a = (\sqrt[b]{x})^a$

Chapter 7

1. $(a - b)(a + b) = a^2 + ab - ab - b^2 = a^2 - b^2$
2. $4y^2 - x^2$
3. $4y^2 + 2xy - 2x^2$
4. $a^2 + 8a + 16$
5. $a^2b^2c^2 - 3abc + 2$
6. $PV + a/V - bP - ab/V^2$
7. $wLT/p - wL^2/P$
8. $s^4 - c^4$
9. $x^2 - 2kx^3 + k^2x^4$
10. $x^2 - x + 1/4$
11. $x^2 - xb/a + b^2/4a^2$
12. $x^2 - x - y^2 - y$
13. $x^2 - 2xy - x + y^2 + y$
14. $x^2 - x - y^2 + y$
15. $x^2 + y^2 + 1 + 2xy + 2x + 2y$
16. $x^2 + 2xy + y^2 - 1$
17. $x^2 - 2x - y^2 + 1$
18. $x^4 + 2x^3 + x^2$
19. $x^5 + 2x^4 + 2x^3 + x^2$
20. $x^5 + 2x^4 + 3x^3 + 2x^2 + x$
21. $x^4 + 2x^3 + 3x^2 + 2x + 1$
22. $6x^2 + 5x + 3ax + a + 1$
23. $6x^2 + 2x + 12ax + 2a + 6a^2 - 4$
24. $1 + 2x + x^2$
25. $1 + 3x + 3x^2 + x^3$
26. $1 + 4x + 6x^2 + 4x^3 + x^4$
27. $1 + 5x + 10x^2 + 10x^3 + 5x^4 + x^5$
28. If x is small, then $(1 + x)^n = 1 + nx$ (approximately).
29. $x^2 + y^2 + [-2h]x + [-2k]y + [h^2 + k^2 - r^2] = 0$
30. $x^2 - y^2 - 1 = 0$
31. $(1/16)x^2 + (1/9)y^2 - 1 = 0$
32. $[x^2 - 6x + 9]/16 + [y^2 + 6y + 9]/9 = 1$; $\frac{x^2}{16} - \frac{3}{8}x + \frac{y^2}{9} + \frac{2}{3}y + \frac{9}{16} = 0$
33. $(1/4)\,x^2 - y = 0$
34. $(1/4)\,x^2 + 5x - y + 30 = 0$
35. $155^2 - 154^2 = (155 - 154)(155 + 154) = 309$
36. $a^3 - b^3$
37. $a^3 + b^3$

Chapter 8

1. 0
2. 1
3. 4
4. a^2
5. q^2
6. $(a + b)^2 = a^2 + 2ab + b^2$
7. $(x + h)^2 = x^2 + 2hx + h^2$
8. $(x + 2)^2 = x^2 + 4x + 4$
9. 1
10. $x^2 + 2xh + h^2 - x^2 = 2xh + h^2$
11. 2
12. 5
13. 1
14. 17
15. $a^2 + 2a + 2$
16. $x^2 + 2xh + h^2 + 2x + 2h + 2$
17. $2xh + 2h + h^2$
18. 2 **19.** $\sqrt{7} = 2.646$
20. $\sqrt{3ab}$
21. $bc\sqrt{ac}$
22. $\sqrt{1 + 15} = \sqrt{16} = 4$
23. $\sqrt{1 + 10} = \sqrt{11} = 3.317$
24. 2.449
25. 1.414

26. 1.225
27. 1.049
28. 1.00499
29. 1.0004998
30. 8.5
31. 6
32. 3.5
33. 1.5
34. 1.25
35. 1.05
36. 1.005
37. 1.0005
38. For small values of x, $\sqrt{1 + x}$ is approximately the same as $1 + x/2$.
39. x must be positive.
40. Odd function.
41. Even function.
42. Odd function.
43. Neither.
44. Even function.
45. Odd function.

Chapter 9

1. $\sqrt{6^2 + 8^2} = 10$; $(y - 8)/(x - 6) = 1\ 1/3$
2. $\sqrt{(6 - 2)^2 + (10 - 3)^2} = 8.062$; $(y - 10)/(x - 6) = 1\ 3/4$
3. 2.236; $(y - 1)/(x - 1) = 1/2$
4. 7.211; $(y - 10)/(x - 2) = 2/3$
5. 6.083; $(y - 17)/(x - 8) = -6$
6. 17.464; $(y - 4)/(x - 16) = 7/16$
7. 8; $x = 2$
8. 7; $y = 10$
9.

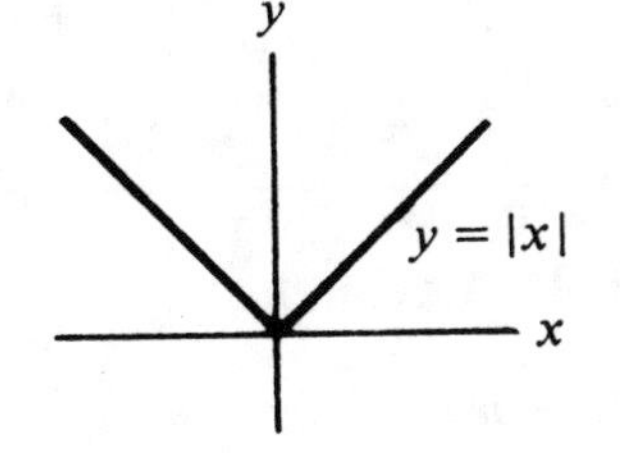

10.

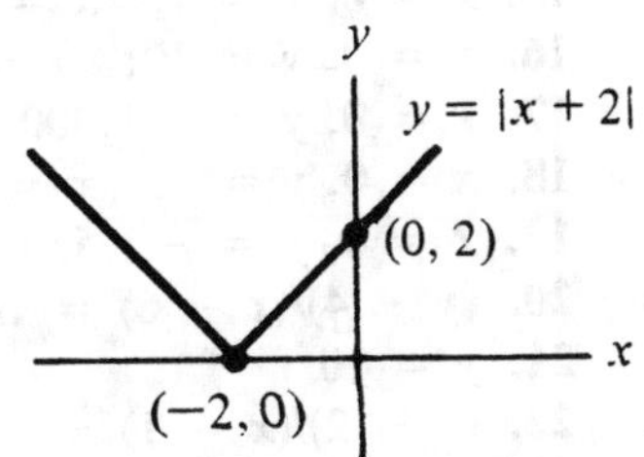

11.

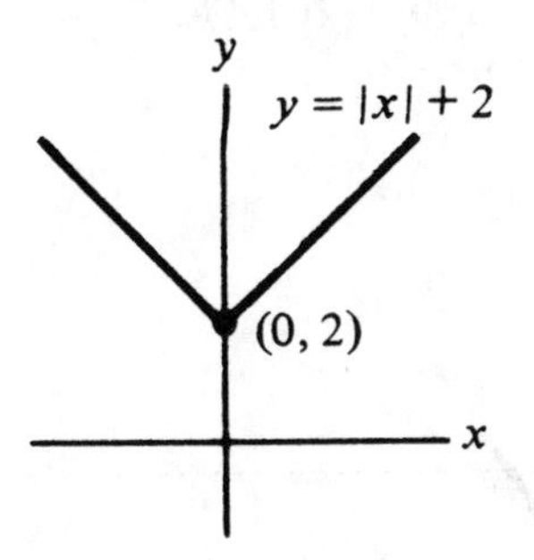

12.

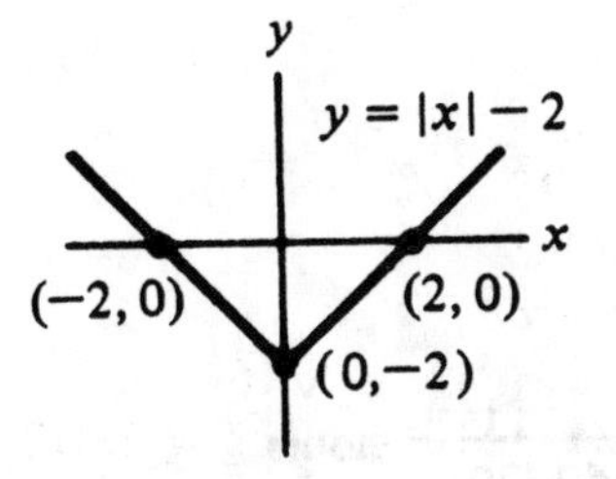

13.

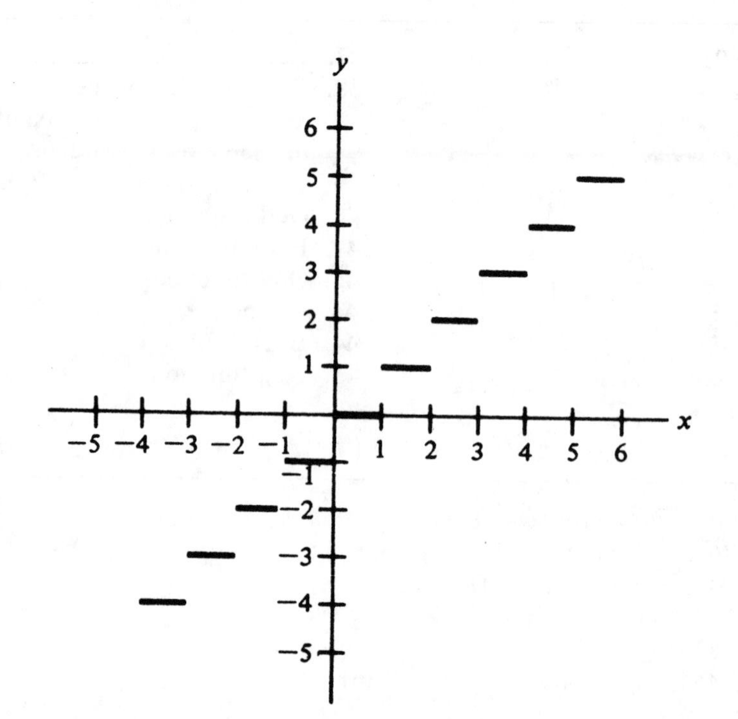

14. $x = 0, y = 0$
15. $x = 0, y = 1; \quad x = 1, y = 0$
16. $x = 0, y = 10; \quad x = 5, y = 0$
17. $x = 0, y = -1{,}000; \quad x = 10, y = 0$
18. $x = 0, y = 3; \quad x = 4, y = 0$
19. $x = 0, y = -1.539; \quad x = 17.135, y = 0$
20. $(y - 4)/(x - 6) = 1/2; \quad y = 1/2\, x + 1$
21. $y = 10$
22. $(y - 2)/(x - 1) = m; \quad y = mx + (2 - m)$
23. $(y - 5)/(x - a) = m; \quad y = mx + (5 - ma)$
24. $(y - 1{,}030)/(x - 1{,}025) = 100; \quad y = 100x - 101{,}470$
25. $(y + 6)/(x + 5) = m; \quad y = mx + (5m - 6)$
26.

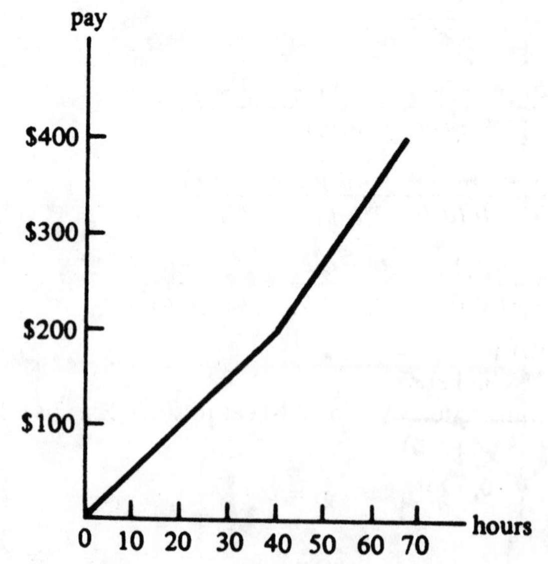

Chapter 10

1. $x = 4 \quad y = 18$
2. $x = 3\ 1/3 \quad y = 16\ 2/3$
3. No solution
4. $x = -0.384 \quad y = 9.23$
5. $x = 1 \quad y = 1$
6. $x = -1/3 \quad y = 3$
7. $x = 8 \quad y = -1$
8. $x = -0.727 \quad y = 2.182$
9. No solution
10. $x = 1.609 \quad y = -0.496$
11. $x = 1 \quad y = 0$
12. $x = 1/2 \quad y = 1/2$
13. $x = 2 \quad y = 1/2$ or $x = -2, y = -1/2$
14. $x = \sqrt{ab}, y = \sqrt{ab}/b$
15. $x = 10/3, \quad y = 10/3$
16. $x = 0 \quad y = 0$
17. $x = 0, \quad y = 0$
18. $x = \dfrac{b_2 - b_1}{m_1 - m_2}, y = \dfrac{m_1 b_2 - m_2 b_1}{m_1 - m_2}$
19. No solutions.
20. No solutions.
21. -40
22. $Y = [a + I + G]/(1 - b), C = a + b[a + I + G]/(1 - b)$
23. $Y = 350, C = 290$
24. $Y = 100, r = 5$
25. $P = (D_0 - S_0)/(s + d)$
26. $P = 100$
27. We know that

$$x = \frac{b_1c_2 - b_2c_1}{a_2b_1 - a_1b_2}. \; y = \frac{c_1 - a_1x}{b_1}$$

$$= \frac{c_1}{b_1} - \frac{a_1}{b_1}\left(\frac{b_1c_2 - b_2c_1}{a_2b_1 - a_1b_2}\right)$$

$$= \frac{c_1(a_2b_1 - a_1b_2) - a_1(b_1c_2 - b_2c_1)}{b_1(a_2b_1 - a_1b_2)}$$

$$= \frac{a_2b_1c_1 - a_1b_2c_1 - a_1b_1c_2 + a_1b_2c_1}{b_1(a_2b_1 - a_1b_2)}$$

$$= \frac{b_1(a_2c_1 - a_1c_2)}{b_1(a_2b_1 - a_1b_2)} = \frac{a_2c_1 - a_1c_2}{a_2b_1 - a_2b_1} = \frac{a_1c_2 - a_2c_1}{a_1b_2 - a_2b_1}$$

28. No. (Try it.)
29. 12 cows, 21 chickens
30. Here is an illustration of a spreadsheet to solve Exercise 8. Rewrite the two equations as:

$$y_1 = 2 - \frac{x}{4}$$

$$y_2 = \frac{8}{3} + \frac{2x}{3}$$

The slope and y-intercept are shown clearly in this form, and adding the subscripts 1 and 2 allows us to distinguish between the values of y that come from the two different equations. Now enter the following formulas and numbers:

	A	B	C
1	X	Y1	Y2
2	−10	+b5+b4*a2	+c5+c4*a2
3	10	+b5+b4*a3	+c5+c4*a3
4	slope:	−1/4	2/3
5	y intercept:	2	8/3

Note that your spreadsheet screen will show you the results of the formulas, not the formulas themselves. The values 10 and −10 are chosen arbitrarily; we will need to change these values to zoom in on the solution.

Now set up an XY (or scatter) graph. The exact way of doing this will vary with different programs, but the general idea is to use A2 to C3 as the range to graph. A2 to A3 is the horizontal, or X range; B2 to B3 is the first data range and C2 to C3 is the second data range. The data is arranged in columns, and choose the option where the computer will draw a line between the data points. The figure shows three different possibilities where the values in the X column are changed.

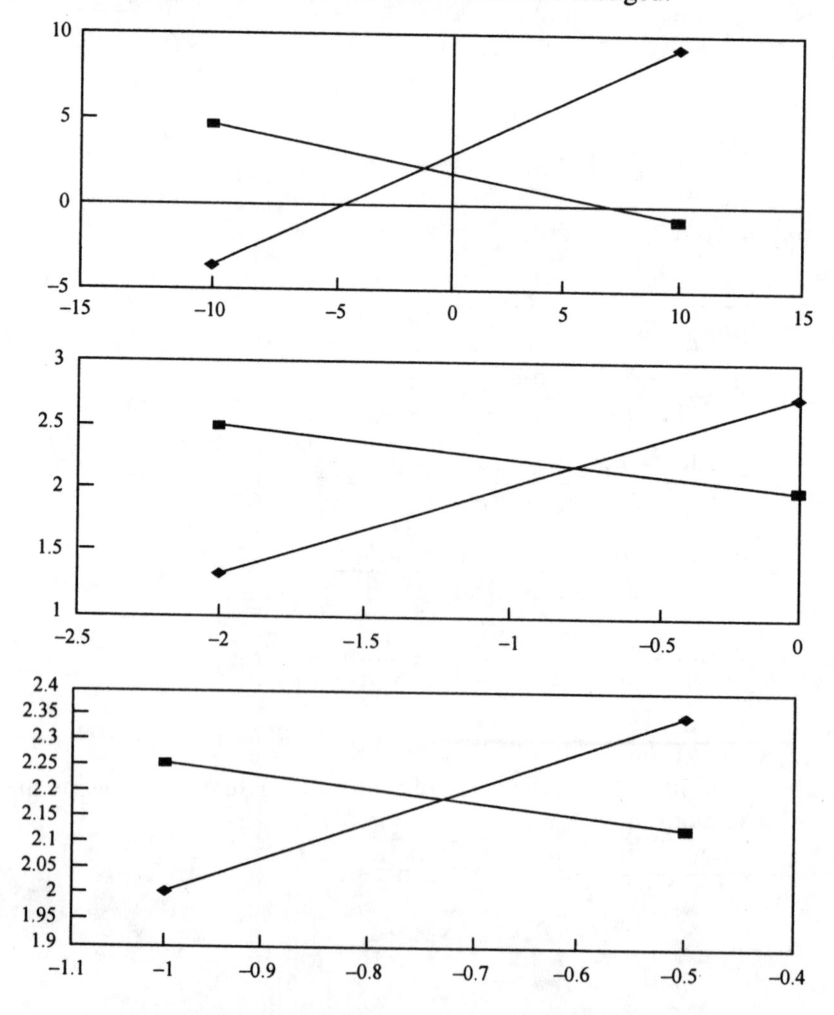

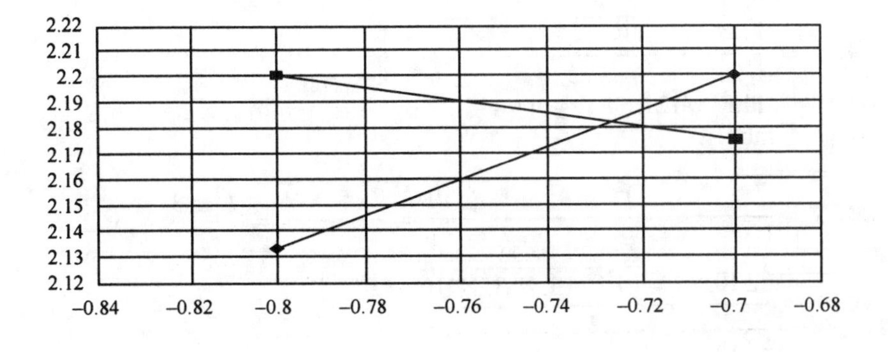

Chapter 11

1. 1 and 1
2. -1 and 1
3. 3 and 2
4. 5 and 4
5. 8 and 1
6. $3x^2 + 2x + 4 = 0$
7. $4x^2 + 7x - 5 = 0$
8. $x^2 + 9x + 10 = 0$
9. $3x^2 - 4x - 6 = 0$
10. $6x^2 - 2x + 6 = 0$
11. $x = \pm 3$
12. $x = \pm 4$
13. $x = 4, x = 0$
14. $x = 0, 4/5$
15. $x = 0, -3/4$
16. $x = -1$
17. $x = 7, 4.$
18. $x = 11, 10$
19. $x = 19, 1$
20. $x = 2, -3$
21. $x = -1/2, -1$
22. $x = 7, -6$
23. $x = 1/2$
24. $x = 1, 1/2$
25. $x = -1/3, -3$
26. $x - 4$
27. $x - 3$
28. $x + 2$
29. $x + 2$
30. Yes.
31. No.
32. No.
33. Yes.
34. It can be transformed into a polynomial.

35.

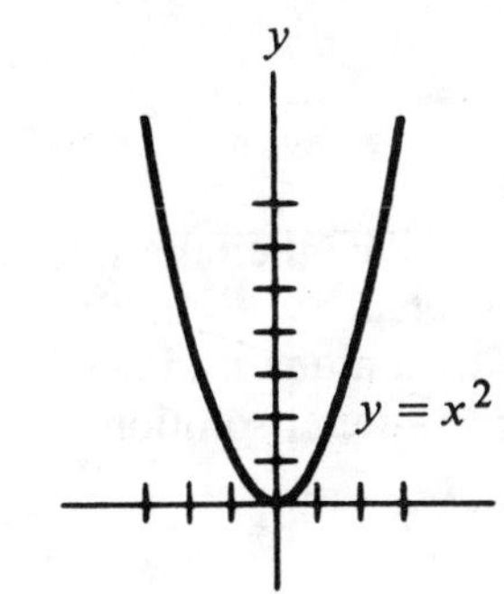

36.

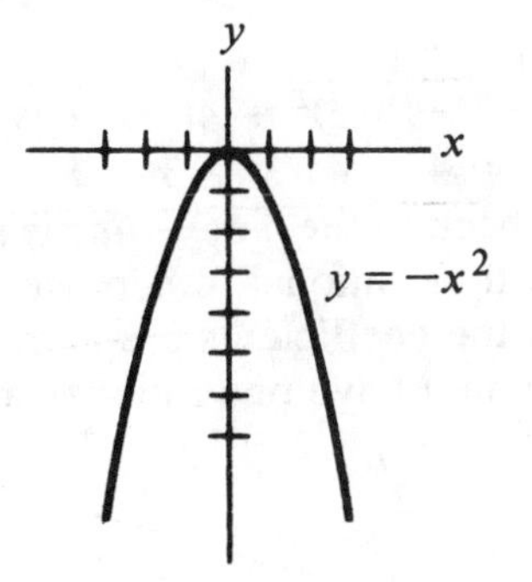

37.

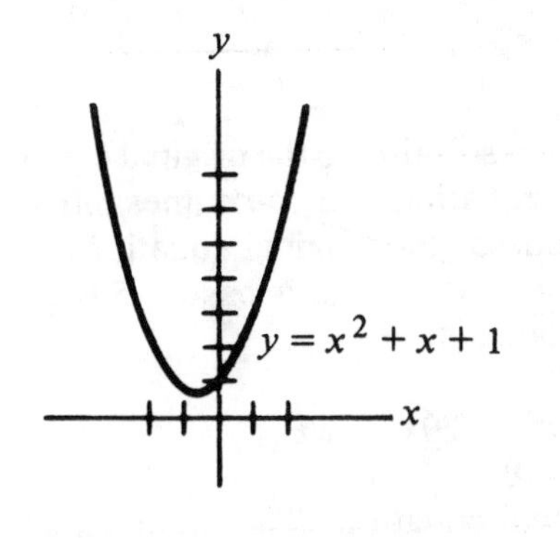

38.

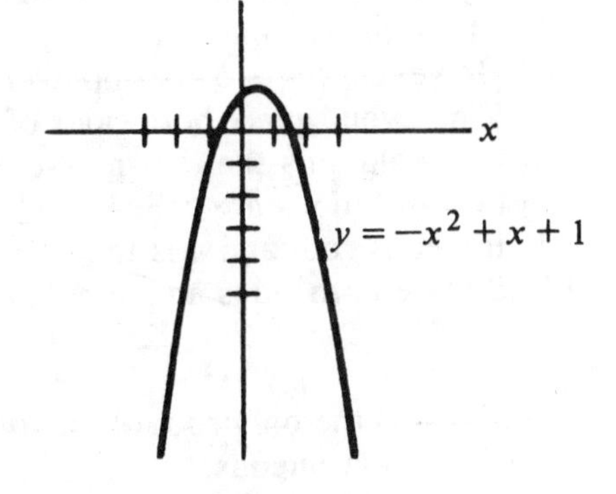

39. $x = \dfrac{1 \pm \sqrt{(-1)^2 - 4 \times 1 \times (-30)}}{2}$

$= \dfrac{1 \pm \sqrt{121}}{2} = \dfrac{1 \pm 11}{2} = 6, -5$

40. $x = \dfrac{7 \pm \sqrt{(-7)^2 - 4 \times 1 \times 10}}{2} = \dfrac{7 \pm \sqrt{9}}{2} = \dfrac{7 \pm 3}{2} = 5, 2$

41. $x = \dfrac{-10 \pm \sqrt{10^2 - 4 \times 1 \times 16}}{2}$

$= \dfrac{-10 \pm \sqrt{36}}{2} = \dfrac{-10 \pm 6}{2} = -2, -8$

42. $x = \dfrac{12 \pm \sqrt{(-12)^2 - 4 \times 1 \times 11}}{2}$

$= \dfrac{12 \pm \sqrt{100}}{2} = \dfrac{12 \pm 10}{2} = 11, 1$

43. $x = \dfrac{-3 \pm \sqrt{3^2 - 4 \times 1 \times (-18)}}{2}$

$= \dfrac{-3 \pm \sqrt{81}}{2} = \dfrac{-3 \pm 9}{2} = 3, -6$

44. $x = (4 \pm \sqrt{56})/4 = 2.871, -0.871$
45. $x = (1 \pm \sqrt{5})/2 = 1.618, -0.618$
46. $x = (-1 \pm \sqrt{5})/2 = 0.618, -1.618$
47. $x = (1 \pm \sqrt{13})/6 = 0.768, -0.434$
48. $x = (-10 \pm \sqrt{60})/20 = -0.113, -0.887$
49. $x = 0, -1$
50. $x = -1/2$
51. $x = 7$
52. $x = -1$
53. $(-2 + \sqrt{3})^2 + 4(-2 + \sqrt{3}) + 1$
$= 4 - 4\sqrt{3} + 3 - 8 + 4\sqrt{3} + 1 = 0$
54. Check to see if $b^2 - 4ac$ is a perfect square. If it is, then the solutions will be rational, otherwise they will be irrational.
55. If the coefficients are all rational numbers then a quadratic equation cannot have one rational solution and one irrational solution.
56. $x = -3, -2$
57. $x = -2$
58. $x = 1, -1$
59. $x = 3, 2$
60. Square both sides: $20 - x = x^2$.
This equation has two solutions: $x = 4$, $x = -5$.
However, $x = -5$ does not work as a solution to the original equation. When you square both sides of an equation you sometimes introduce new solutions that are not solutions of the original equation. These false solutions are called *extraneous roots*. In this case -5 is an extraneous root and 4 is the only true root.
61. Square both sides and simplify:

$$x^2 + 2x + 1 = 4x^2 - 20x + 25$$
$$3x^2 - 22x + 24 = 0$$

$x = 6$ is the only root of the original equation. The other root $x = 1\ 1/3$ is extraneous.

62. $y^2 = 4$ or 9, so $y = \pm 2$ or ± 3.
63. $y^3 = 8$ or 1, so $y = +2$ or $+ 1$.
64. $(x - 3)(x - 5) = x^2 - 8x + 15 = 0$
65. $x^2 - 23x + 132 = 0$
66. $x^2 + x - 20 = 0$
67. $x^2 + 13x + 42 = 0$
68. $x^2 - 4x - 1 = 0$
69. $-(\frac{1}{2})gt^2 + v_0t + (h_0 - h) = 0$

$$t = \frac{-v_0 \pm \sqrt{v_0^2 + 4(\frac{1}{2})\, g(h_0 - h)}}{-g} = v_0/g \pm \sqrt{v_0^2 + 2g(h_0 - h)}/g$$

70. $t = 3.632$; $t = 0.4495$.
71. If $b = 0$, the solution is $\pm\sqrt{-c/a}$.
If $c = 0$, the solutions are $x = 0, -b/a$.
If $a = 0$, the quadratic formula does not work because the equation is no longer a quadratic equation.
72. 12 by 8.
73. Your answer contains a square root of a negative number; see Chapter 20.
74. You will receive a total of $x(13 - x)$ items of the gift you start receiving on day x. Write the quadratic equation: $13x - x^2 = 42$; $x^2 - 13x + 42 = 0$. Factor: $(x - 6)(x - 7) = 0$, so $x = 6$ or $x = 7$. Therefore, you will receive 42 geese a'layin (starting on day 6) and 42 swans a'swimming (starting on day 7).
75. Here is a spreadsheet that graphs the quadratic function $ax^2 + bx + c$. The numbers in this example show Exercise 44; you can change the values of a, b, and c in cells D1 to D3 if you want to graph a different quadratic function.
Enter these formulas and numbers as shown:

	A	B	C	D
1	low value of X:	–10	a=	–2
2	high value of X:	10	b=	–4
3	step value:	+(b2–b1)/100	c=	–5
4				
5	X	Y		
6	+B1	+d1*a6^2+d2*a6+d3		
7	+A6+$B$3	+d1*a7^2+d2*a7+d3		
8	+A7+$B$3	+d1*a8^2+d2*a8+d3		
9	+A8+$B$3	+d1*a9^2+d2*a9+d3		

You will have to use the copy command to copy the formulas in column A and B so they fill all the cells from row 6 to row 106. Set up an XY (or scatter) graph, using A6 to A106 as the X range and B6 to B106 as the Y range. The graph will always show 101 data points, but you can adjust the low and high value of X, allowing you to zoom in on one of the roots. The figures show some examples.

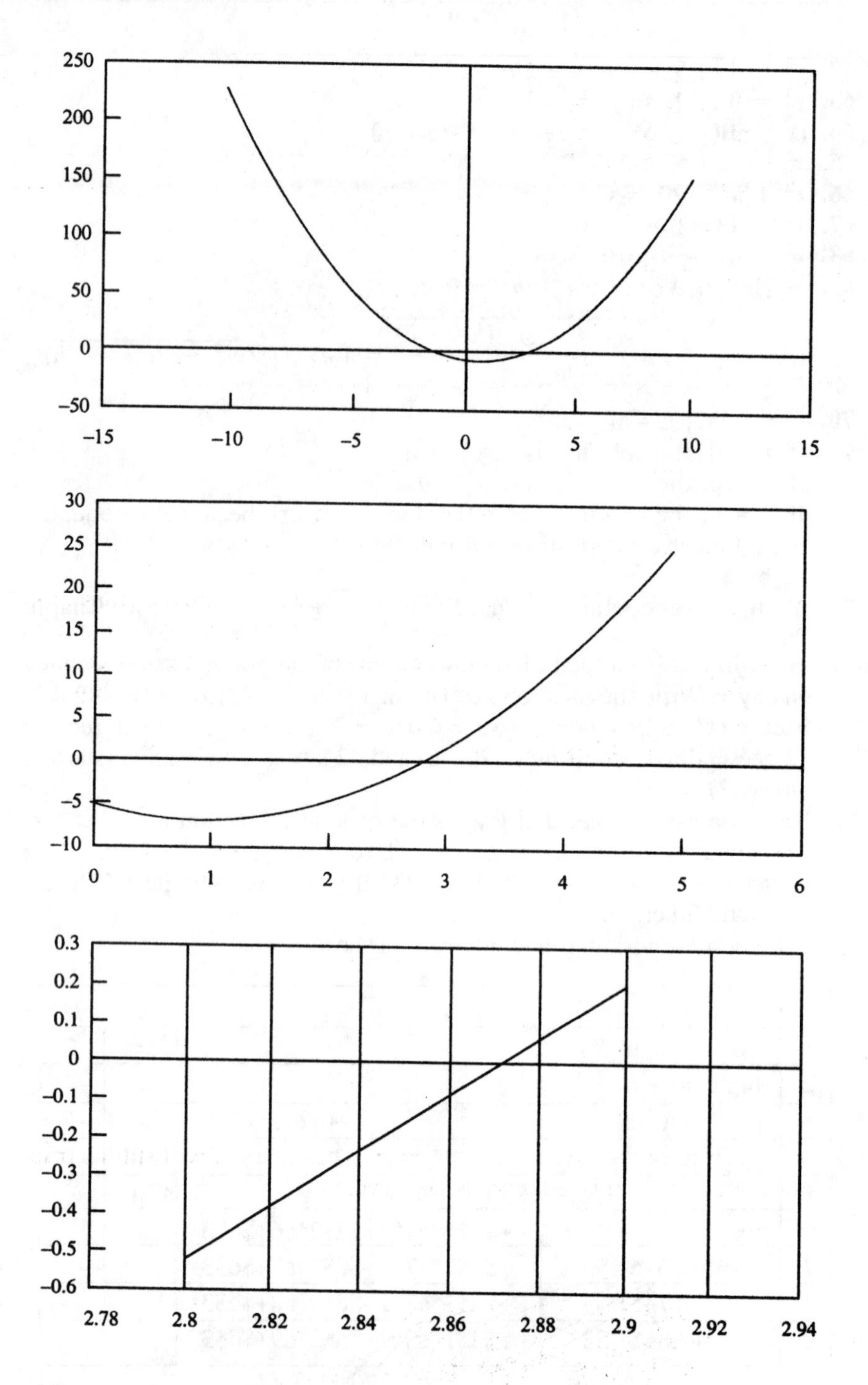

Chapter 12

1. $C = 31.42$, $A = 78.54$
2. $C = 6.28$, $A = 3.14$
3. $C = 62.83$, $A = 314.16$
4. $C = 35.44$, $A = 99.93$
5. $C = 24{,}881$, $A = 4.9 \times 10^7$
6. $C = 5.8 \times 10^8$, $A = 2.7 \times 10^{16}$
7. circumference $= 628{,}318{,}000$ light years (approximately)
   area $= 3.14159 \times 10^{16}$ square light years
8. $r = 1$
9. Diameter $= 1/\pi = 0.318$; $r = 0.159$
10. $r = \sqrt{1/4\pi} = 0.282$
11. $r = 2$
12. $r = 1.59$
13. $22/7 - \pi = 0.00126$ (approximately)

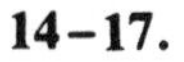
14–17.

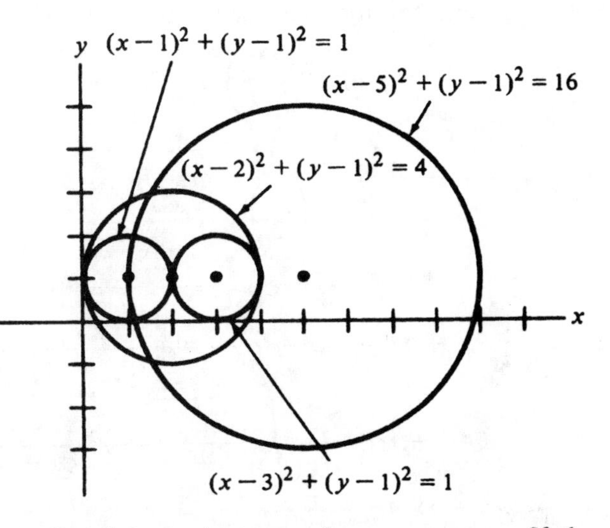

18. Calculate the distance between the two centers. If the distance is greater than the sum of the two radii, then clearly the circles cannot intersect. In this case the distance between the two centers is 20 and each radius is 10, so the two circles will intersect at just one point.

19. Intersect at two points.

20. No intersection.

21. No intersection.

22. $x^2 + (-2h)x + y^2 + (-2k)y + (h^2 + k^2 - r^2) = 0$

Let the point (h, k) represent the center of the circle. By comparing each equation to the answer for Exercise 22, we can identify h, k, and r.

	h	k	r
23.	1	1	1
24.	0	5	5
25.	−2	1	3
26.	10	−3	$\sqrt{2}$
27.	−2	−1	$\sqrt{3}$
28.	0	0	7

29. The equation of the circle is $x^2 + y^2 = r^2$. The equation will still be true if either x or y is replaced with its negative.

30.

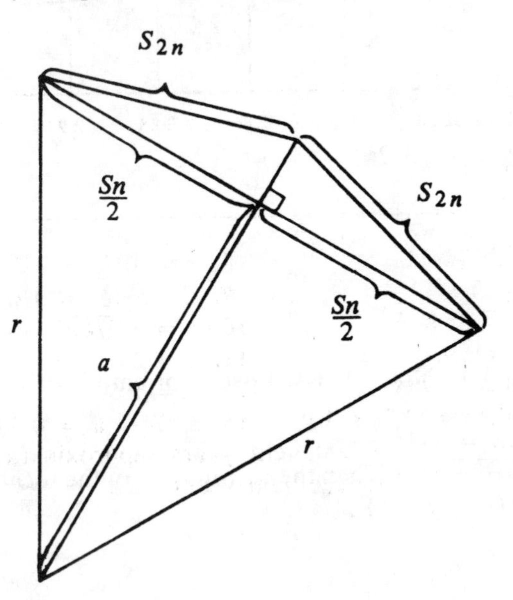

Using the Pythagorean theorem for the two right triangles:

$$(r - a)^2 + \left(\frac{S_n}{2}\right)^2 = S_{2n}{}^2$$

and

$$a^2 + \left(\frac{S_n}{2}\right)^2 = r^2$$

Therefore,

$$S_{2n}{}^2 = \frac{1}{4} S_n{}^2 + r^2 - 2ar + a^2$$

$$= \frac{1}{4} Sn^2 + r^2 - 2r\sqrt{r^2 - \frac{1}{4} Sn^2} + r^2 - \frac{1}{4} Sn^2$$

$$= 2r^2 - r\sqrt{4r^2 - S_n{}^2}$$

31. The area of each triangle will be (1/2) (base) (altitude). As n becomes large, the base of each triangle will come close to $2\pi r/n$, so the area is

$$\frac{1}{2}(2\pi r/n)r = \pi \frac{r^2}{n}$$

The total area of all of the triangles will therefore be πr^2. Note that, as n becomes large, the area of all of the triangles comes close to the area of the whole circle.

32. $x^2/a^2 + y^2/b^2 = 1$, $y^2/b^2 = 1 - x^2/a^2$, $y^2 = b^2(1 - x^2/a^2)$, $y = \pm b\sqrt{1 - x^2/a^2}$

33. $b^2 + e^2a^2 = a^2$

$$e^2a^2 = a^2 - b^2$$

$$e = \frac{\sqrt{a^2 - b^2}}{a}$$

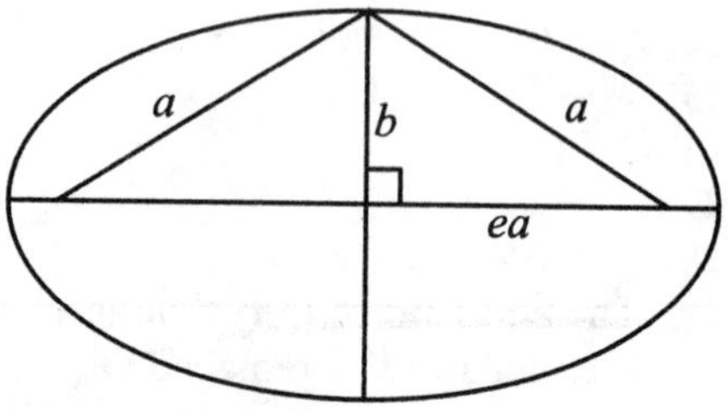

34. $A = 47.12$, $e = .8$

35. $A = 31.42$, $e = .995$

36. $A = 29845$, $e = .312$

37. $e = .094$

38. $e = .048$

39. $e = .056$

40. $e = .250$

41. $a = 10$, $b = 9$

42. $a = 2$, $b = 1$

43. $a = 11$, $b = 6$ (Note: This ellipse is oriented so that its major axis is along the y axis.)

44. $a = \sqrt{28}$; $b = \sqrt{11}$

45. Let $e = \sqrt{a^2 - b^2}/a$. Then the coordinates of the focal points are $(h + ea, k)$ and $(h - ea, k)$.

46. (0, 1/4); $y = -1/4$
47. (0, 1/40); $y = -1/40$
48. (0, $1/4c$); $y = -1/4c$
49. (0, 4); $y = -4$
50. (−3, 6); $y = 4$
51. (3, 8 1/4); $y = 7\ 3/4$
52. (1/4, 0); $x = -1/4$
53. (1, 0); $x = -1$
54. $(h, k + 1/4)$; $y = k - 1/4$
55. $(h, k + 1/4b)$; $y = k - 1/4b$
56. First, find the coordinates of the vertex (h, k). In this case $h = 0$ and $k = 4$. Then, find the distance from the vertex to the focus (call that distance a.) The equation of the parabola is

$$(y - k) = \frac{1}{4}a(x - h)^2$$

In this case the equation is $(y - 4) = (1/16)x^2$
57. $(y - 5) = (1/4)(x - 10)^2$
58. $x - 14.5 = (1/10)(y - 9)^2$
59. $y = (1/40)x^2$
60.

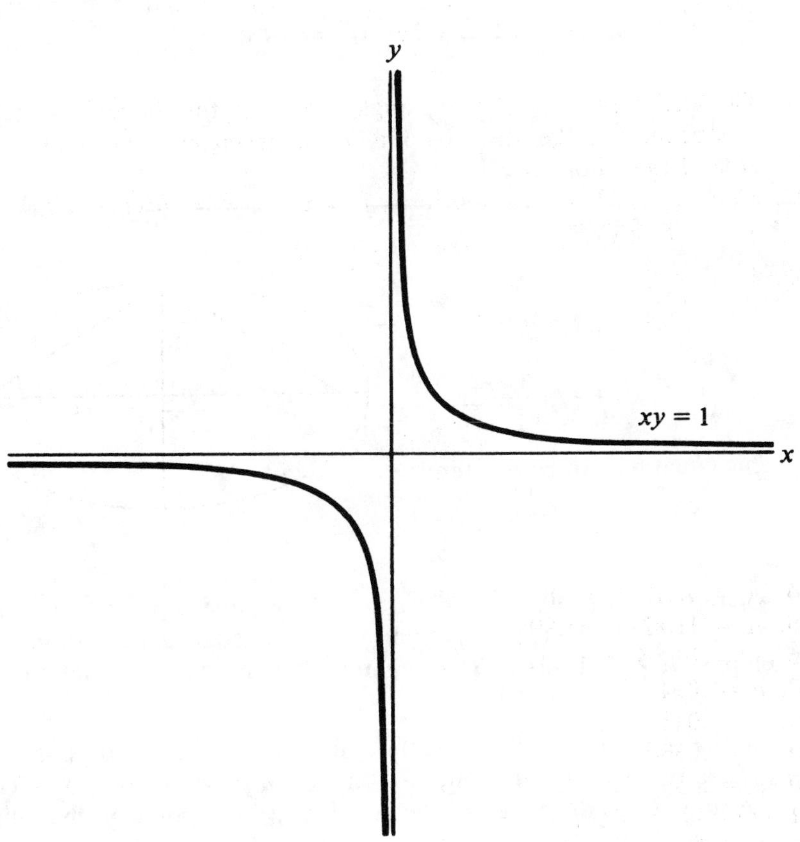

61.

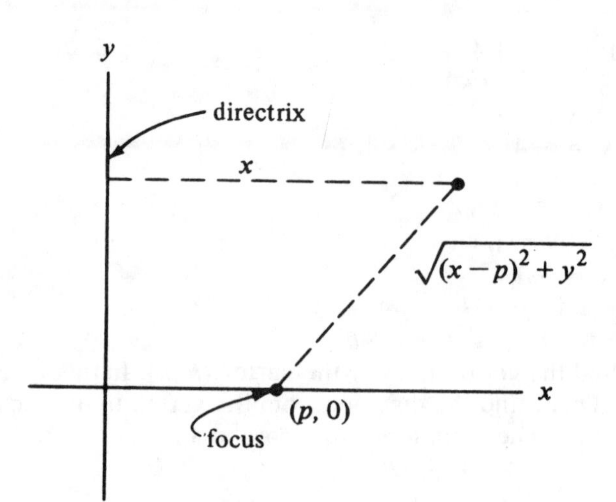

$$\frac{\sqrt{(x-p)^2+y^2}}{x} = e$$

$$x^2(1-e^2) - 2px + y^2 + p^2 = 0$$

$$x^2 - \frac{2px}{1-e^2} + \frac{p^2+y^2}{(1-e^2)} = 0$$

Completing the square for x:

$$x^2 - \frac{2px}{1-e^2} + \frac{p^2}{(1-e^2)^2} - \frac{p^2}{(1-e^2)^2}$$

$$+ \frac{y^2}{1-e^2} + \frac{p^2}{1-e^2} = 0$$

$$\left(x - \frac{p}{1-e^2}\right)^2 + \frac{y^2}{1-e^2} = \frac{e^2p^2}{(1-e^2)^2}$$

This equation can be written:

$$\frac{(x-h)^2}{A} + \frac{y^2}{B} = 1$$

where $h = p/(1-e^2)$, $A = e^2p^2/(1-e^2)^2$, and $B = e^2p^2/(1-e^2)$.

If $e < 1$, then B is positive and this is the standard equation of an ellipse. If $e > 1$, then B is negative and this is the equation of a hyperbola.

62. Put the center of the ellipse at the origin and put the major axis along the x axis. Let a = semimajor axis and let b = semiminor axis. Then the sum of the distances to the two foci from a point on the ellipse is $2a$. Let $f = \sqrt{a^2-b^2}$. Then the coordinates of the two foci are $(f, 0)$ and $(-f, 0)$. The sum of the distances from a point (x, y) to the two foci is

$$\sqrt{(x+f)^2+y^2} + \sqrt{(x-f)^2+y^2} = 2a$$

$$\sqrt{(x+f)^2+y^2} = 2a - \sqrt{(x-f)^2+y^2}$$

Square both sides:

$$(x + f)^2 + y^2 = 4a^2 - 4a\sqrt{(x - f)^2 + y^2} + (x - f)^2 + y^2$$

$$x^2 + 2xf + f^2 = 4a^2 - 4a\sqrt{(x-f)^2 + y^2} + x^2 - 2xf + f^2$$

$$4xf - 4a^2 = -4a\sqrt{(x - f)^2 + y^2}$$

$$a^2 - xf = a\sqrt{(x - f)^2 + y^2}$$

Square both sides again:

$$a^4 - 2a^2xf + x^2f^2 = a^2(x^2 - 2xf + f^2 + y^2)$$

$$a^4 - 2a^2xf + x^2f^2 = a^2x^2 - 2a^2xf + a^2f^2 + a^2y^2$$

$$x^2(a^2 - f^2) + a^2y^2 = a^4 - a^2f^2$$

Since $f^2 = a^2 - b^2$, $a^2 - f^2 = b^2$:

$$b^2x^2 + a^2y^2 = a^4 - a^2(a^2 - b^2)$$

$$b^2x^2 + a^2y^2 = a^4 - a^4 + a^2b^2$$

$$b^2x^2 + a^2y^2 = a^2b^2$$

$$\frac{x^2}{a^2} + \frac{y^2}{b^2} = 1$$

63–66.

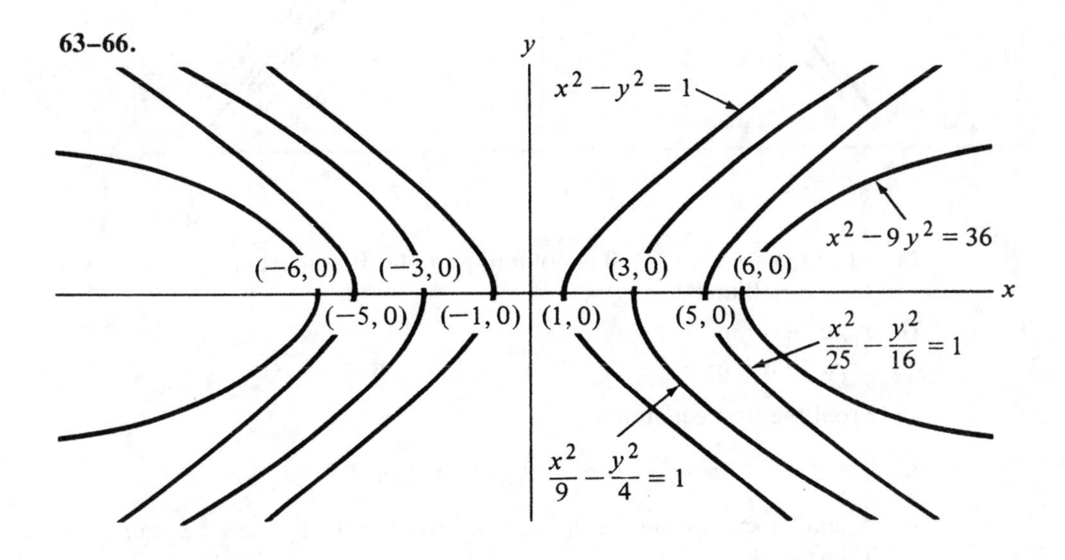

67–71.

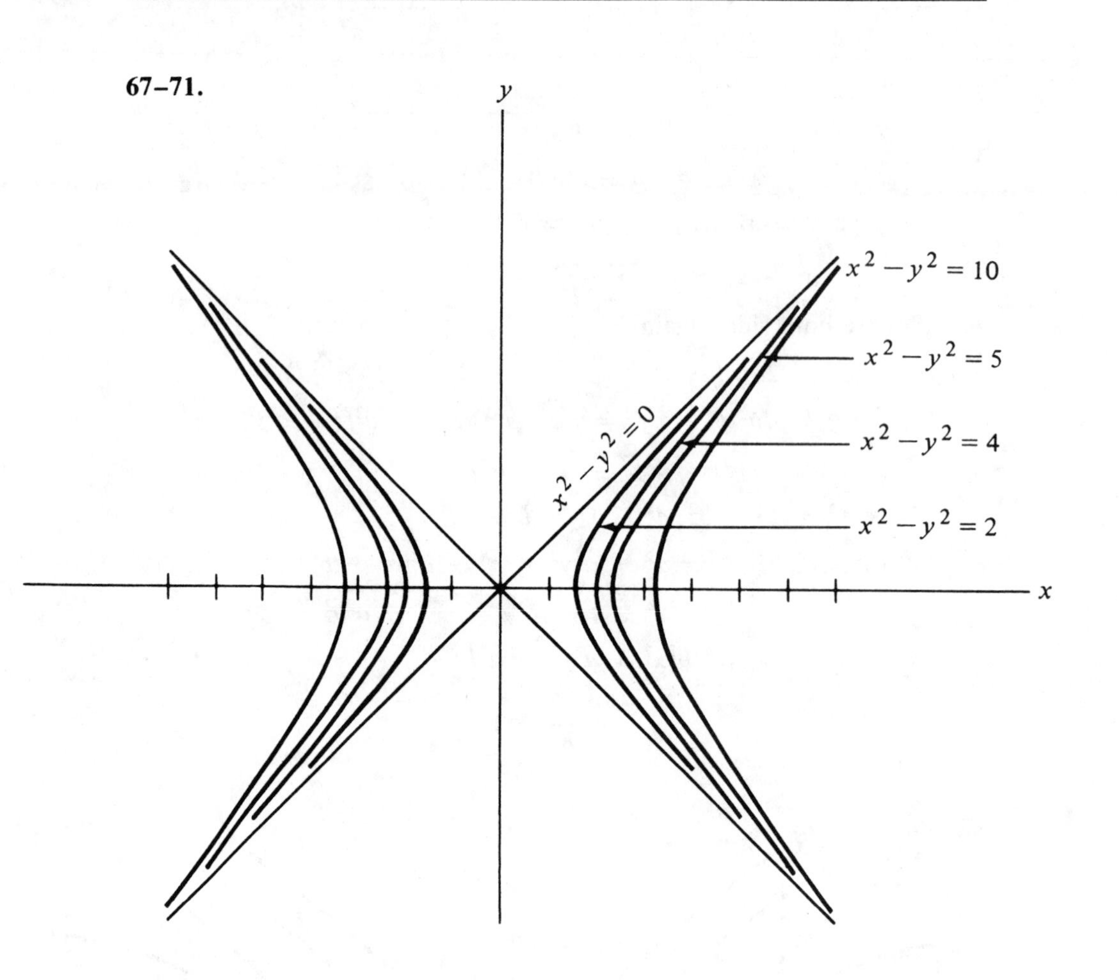

71. The curve $x^2 - y^2 = 0$ is not a hyperbola. It consists of the two lines $x - y = 0$ and $x + y = 0$.

72. $x_1^2/a^2 + y_1^2/b^2 = 1$
$x_2^2/a^2 + y_2^2/b^2 = 1$

From the first equation:

$$b^2 = y_1^2/(1 - x_1^2/a^2)$$

A similar expression for b^2 can be found from the second equation. Then:

$$\frac{y_2^2}{\dfrac{a^2 - x_2^2}{a^2}} = \frac{y_1^2}{\dfrac{a^2 - x_1^2}{a^2}}$$

$$y_2^2(a^2 - x_1^2) = y_1^2(a^2 - x_2^2)$$

Solve the above equation for a:

$$a = \sqrt{\left|\frac{y_2^2x_1^2 - y_1^2x_2^2}{y_2^2 - y_1^2}\right|}$$

73. $Ax^2 + Bx + Cy^2 + Dy + E = 0$

$x = x' - \frac{B}{2A} \quad y = y' - \frac{D}{2C}$

Substitute the expressions for x and y into the equation:

$$A\left(x' - \frac{B}{2A}\right)^2 + B\left(x' - \frac{B}{2A}\right) + C\left(y' - \frac{D}{2C}\right)^2 + D\left(y' - \frac{D}{2C}\right) + E = 0$$

$$A\left(x'^2 - 2x'\frac{B}{2A} + \frac{B^2}{4A^2}\right) + Bx' - \frac{B^2}{2A} + C\left(y'^2 - 2y'\frac{D}{2C} + \frac{D^2}{4C^2}\right) + Dy' - \frac{D^2}{2C} + E = 0$$

$$Ax'^2 - Bx' + \frac{B^2}{4A} + Bx' - \frac{B^2}{2A} + Cy'^2 - Dy' + \frac{D^2}{4C} + Dy' - \frac{D^2}{2C} + E = 0$$

$$Ax'^2 - \frac{B^2}{4A} + Cy'^2 - \frac{D^2}{4C} + E = 0$$

$$\text{Let } K = \frac{B^2}{4A} + \frac{D^2}{4C} - E$$

Then the equation becomes:

$$Ax'^2 + Cy'^2 = K$$

If $A = C$, then this will be the equation of a circle, centered at the origin in the (x', y') coordinate system. If A and C have the same sign, the equation represents an ellipse; if they have opposite signs, it will be a hyperbola. (There are also some other possibilities. Try to determine what values for A, C, and K will make the solution a single point, or two lines, or that cause there to be no solution.)

Chapter 13

1. $x = 0, 1, -1$
2. $x = 0, \pm\sqrt{2}$
3. $x = 0, -1, 5/2$
4. $x = 0, x = -2/3 \pm \sqrt{34}/3$
5. $y = 4, 9$
6. $x = \pm 2, \pm 3$
7. $x = \pm\sqrt{7}$
8. $x = 2$
9. $x = 2, 1$
10. $x = 0, 4, 5$
11. $x = 0, 7, -5$
12. $(x-1)(x-2)(x-3) = x^3 - 6x^2 + 11x - 6 = 0$
13. $x^3 - 7x^2 - 6x + 72 = 0$
14. $x^3 - 10x^2 - 13x + 22 = 0$
15. $x^3 - 8x^2 + 17x - 4 = 0$
16. $x^3 + 2x^2 - 3x = 0$
17. $x + 3 \quad R\ 5$ (R means remainder)

18. Long division of $2x^4 + 5x^3 + 7x^2 + 5x$ by $(2x^2 + 3x + 4)$:

$$\begin{array}{r} x^2 + x \\ (2x^2 + 3x + 4)\overline{)2x^4 + 5x^3 + 7x^2 + 5x} \\ \underline{2x^4 + 3x^3 + 4x^2} \\ 2x^3 + 3x^2 + 5x \\ \underline{2x^3 + 3x^2 + 4x} \\ x \end{array}$$

result: $x^2 + x\ R\ x$

19. $5x\ R\ 1$
20. $5x^2 + 4x\ R\ 1$
21. $3x^2 + 2x + 5 \quad R\ 4x - 1$
22. $x \quad R\ 2$
23. $x - 2 \quad R\ -3$
24. $x + 5 \quad R\ 21$
25. $3x - 4 \quad R\ 13$
26. $6x^2 + 12x + 41 \quad R\ 133$
27. $3x^2 + 20x + 125 \quad R\ 754$
28. $2x^2 + 8x \quad R\ -9$
29. $5x^2 + 28x + 106 \quad R\ 430$
30. $x^2 - 3x - 1$ (no remainder)
31. $3x^2 + 5x + 1$
32. $6x^2 - 8x - 9$
33. $8x^2 - 4x + 12$
34. $1 + x$
35. $1 + x + x^2$
36. $1 + x + x^2 + x^3$
37. $1 + x + x^2 + x^3 + x^4$
38. Judging from the pattern of the previous four answers, we can guess that

$$\frac{x^n - 1}{x - 1} = 1 + x + x^2 + x^3 + \cdots + x^{n-1}$$

and that guess turns out to be right. (See page 189.)

39. When the coefficient of the highest power of x is 1, then the rational roots (if any) must be factors of the last term. The factors of 50 are 1, 2, 5, 10, 25, and 50. 1 turns out to be a solution, so we can perform the synthetic division

$$(x^3 - 16x^2 + 65x - 50)/(x - 1) = x^2 - 15x + 50$$

We can find that 10 and 5 are the two solutions to the equation $x^2 - 15x + 50 = 0$, so the three solutions to our original equation are 1, 5, and 10.

40. $x = 2, 3, -2$
41. $x = -2$
42. $x = 3, x = -1$
43. $x = 4, x = 8, x = 12$
44. $x = 3, x = -3, x = -1$
45. $x = 1/2, x = -1/3, x = -1/2$
46. $x = 7, x = 5, x = -7$
47. $x = \pm\sqrt{21}$
48. $x = \pm\sqrt{7}$
49. $x = 0, x = -1/2 \pm \sqrt{17}/2$
50. $x = -\sqrt{3}, x = 5$

51. $x = -\sqrt{8} - 2, x = -6$
52. If $f(x) = 5x + 6$, then $f(2) = 16$ and $(5x + 6)/(x - 2) = 5$ R 16
53. $f(x) = x^2 + x + 5$; $f(-5) = 25$; $(x^2 + x + 5)/(x + 5) = x - 4$ R 25
54. $f(x) = 3x^2 - 16x + 7$; $f(7) = 42$; $(3x^2 - 16x + 7)/(x - 7) = 3x + 5$ R 42.
55. $f(x) = x^2 - x - 12$; $f(4) = 0$; $(x^2 - x - 12)/(x - 4) = x + 3$ R 0
56.
```
10  REM  THIS PROGRAM DRAWS A GRAPH OF A
20  REM  POLYNOMIAL. THE USER CAN CHOOSE
30  REM  THE SCALE OF THE DRAWING AND THE
40  REM  COORDINATES OF THE CENTER OF THE SCREEN.
50 SH = 160   : REM SCREEN HEIGHT
60 S2 = SH/2  : REM HALF OF SCREEN HEIGHT
70 SW = 280   : REM SCREEN WIDTH
80 INPUT "DEGREE OF POLYNOMIAL:";D
90 FOR I=0 TO D
100    PRINT "INPUT COEFFICIENT FOR X^";I;:INPUT A(I)
110 NEXT I
120 PRINT "DO YOU WANT AXES SHOWN?"
130 INPUT "Y OR N ";QA$
140 REM  SET INITIAL SPECIFICATIONS
150 XC = 0 : YC = 0
160 REM  THE CENTER IS AT (0,0)
170 W = 20 : H = 16
180 REM W IS WIDTH, H IS HEIGHT
190 REM *************************
200 REM INPUT SPECIFICATIONS
210 CLS
220 LOCATE 21,1
230 PRINT "W:";W;"  H:";H;
240 PRINT " CENTER:";XC;",";YC;
250 PRINT "   ?"
260 X$=INKEY$: IF X$="" THEN GOTO 260
270 IF X$="G" THEN GOSUB 450 : GOTO 220
280 IF X$="W" THEN GOTO 360
290 IF X$="H" THEN GOTO 380
300 IF X$="X" THEN GOTO 400
310 IF X$="Y" THEN GOTO 420
320 IF X$="S" THEN END
330 GOTO 150
340 REM ****************************
350 REM  CHANGE SPECIFICATIONS
360 INPUT "NEW WIDTH:";W
370 GOTO 200
380 INPUT "NEW HEIGHT:";H
390 GOTO 200
400 INPUT "NEW X CENTER:";XC
410 GOTO 200
420 INPUT "NEW Y CENTER:";YC
430 GOTO 200
440 REM ****************************
450 REM  DRAW DIAGRAM
460 SCREEN 1 ' TURN ON GRAPHICS
470 IF QA$ = "Y" THEN GOSUB 650 : REM PLOT AXES
480 XS = XC - W/2 : REM STARTING X VALUE
490 DX = W/SW : REM X INCREMENT
```

```
500 X2 = 0
510 FOR X = XS TO (XS+W) STEP DX
520 X2 = X2 + 1
530 GOSUB 750 : REM CALCULATE Y
540 Y2 = INT(S2 - SH*(Y-YC)/H)
550 REM  NOW CHECK TO MAKE SURE WE'RE
560 REM  NOT OVER THE EDGE
570 IF X2<1 THEN GOTO 620
580 IF X2>=SW THEN GOTO 620
590 IF Y2<1 THEN GOTO 620
600 IF Y2>=SH THEN GOTO 620
610 PSET (X2,Y2)
620 NEXT X
630 RETURN
640 REM *********************************
650 REM  PLOT AXES
660 XA = SW * (W/2 - XC)/W
670 XA = INT(XA + .5)
680 YA = SH - (SH * (H/2 - YC)/H)
690 YA = INT(YA + .5)
700 IF (XA<1) OR (XA>=SW) THEN GOTO 720
710 LINE (XA,0)-(XA,SH-1)
720 IF (YA<1) OR (YA >= SH) THEN GOTO 740
730 LINE (0,YA)-(SW-1,YA)
740 RETURN
750 REM SUBROUTINE TO CALCULATE VALUE OF Y
760 Y = 0
770 FOR I = 0 TO D
780   Y = Y + A(I)*X^I
790 NEXT I
800 RETURN
```

57.

```
1  REM  POLYNOMIAL DIVISION PROGRAM
5  DIM NUMER(20),DENOM(20),QLIST(20),ATEMP(20)
10 INPUT "DEGREE OF NUMERATOR:";NMAX
20 FOR I = 0 TO NMAX
30   PRINT "ENTER COEFFICIENT OF X^";I;":  ";:INPUT NUMER(I)
40 NEXT I
50 INPUT "DEGREE OF DENOMINATOR:";DMAX
60 FOR I = 0 TO DMAX
70   PRINT "ENTER COEFFICIENT OF X^";I;":  ";:INPUT DENOM(I)
80 NEXT I
90 GOSUB 1100  ' PERFORM DIVISION
100 PRINT "QUOTIENT"
110 FOR I = 0 TO QMAX : Z(I) = QLIST(I) : MAX = QMAX : NEXT I
120 GOSUB 1500  'POLYNOMIAL OUTPUT
130 PRINT "REMAINDER"
140 FOR I = 0 TO NMAX : Z(I) = ATEMP(I) : MAX = NMAX : NEXT I
150 GOSUB 1500
160 END
1099 '
1100  REM ---  POLYNOMIAL DIVISION SUBROUTINE
1130 QMAX=NMAX-DMAX
1135 FOR I=0 TO NMAX:ATEMP(I)=NUMER(I):QLIST(I)=0: NEXT I
1140 NINDEX=NMAX
1145 QINDEX=QMAX
1150 FOR J1=1 TO (NMAX-DMAX+1)
1155   QUOT=ATEMP(NINDEX)/DENOM(DMAX)
```

```
1160    QLIST(QINDEX)=QUOT
1165    FOR I2=NINDEX TO (NINDEX-DMAX) STEP -1
1170      ATEMP(I2)=ATEMP(I2)-QUOT*DENOM(I2-NINDEX+DMAX)
1175    NEXT I2
1180    NINDEX=NINDEX-1
1185    QINDEX=QINDEX-1
1190 NEXT J1
1297 '
1500 REM POLYNOMIAL OUTPUT
1510 FOR I = MAX TO 1 STEP - 1
1520    IF Z(I) = 0 THEN GOTO 1550
1530    IF Z(I)>0 THEN PRINT " + ";
1535    IF ABS(Z(I))<>1 THEN PRINT Z(I);
1540    PRINT "X";
1545    IF I>1 THEN PRINT "^";I;
1550 NEXT I
1580 IF Z(0)>0 THEN PRINT " + ";Z(0);
1590 IF Z(0)<0 THEN PRINT Z(0);
1600 PRINT
1610 RETURN
```

58.

```
1  REM  POLYNOMIAL EQUATION SOLUTION PROGRAM
2  REM   THIS PROGRAM USES NEWTON'S METHOD TO FIND
3  REM   A SOLUTION FOR A POLYNOMIAL EQUATION OF DEGREE 4
10 DIM A(4)
20 FOR I = 0 TO 4
30   PRINT "ENTER COEFFICIENT OF X^";I;": ";:INPUT A(I)
40 NEXT I
50 INPUT "INITIAL GUESS FOR SOLUTION:";X
60 GOSUB 200  'CALCULATE Y = F(X)
65   IF ABS(Y)<.0001 THEN GOTO 100
70 GOSUB 300  'CALCULATE DENOMINATOR
80 X = X - Y/D
90 GOTO 60
100  REM -- SOLUTION FOUND --
110  PRINT "SOLUTION:";X
120  END
200  REM CALCULATE Y = F(X)
210  Y = 0
220  FOR I = 0 TO 4
230    Y = Y + A(I)*X^I
240  NEXT I
250  RETURN
300  REM CALCULATE DENOMINATOR
310  D = 0
320  FOR I = 0 TO 3
330    D = D + A(I+1)*(I+1)*X^I
340  NEXT I
350  RETURN
360  REM  WARNING -- SOMETIMES THIS PROGRAM WILL GO AROUND
370  REM  IN AN ENDLESS LOOP
380  REM  IN THAT CASE, INTERRUPT THE PROGRAM AND TRY AGAIN
390  REM  WITH ANOTHER INITIAL GUESS
400  REM  HOWEVER, IF THERE ARE NO REAL SOLUTIONS, THEN
410  REM  THE PROGRAM WILL NEVER FIND A SOLUTION
```

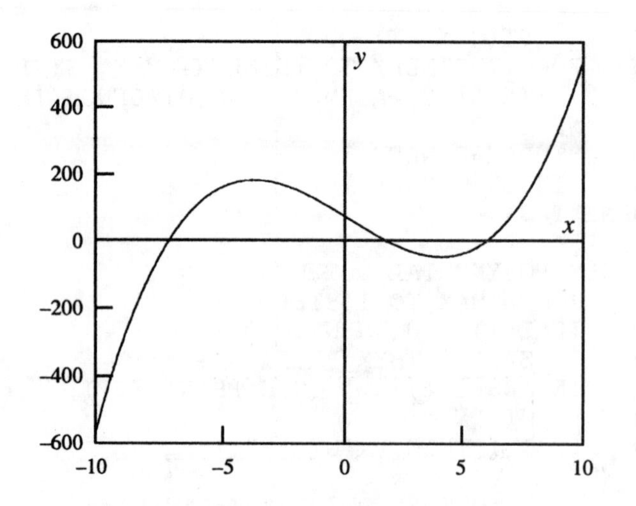

59. $x^3 - x^2 - 44x + 84$; 3 solutions: $x = -7, 2, 6$

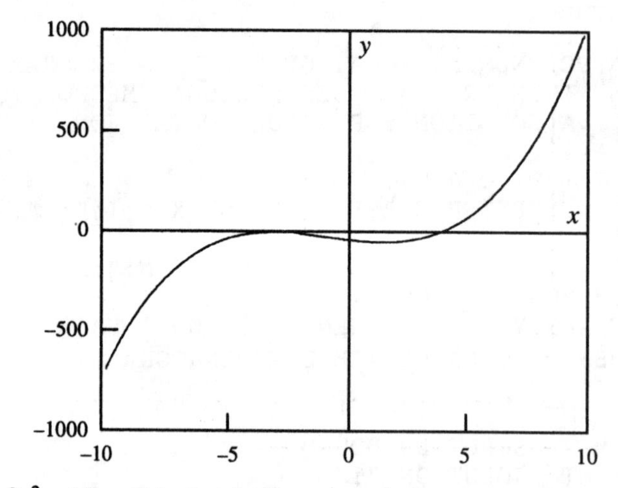

60. $x^3 + 2x^2 - 15x - 36$; 2 solutions: $x = -3, 4$

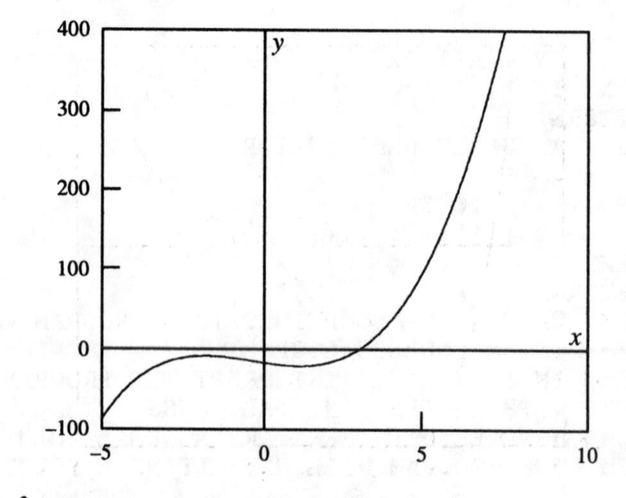

61. $x^3 + x^2 - 6x - 18$; 1 solution: $x = 3$

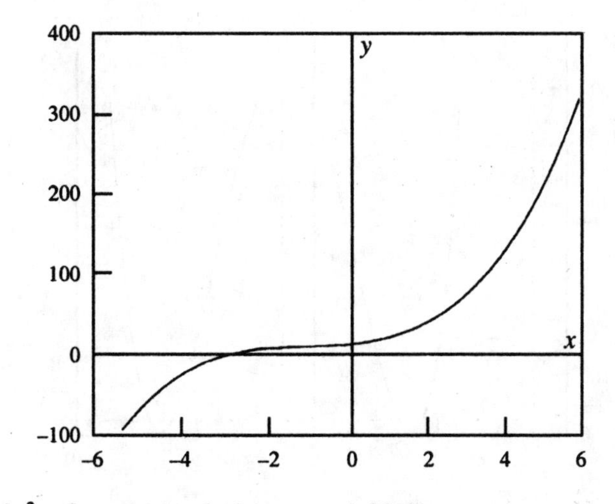

62. $x^3 + 2x^2 + 3x + 4$; 1 solution: $x = -1.6506$

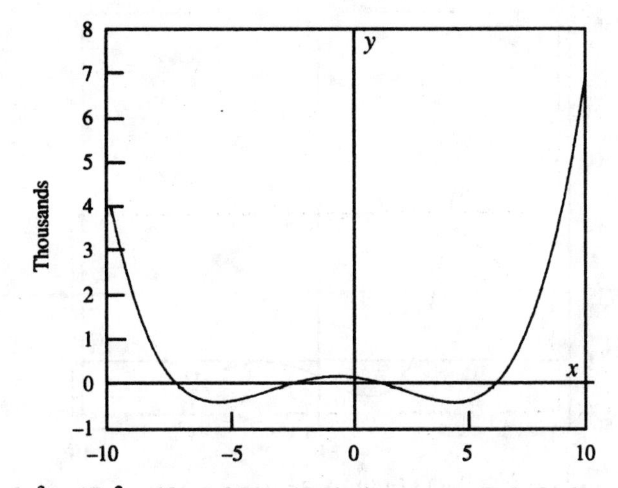

63. $x^4 + 2x^3 - 47x^2 - 48x + 252$; 4 solutions: $x = -7, 3, 2, 6$

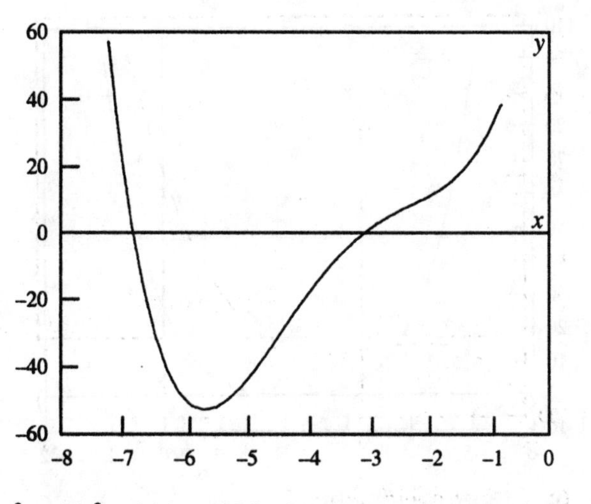

64. $x^4 + 14x^3 + 67x^2 + 144x + 126$; 2 solutions: $x = -7, -3$

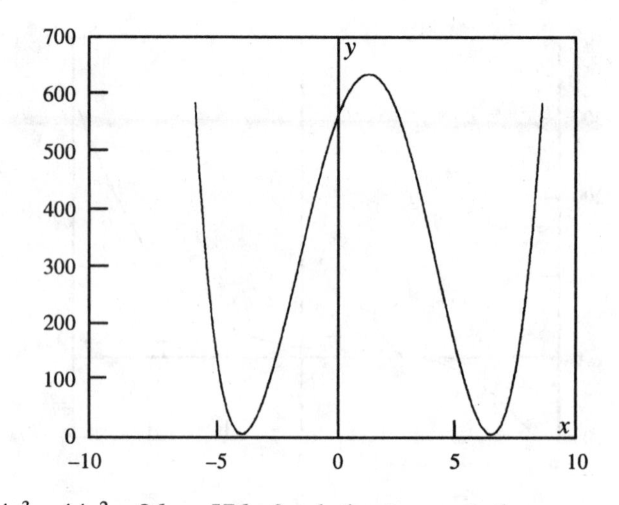

65. $x^4 - 4x^3 - 44x^2 + 96x + 576$; 2 solutions: $x = -4, 6$

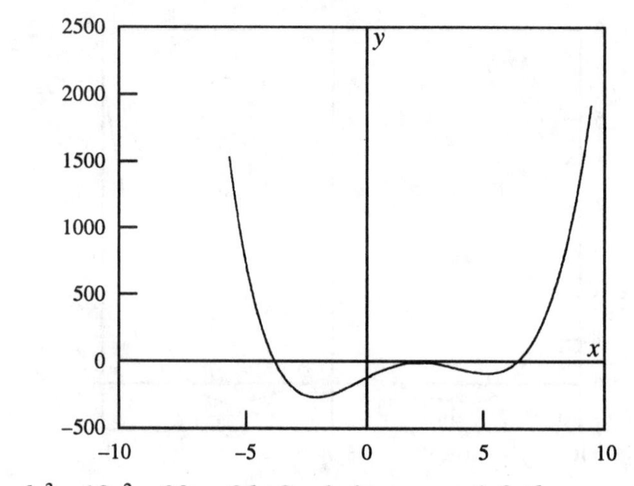

66. $x^4 - 6x^3 - 12x^2 + 88x - 96$; 3 solutions: $x = -4, 2, 6$

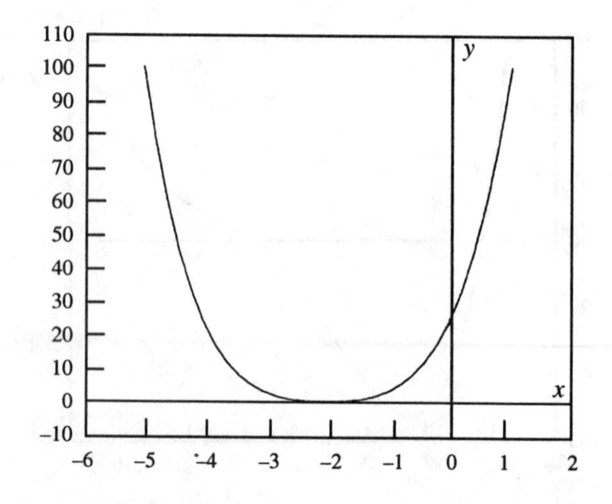

67. $x^4 + 8x^3 + 26x^2 + 40x + 24$; 1 solution: $x = -2$

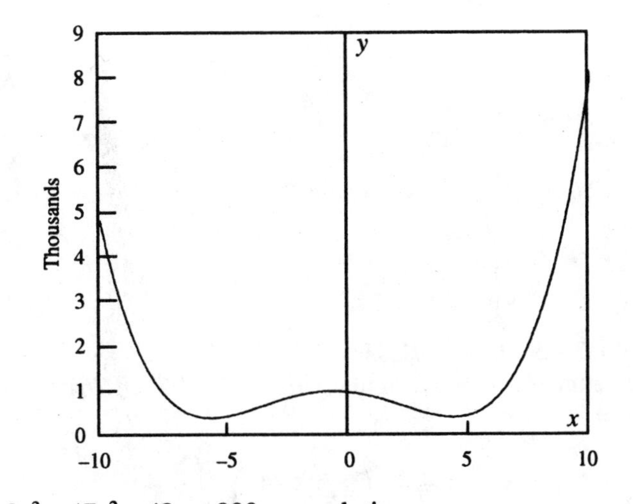

68. $x^4 + 2x^3 - 47x^2 - 48x + 900$; no solutions

Chapter 14

1. $\sum_{i=10}^{30} i$ **2.** $\sum_{i=1}^{25} (2i - 1)$ **3.** $\sum_{i=10}^{30} 2i$ **4.** $\sum_{i=1}^{99} i^2$

5. This property follows directly from the distributive property.
6. 165
7. 220
8. 550
9. $5.000\,005 \times 10^{11}$
10. 1,050
11. 506
12. 104.5
13. 9 (Note that the formula works even if the first term is negative.)
14. 4
15. 1/3
16. 2/9
17. 2/11
18. 5
19. 3
20. 111, 111, 110
21. 5.8019
22. 34.719
23. 254
24. 1,092
25. 0 (If $r = -1$, then the sum of the geometric series will be 0 if n is even, and a if n is odd.
26. 1
27. -182
28. -129
29. $2n$
30. $20 + 5n$
31. $12n$
32. $53 + 3n$
33. 2^{6-n}

34. 3^n
35. $100(1.05)^{n-1}$
36. $d = 1$
37. $d = 3$
38. $d = 5$
39. $d = 30$
40. $a = 80$
41. $a = 10$
42. $a = 1/9$
43. $a = 1$
44. $PV = 5/1.05 + 5/1.05^2 + 5/1.05^3 + \ldots + 5/1.05^{11}$
Use the geometric series formula with $a = 5/1.05 = 4.7619$; $r = 1/1.05 = .9524$; $n = 11$.

$$PV = 4.7619\left(\frac{1-.9524^{11}}{1-.9524}\right) = 41.53$$

45. 100
46. $S_{n+1} = (n + 1)a + (1/2)(n + 1)nd$
47. $S_{n+1} = a(r^{n+1} - 1)/(r - 1)$
48. First, we'll calculate the distance that the dog travels during its first round trip to the finish line and back. Let x represent the distance that the dog travels in that time. The distance that you travel is $2d - x$, where d is the total distance (in this case 4). The time it takes you to walk a distance $(2d - x)$ is $(2d - x)/5$, and the time it takes the dog to travel a distance x is $x/9$. These two times must be equal, so we can solve for x: $(2d - x)/5 = x/9$, so $x = (9/7)d = 36/7$.
When the dog starts the next round trip it will be only $(2/7)d$ miles away from the finish line. We can repeat exactly the same calculation to show that the next round trip is of length $(2/7)(36/7)$. The third round trip has length $(2/7)^2(36/7)$, and so on. The total distance the dog travels is $(36/7)[1 + 2/7 + (2/7)^2 + (2/7)^3 + \cdots] = (36/7)[1/(1 - 2/7)] = 36/5$.
Here is the easy way. It will take you 4/5 hours to walk 4 miles. Since the dog is running at a constant speed of 9 miles per hour for 4/5 hours, the total distance will be $9 \times (4/5) = 36/5$.
49. 250
50. Let L = the last term of the arithmetic series. Then $L = a + (n - 1)d$, from which we get $n = [(L - a)/d] + 1$. $S_n = na + (\frac{1}{2})\, n(n - 1)d$; substituting the expression for n and simplifying gives $S_n = (\frac{1}{2d})(L^2 - a^2 + ad + dL) = (\frac{1}{2d})(L + a)(L - a + d)$.
51. You receive $13x - x^2$ items for the gift you start receiving on day x. Therefore, the total number of items is:

$$\Sigma_{x=1}^{12}(13x - x^2) = 13\Sigma_{x=1}^{12} x - \Sigma_{x=1}^{12} x^2$$

$$= 13\left(\frac{12}{2}\right)(12 + 1) - \left(\frac{12}{6}\right)(12 + 1)(2 \times 12 + 1)$$

$$= 13 \times 6 \times 13 - 2 \times 13 \times 25 = 1014 - 650 = 364$$

Chapter 15

1. $n(n-1)! = n(n-1)(n-2)(n-3) \times \cdots \times (3)(2)(1) = n!$

2. $\binom{n}{j} = \frac{n!}{j!(n-j)!} = \frac{n!}{(n-j)!\,j!} = \binom{n}{n-j}$

3. $\binom{n}{0} = \frac{n!}{0!n!} = 1$

4. $\binom{n}{1} = n$

5. $\binom{n}{n-1} = n$

6. $\binom{n}{n} = 1$

7. $$\frac{n!}{j!(n-j)!} + \frac{n!}{(j+1)!(n-j-1)!}$$

$$= \frac{n!(j+1)}{(j+1)j!(n-j)!} + \frac{n!(n-j)}{(j+1)!(n-j)(n-j-1)!}$$

$$= \frac{n!(j+1)}{(j+1)!(n-j)!} + \frac{n!(n-j)}{(j+1)!(n-j)!}$$

$$= \frac{n!(j + 1 + n - j)}{(j+1)!(n-j)!}$$

$$= \frac{n!(n+1)}{(j+1)!(n-j)!} = \frac{(n+1)!}{(j+1)!(n-j)!} = \binom{n+1}{j+1}$$

8. row 7: 1 7 21 35 35 21 7 1
row 8: 1 8 28 56 70 56 28 8 1
row 9: 1 9 36 84 126 126 84 36 9 1

9. $(a+b)^n = \sum_{i=0}^{n} \binom{n}{i} a^{n-i} b^i$

10. The sum of the elements in row j is 2^j.

11. no heads: $\binom{4}{0} = 1$
one head: $\binom{4}{1} = 4$
two heads: $\binom{4}{2} = 6$
three heads: 4
four heads: 1

12. no heads: $\binom{6}{0} = 1$
one head: $\binom{6}{1} = 6$
two heads: $\binom{6}{2} = 15$
three heads: 20
four heads: 15
five heads: 6
six heads: 1

13. $\binom{n}{j}$

14. $\binom{52}{13} = 6.35 \times 10^{11}$ $\binom{52}{7} = 133{,}784{,}560$

$\binom{52}{4} = 270{,}725$ $\binom{52}{3} = 22{,}100$

15. $\binom{20}{5} = 15{,}504$

16. $1 + 5x + 10x^2 + 10x^3 + 5x^4 + x^5$

17. $1 + nx + \binom{n}{2} x^2 + \binom{n}{3} x^3 + \cdots + \binom{n}{1} x^{n-1} + x^n$

18. $(10 + 1)^4 = 10^4 + 4 \times 10^3 + 6 \times 10^2 + 4 \times 10 + 1$
$= 14{,}641$

19. $(10 + 1)^8 = 10^8 + 8 \times 10^7 + 28 \times 10^6 + 56 \times 10^5$
$+ 70 \times 10^4 + 56 \times 10^3 + 28 \times 10^2 + 8 \times 10 + 1$
$= 214{,}358{,}881$

20. $a^5 + 5a^4b + 10a^3b^2 + 10a^2b^3 + 5ab^4 + b^5 - a^3 - 3a^2b - 3ab^2 - b^3$

21. $a^6 + 12a^5 + 60a^4 + 160a^3 + 240a^2 + 192a + 64$

22. $a^3 + 9a^2 + 27a + 27$

23. $a^7 + 70a^6 + 2{,}100\ a^5 + 35{,}000\ a^4 + 350{,}000\ a^3 + 2{,}100{,}000\ a^2 + 7{,}000{,}000\ a + 10{,}000{,}000$

24. $x^6 - 6x^5 + 15x^4 - 20x^3 + 15x^2 - 6x + 1$

25. $x^6 - 12x^5 + 60x^4 - 160x^3 + 240x^2 - 192x + 64$

26. $a^n - na^{n-1}b + \binom{n}{2} a^{n-2}b^2 - \binom{n}{3} a^{n-3}b^3 + \cdots$

27. $1 \times 2^5 + 5 \times 2^5 + 10 \times 2^5 + 10 \times 2^5 + 5 \times 2^5 + 1 \times 2^5 = 1{,}024$

28. $32x^5 + 80x^4 + 80x^3 + 40x^2 + 10x + 1$

29. $243a^5 + 1{,}620a^4b + 4{,}320a^3b^2 + 5{,}760a^2b^3 + 3{,}840ab^4 + 1{,}024b^5$

30. $\left[a^n + na^{n-1}h + \binom{n}{2} a^{n-2}h^2 + \cdots + h^n - a^n\right]/h$

$= na^{n-1} + \binom{n}{2} a^{n-2}h + \cdots + nah^{n-2} + h^{n-1}$

31. As h becomes very small, this expression comes close to na^{n-1}. (This expression is very important in calculus.)

32.
```
10 INPUT N
20 X = N
30 GOSUB 1000
40 PRINT Z
50 END
1000 REM  THIS SUBROUTINE CALCULATES X FACTORIAL (X!)
1010 Z = 1
1020 FOR I = 1 TO X
1030   Z = Z * I
1040 NEXT I
1050 RETURN
```

33.
```
1 REM    THIS PROGRAM CALCULATES THE NUMBER OF COMBINATIONS
2 REM    OF N THINGS TAKEN J AT A TIME
10 INPUT "N, J:";N,J
20 X = N
30 GOSUB 1000
40 C = Z       : REM  Z IS NOW N!
50 X = J
60 GOSUB 1000
70 C = C/Z     : REM  Z IS NOW J!
80 X = N - J
90 GOSUB 1000
100 C = C/Z    : REM  Z IS NOW (N-J)!
110 PRINT C
120 END
1000 REM  THIS SUBROUTINE CALCULATES X FACTORIAL (X!)
1010 Z = 1
1020 FOR I = 1 TO X
1030    Z = Z * I
1040 NEXT I
1050 RETURN
```

34.
```
1 REM    THIS PROGRAM CALCULATES THE NUMBER OF COMBINATIONS
2 REM    OF N THINGS TAKEN J AT A TIME
10 INPUT "N:";N
20 INPUT "J:";J
30 B = J
40 IF J > (N-J) THEN B = N - J
50 Z = 1
60 FOR I = 1 TO B
70     Z = Z * (N-I+1)/I
80 NEXT I
90 PRINT Z
100 END
```

Chapter 16

1. The formula is true for $n = 1$. Now assume $S_n = (\frac{1}{2})\, n(n + 1)$. Then $S_{n+1} = (\frac{1}{2})\, n(n + 1) + (n + 1) = (\frac{1}{2})\,(n + 1)\,(n + 2)$.
2. The formula is true for $n = 2$. Assume $S_n = (\frac{1}{2})n(n/2 + 1)$. Then $S_{n+2} = (\frac{1}{2})n(n/2 + 1) + (n + 2) = (\frac{1}{2})(n + 2)[(n + 2)/2 + 1]$.

3. Show that the formula works for $n = 1$:

$$1^2 = 1 = \frac{1}{6}(1 + 1)(2 + 1)$$

Now, assume that the formula is true for $n = j$:

$$S_j = 1^2 + 2^2 + \cdots + j^2 = \frac{j}{6}(j + 1)(2j + 1)$$

Then

$$S_{j+1} = S_j + (j + 1)^2 = \frac{j}{6}(j + 1)(2j + 1) + (j + 1)^2$$

$$= \frac{(j + 1)}{6}[j(2j + 1) + 6(j + 1)]$$

$$= \frac{(j + 1)}{6}[2j^2 + 7j + 6]$$

$$= \frac{(j + 1)}{6}[(j + 1) + 1][2(j + 1) + 1]$$

Therefore, the formula works.

4. Show that the formula works for $n = 1$:

$$1^3 = 1 = \frac{1}{4}2^2$$

Now, assume that the formula is true for $n = j$:

$$S_j = 1^3 + 2^3 + \cdots + j^3 = \frac{j^2}{4}(j + 1)^2$$

Find a formula for S_{j+1}:

$$S_{j+1} = S_j + (j + 1)^3 = \frac{j^2}{4}(j + 1)^2 + (j + 1)^3$$

$$= \frac{(j + 1)^2}{4}[j^2 + 4(j + 1)]$$

$$= \frac{(j + 1)^2}{4}[(j + 1) + 1]^2$$

Therefore, the formula works.

5. The formula is true for $n = 1$. Assume $S_{2n-1} = n^2$. Then $S_{2(n+1)-1} = n^2 + 2(n + 1) - 1 = (n + 1)^2$.

6. The formula is true for $n = 1$. Assume $S_n = n/(n + 1)$. Then $S_{n+1} = n/(n + 1) + 1/[(n + 1)(n + 2)] = (n + 1)/(n + 2)$.

7. 1 is odd, and 1^2 is odd. Assume n is odd and n^2 is odd. The next *odd* number after n is $n + 2$. $(n + 2)^2 = n^2 + 4n + 4 = n^2 + 2(2n + 2)$. Since this is the sum of an odd and an even number, it follows that $(n + 2)^2$ is odd.

8. 2 is even and 2^2 is even. Assume n is even and n^2 is even. Then $(n + 2)^2 = n^2 + 4n + 4 = n^2 + 2(2n + 2)$. Since this is the sum of two even numbers, it follows that $(n + 2)^2$ is even.

Chapter 17

1. 2
2. 5
3. 32
4. 1.1761
5. 1.3979
6. 3.6990
7. $-2 + 0.8129 = -1.1871$
8. 2.8751
9. 10.9777
10. 3.9294
12. $\log_2 32 = 5$
13. $\log_3 243 = 5$
14. $\log_3 81 = 4$
15. $\log_{16} 4 = 1/2$
16. $\log_2 1.41 = 0.5$ (approximately)
17. 8
18. 10
19. 4

20. 5
21. 4
22. 6
23. 6
24. −4
25. −2
26. −4
27. 4,094
28. Since $1^a = 1$ for all values of a, you cannot find a base 1 logarithm for any number other than 1. Negative bases would create problems because when a negative number is raised to a fractional power the result might not be a real number.
29. $\frac{1}{2} \log x$
30. $\frac{1}{2} \log(x + y)$
31. $n \log x$
32. $\log (x - 4) + \log (x + 5)$
33. $\log (a + b) - \log (c + d)$
34. $\log (x - 5) - \log (x - 29)$
35. $\log (x + 15) + \log (2x + 3) - \log (x + 16)$
36. $3 \log (x + 4) + \log (a + b) - 2 \log (x - 1) - \log (a^2 + b^2)$
37. $a \log x + b \log y$
38. From the change of base formula, we know that $\ln x = \log x/0.4343$.
Using the results from the table in the chapter for $\log x$, we can calculate:

x	$\ln x$
1	0
2	0.693
3	1.099
4	1.386
5	1.609
6	1.792
7	1.946
8	2.079
9	2.197
10	2.303

39. If L is the relative loudness, then the decibel measure (d) can be found from the formula $D = 10 \log L$. If L is 1,000, then D is 30.
40. 70
41. 3.01
42. 4.77
43. 6.99
44. 26.99
45. Use the formula $L = 10^{D/10}$. If D is 90, then $L = 1 \times 10^9$.
46. 1×10^{10}
47. 1.585
48. 3.162
49. 5×10^6
50. 3.16×10^6
51. 3.16×10^3
52. The formula is $m = 6 - 2.5 \log b$, where m is the magnitude and b is the relative brightness. If $b = 1/100$, then $m = 11$.
53. 16
54. 3.5
55. 4.25
56. 1.75

57. Use the formula $b = 10^{2/5(6 - m)}$, which gives 6.3×10^{12}
58. 1.6×10^{7}
59. 14,500
60. 3,300
61. 2,500
62. 360
63. 2.8×10^{-4}
64. 910
65. 250
66. 40
67. 31
68. 10
69. The relative strength of an earthquake is 10^{m-5}, where m is the magnitude.

A magnitude 8 earthquake is 1,000 times stronger than magnitude 5.

70. 10,000 times stronger
71. 3.16 times stronger
72. 199.5 times stronger
73. 63.1 times stronger
74. To prove the change of base formula, start with the expression

$$x = b^{\log_b x}$$

Also,

$$b = a^{\log_a b}$$

Therefore,

$$x = [a^{\log_a b}]^{\log_b x}$$

$$x = a^{\log_a b \, \log_b x}$$

$$\log_a x = \log_a b \, \log_b x$$

$$\log_b x = \frac{\log_a x}{\log_a b}$$

75. Suppose you have the equation $x = a + b$. You can find b by subtraction: $b = x - a$. Because addition is commutative, you can also find a by subtraction: $a = x - b$. However, exponentiation is not commutative, so you need two inverse operations for exponentiation. If you have the equation $x = a^b$, then you can find a by taking a root: $a = \sqrt[b]{x}$, but you must find b by taking a logarithm: $b = \log_a x$.

76.

```
1 REM  THIS PROGRAM CALCULATES AN APPROXIMATION FOR THE
2 REM  NATURAL LOGARITHM
10 INPUT K
11 FS = 1
15 IF K>2 THEN K=1/K : FS = -1
20 X = K - 1
30 Z = 0
40 S = 1
70 FOR I = 1 TO 100
```

```
80    Z = Z + S*X^I/I
90    S = -S
100 NEXT I
110 PRINT FS*Z
120 END
```

Chapter 18

1. $a = 10, b = 5, c = 6$
2. $a = 7, b = 12, c = 6$
3. $a = 52, b = 62, c = 36$
4. $a = 6, b = 4, c = 3$
5. $x = 1.9225, y = 0.4455, z = -0.7651$
6. $x = 11, y = 2, z = 6$
7. $a = 0.7906, b = 80.8837, c = 18.3255$
8. $Y = 150, C = 85, T = 57.5, I = 35$
9. $Y = [C_0 - bT_0 + I_0 + G]/[1 - b(1 - t) - d]$
10. Let's let $ax^2 + bx + c$ represent the polynomial we're looking for. We need to find the values of a, b, and c. We have three equations, since we know the polynomial must pass through the three points that are given:

$$a \times 1^2 + 1 \times b + c = 7$$

$$a \times 2^2 + 2b + c = 8$$

$$a \times 3^2 + 3b + c = 13$$

We can solve this system to find that $a = 2, b = -5, c = 10$.

11. $x^2 + 4x - 5$
12. $3x^2 + (1/2)x + 4\ 1/2$
13. $2x^2 - x - 1$
14. $5x^2 + 7x - 11$
15. NaCl (salt)
16. H_2O (water)
17. NH_3 (ammonia)
18. CH_4
19. SO_2
20. $KClO_3$
21. $K_2Cr_2O_7$
22. KOH
23. K_2SO_4
24. C_2H_6O
25. $C_3H_8O_3$
26. C_3H_6O
27. $x = 5\sqrt{2}/2, \quad y = 5\sqrt{2}/2$ or $x = -5\sqrt{2}/2, y = -5\sqrt{2}/2$
28. $x = 4, \quad y = 0$ or $x = -12/5, y = -16/5$
29. $x = -0.441, y = 3.051$, or $x = 0.566, \quad y = 1.793$
30. center $(-1, -1)$, radius 25
31. center $(12, 15)$, radius 13
32. center $(13\frac{1}{8}, 13\frac{1}{8})$, radius 13.63
33. center $(-27.346, -11.5769)$, radius 40.27

Chapter 19

1. $\begin{pmatrix} 19 & 22 \\ 43 & 50 \end{pmatrix}$

2. $\begin{pmatrix} 70 & 82 \\ 45 & 51 \end{pmatrix}$

3. $\begin{pmatrix} 132 & 92 \\ 92 & 62 \end{pmatrix}$

4. $\begin{pmatrix} 70 & 42 & 77 \\ 80 & 76 & 112 \\ 70 & 119 & 231 \end{pmatrix}$

5. $\begin{pmatrix} 0 & 2 & 0 \\ 1 & 0 & 1 \\ 0 & 2 & 0 \end{pmatrix}$

6. $\begin{pmatrix} 30 & 33 & 36 \\ 36 & 39 & 33 \\ 36 & 30 & 30 \end{pmatrix}$

7. They cannot be multiplied.

8. $\begin{pmatrix} 19 \\ 7 \end{pmatrix}$

9. $\begin{pmatrix} 27 & 35 & 15 \\ 69 & 25 & 19 \end{pmatrix}$

10. They cannot be multiplied.

11. They cannot be multiplied.

12. $\begin{pmatrix} 11 \\ 32 \\ 27 \end{pmatrix}$

13. $\begin{pmatrix} 0 & 0 & 0 \\ 1 & 2 & 3 \\ 11 & 12 & 13 \end{pmatrix}$

14. $\begin{pmatrix} 0 & 0 & 0 \\ 0 & 0 & 0 \\ 1 & 2 & 3 \end{pmatrix}$

15. $\begin{pmatrix} 0 & 0 & 0 \\ 0 & 0 & 0 \\ 0 & 0 & 0 \end{pmatrix}$

16. $\begin{pmatrix} 0 & 0 & 0 \\ a & b & c \\ d & e & f \end{pmatrix}$

17. $\begin{pmatrix} 0 & 0 & 0 \\ 0 & 0 & 0 \\ 1 & 0 & 0 \end{pmatrix}$

18. $\begin{pmatrix} a_1c_1 + b_1c_2 & a_1d_1 + b_1d_2 \\ a_2c_1 + b_2c_2 & a_2d_1 + b_2d_2 \end{pmatrix}$

19. $\begin{pmatrix} 1 + \frac{b}{c} & \frac{a}{b} + \frac{b}{d} \\ \frac{c}{a} + \frac{d}{c} & \frac{c}{b} + 1 \end{pmatrix}$

20.

$$\begin{matrix} \sum_{k=1}^{n} a_{1k}b_{k1} & \sum_{k=1}^{n} a_{1k}b_{k2} & \sum_{k=1}^{n} a_{1k}b_{k3} & \cdots & \sum_{k=1}^{n} a_{1k}b_{kp} \\ \sum_{k=1}^{n} a_{2k}b_{k1} & \sum_{k=1}^{n} a_{2k}b_{k2} & \sum_{k=1}^{n} a_{2k}b_{k3} & \cdots & \sum_{k=1}^{n} a_{2k}b_{kp} \\ \vdots & & & & \vdots \\ \sum_{k=1}^{n} a_{mk}b_{k1} & \sum_{k=1}^{n} a_{mk}b_{k2} & \sum_{k=1}^{n} a_{mk}b_{k3} & \cdots & \sum_{k=1}^{n} a_{mk}b_{kp} \end{matrix}$$

21. $x = 6, y = 10, z = 4$

22. $x = 3, y = 2, z = 1$

23. $x = .5064, y = .8498, z = 1.6309$

24. $x = .9032, y = 3.2903, z = 1.3548$

25. $x = 2.1722, y = 0.9237, z = -0.0422$

26. $x = -4.8, y \doteq 31.2, z = 54.6$

27. $x = 7, y = 9, z = 6$

28. $x = 0.6033, y = 0.1135, z = 0.0223$

29. 0

30. −18

31. −335

32. 45

33. 0

34. 0

35. −2

36. 2

37. 0

38. −39

39. 18

40. 52

41. 660

42. 262

43. −18

44. 0

45. $\begin{vmatrix} a & b \\ c & d \end{vmatrix} = ad - bc$

$\begin{vmatrix} b & a \\ d & c \end{vmatrix} = bc - ad$

46. $\begin{vmatrix} a & a \\ b & b \end{vmatrix} = ab - ab = 0$

47. $\begin{vmatrix} a & b \\ c & d \end{vmatrix} = ad - bc$

$$\begin{vmatrix} (a + kb) & b \\ (c + kd) & d \end{vmatrix} = (a + kb)d - (c + kd)b$$
$$= ad + kbd - cb - kbd = ad - bc$$

48. $\begin{vmatrix} ka & b \\ kc & d \end{vmatrix} = kad - kbc = k(ad - bc)$

49. Let $A = \begin{pmatrix} a & b \\ c & d \end{pmatrix}$ $\quad$ $\det(A) = ad - bc$

Let $B = \begin{pmatrix} e & f \\ g & h \end{pmatrix}$ $\quad$ $\det(B) = eh - gf$

$$\det(A)\det(B) = (ad - bc)(eh - gf)$$
$$= adeh - adfg - bceh + bcgf$$

$$AB = \begin{pmatrix} ae + bg & af + bh \\ ce + dg & cf + dh \end{pmatrix}$$

$$\det(AB) = (ae + bg)(cf + dh) - (ce + dg)(af + bh)$$
$$= acef + adeh + bcfg + bdgh - acef - bceh$$
$$- adfg - bdgh$$
$$= adeh + bcfg - bceh - adfg$$

50. $\begin{pmatrix} 1 & 0 & 0 \\ 0 & 1 & 0 \\ 0 & 0 & 1 \end{pmatrix} \begin{pmatrix} a & b & c \\ d & e & f \\ g & h & i \end{pmatrix} = \begin{pmatrix} 1 \cdot a + 0 \cdot d + 0 \cdot g & b & c \\ 0 \cdot a + 1 \cdot d + 0 \cdot g & e & f \\ 0 \cdot a + 0 \cdot d + 1 \cdot g & h & i \end{pmatrix}$

(Second and third columns calculated in same way as first.)

51.
```
1 REM  THIS PROGRAM SOLVES A SYSTEM OF N LINEAR EQUATIONS
2 REM  IN N UNKNOWNS, USING MATRIX TRIANGULARIZATION
100  REM READ IN COEFFICIENTS
110  INPUT "Number of Equations:";N
120  DIM M(N,(N+1)),X(N)
130  FOR I=1 TO N
140    PRINT "Enter coefficients for equation ";I
150    FOR J=1 TO N
155       PRINT J;": ";:INPUT M(I,J)
160    NEXT J
170    INPUT "Right hand side constant:";M(I,(N+1))
180  NEXT I
197 '
198 '
199 '
200 REM  BEGIN CALCULATIONS
210 FOR B= 1 TO (N-1)
220   FOR A=N TO (B+1) STEP -1
230     C=-M(A,B)/M(B,B)
240     FOR J=1 TO (N+1)
250        M(A,J)=M(A,J)+C*M(B,J)
260     NEXT J
270   NEXT A
280 NEXT B
299 '
```

```
300 X(N)=M(N,N+1)/M(N,N)
310 FOR K=(N-1) TO 1 STEP -1
320   T = 0
330   FOR J=(K+1) TO N
340       T=T+X(J)*M(K,J)
350   NEXT J
360   X(K)=(M(K,N+1)-T)/M(K,K)
370 NEXT K
397 '
398 '
399 '
400 REM OUTPUT
410 FOR I= 1 TO N
420   PRINT "X";I;"=";X(I)
430 NEXT I
440 END
449 '
450 REM -- NOTE: THIS PROGRAM WILL NOT WORK IF THERE ARE ZEROS
455 REM    IN THE COEFFICIENT MATRIX AT PLACES THAT REQUIRE
460 REM    DIVISION BY ZERO.  HOWEVER, IF THE COEFFICIENT MATRIX
465 REM    CONTAINS ZEROS THEN THE OTHER SOLUTION METHODS, SUCH
470 REM    AS THE SUBSTITUTION METHOD OR CRAMER'S RULE, BECOME EASIER
```

52.

```
1 REM  THIS PROGRAM CALCULATES THE INVERSE OF A
2 REM  MATRIX   - FOR THIS EXAMPLE, N = 3
10 N = 3
20 DIM A(6,3)
30 FOR I = 1 TO N
40   PRINT "ENTER VALUES FOR ROW ";I
50   FOR J = 1 TO N
60     INPUT A(I,J)
70   NEXT J
80 NEXT I
90 REM  PUT IDENTITY MATRIX IN BOTTOM HALF OF A
100 FOR I = N+1 TO 2*N
110   FOR J = 1 TO N
120     A(I,J) = 0
130   NEXT J
140   A(I,(I-N)) = 1
150 NEXT I
200 REM  CALCULATE INVERSE
210 FOR J = 1 TO N
220   Z5 = A(J,J)
230   FOR I = 1 TO 2*N
240     A(I,J) = A(I,J)/Z5
250   NEXT I
255 '
260   FOR K = 1 TO N
270     Z6 = A(J,K)
280     IF K = J THEN GOTO 320
290     FOR I = 1 TO 2*N
300       A(I,K) = A(I,K) - A(I,J)*Z6
310     NEXT I
320   NEXT K
330 NEXT J
335 '
340 PRINT "THE INVERSE IS:"
350 FOR I = N+1 TO 2*N
```

```
360    FOR J = 1 TO N
370       PRINT USING "#####.########";A(I,J);
380    NEXT J
390    PRINT
400 NEXT I
```

53.

```
1 REM  THIS PROGRAM CALCULATES THE DETERMINANT OF A
2 REM  4 BY 4 MATRIX
10 DIM A4(4,4), A3(3,3), A2(2,2)
20 FOR I4 = 1 TO 4 : FOR J4 = 1 TO 4 : INPUT A4(I4,J4)
30    NEXT J4 : NEXT I4
40 GOSUB 1000
50 PRINT "DETERMINANT IS:";Z4
60 END
999 '
1000  REM  CALCULATE DETERMINANT OF 4 BY 4 MATRIX (A4)
1005  Z4 = 0
1006  S4 = 1
1010  FOR I4 = 1 TO 4
1020    H4 = 0
1030    FOR K4 = 1 TO 4
1040      IF K4 = I4 THEN GOTO 1090
1050      H4 = H4 + 1
1060      FOR M4 = 1 TO 3
1070        A3(M4,H4) = A4(M4+1,K4)
1080      NEXT M4
1090    NEXT K4
1100    GOSUB 2000   : REM CALCULATE DETERMINANT OF A3
1110    Z4 = Z4 + S4*A4(1,I4)*Z3
1120    S4 = -S4
1130  NEXT I4
1140  RETURN   : REM Z4 IS THE RESULT
1999  '
2000  REM  CALCULATE DETERMINANT OF 3 BY 3 MATRIX (A3)
2005  Z3 = 0
2006  S3 = 1
2010  FOR I3 = 1 TO 3
2020    H3 = 0
2030    FOR K3 = 1 TO 3
2040      IF K3 = I3 THEN GOTO 2090
2050      H3 = H3 + 1
2060      FOR M3 = 1 TO 2
2070        A2(M3,H3) = A3(M3+1,K3)
2080      NEXT M3
2090    NEXT K3
2100    GOSUB 3000   : REM CALCULATE DETERMINANT OF A2
2110    Z3 = Z3 + S3*A3(1,I3)*Z2
2120    S3 = -S3
2130  NEXT I3
2140  RETURN   : REM Z3 IS THE RESULT
2999  '
3000  REM  CALCULATE DETERMINANT OF 2 BY 2 MATRIX
3010  Z2 = A2(1,1)*A2(2,2) - A2(1,2)*A2(2,1)
3020  RETURN
```

Note that the structure of the subroutine to calculate the determinant of the 3 by 3 matrix is almost the same as the structure of subroutine for the 4 by 4 matrix, so you may be wondering if it is possible to write the program without duplicating that structure.

```
PROGRAM cramer (INPUT,OUTPUT) ;
{This program uses Cramer's rule to solve a system}
{of linear equations by calculating determinants.}
{This is Turbo Pascal}
CONST maxsize=5;
TYPE matrixtype=ARRAY[1..maxsize,1..maxsize] OF REAL;
        {matrixtype is the type for the matrix of
        coefficients}
    rhstype=ARRAY[1..maxsize] OF REAL;
        {rhstype is the type for the matrix of right
        hand side constants}
VAR a:matrixtype; c:rhstype; n,i,j:INTEGER; denominator:
REAL;
FUNCTION determ(n:INTEGER;  a:matrixtype):REAL;
  {this function calculates the determinant of the}
  {n by n matrix a}
  VAR z:REAL;  j,j2,j3,i,signindicator:INTEGER;
  b:matrixtype;
 BEGIN
  IF n=2 THEN determ := a[1,1]*a[2,2]-a[1,2]*a[2,1]
  ELSE BEGIN
       z := 0;  signindicator := 1;
       FOR j := 1 TO n DO
         BEGIN
           j2 := 0;
           FOR j3:=1 TO n DO
             IF j3<>j THEN BEGIN
                        j2 := j2+1;
                        FOR i := 2 TO n DO b[i-1,j2]
                        := a[i,j3]
                     END;
           z := z+signindicator*a[1,j]*determ(n-1,b);
                {this function uses recursion here
                because}
                {the function determ calls itself}
           signindicator := -signindicator
          END;  {for j:=1 to n loop}
        determ := z
      END {ELSE}
END; {end of function determ}

FUNCTION numerator(j,n:INTEGER;a:matrixtype;c:rhstype):
REAL;
  {This function calculates the numerator in the}
  {Cramer's rule expression for variable j}
  VAR a2:matrixtype;  i:INTEGER;
  BEGIN
   a2 := a;
   FOR i := 1 TO n DO a2[i,j] := c[i];
   numerator := determ(n,a2)
  END;

BEGIN {main program block}
  WRITELN('Cramers rule for solving equation systems');
  WRITELN('the variables will be designated
  x1,x2,x3,...xn');
```

```
WRITE('Enter number of equations:'); READLN(n);
WRITELN('Enter coefficients:');
FOR i := 1 TO n DO
  BEGIN
    WRITE('row',i,': ');
    FOR j := 1 TO n DO
      BEGIN WRITE('  '); READ(a[i,j]) END; WRITELN
  END;
WRITELN('Enter right hand side constants:');
FOR i := 1 TO n DO
  BEGIN WRITE(i:4,':'); READLN(c[i]) END;
denominator := determ(n,a);
IF denominator=0 THEN WRITELN('determinant is zero—
no solution') ELSE BEGIN
     FOR j := 1 TO n DO
        BEGIN
          WRITE('x',j:2,':');
          WRITELN((numerator(j,n,a,c)/denominator):
          12:4)
          END
     END
END.
```

Chapter 20

1–6.

imaginary axis

real axis

• $12 + 16i$

• $2 + 2i$

• $2 - 2i$

• $2 - 5i$

$-6i$

• $12 - 8i$

7. $66 + 7i$
8. $31 - 1.08i$
9. $419 + 221i$
10. $-5 + 12i$
11. $-19 + 116i$
12. $99 + 132i$

13. $16 + (1.95)^2 = 19.8025$
14. $41 + 129i$
15. $556 + 321i$
16. $i^8 = 1$
Let n be an integer. Then

$$i^{4n} = 1$$
$$i^{4n+1} = i$$
$$i^{4n+2} = -1$$
$$i^{4n+3} = -i$$

17. i
18. -1
19. 1
20. i
21. -1
22. $-i$
23. $-i$
24. 1
25. 1
26. To find $\sqrt{i}$, we need to find a complex number $(a + bi)$ such that $(a + bi)^2 = i$.

$$a^2 + 2abi - b^2 = i$$

Equate the real parts of both sides, and then the imaginary parts:

$$a^2 - b^2 = 0$$
$$2ab = 1$$
$$a = \frac{1}{\sqrt{2}}, \quad b = \frac{1}{\sqrt{2}}$$

Therefore,

$$\sqrt{i} = \frac{1}{\sqrt{2}} + \frac{1}{\sqrt{2}}i$$

27. $\frac{1}{i} = -i$
28. $2i$
29. $-5 + 12i$
30. $-13 + 84i$
31. $8 - 6i$
32. $22 - 10\sqrt{3}\,i$
33. $(a^2 - b^2) + 2abi$
34. $a^3 + 3a^2 bi - 3ab^2 - b^3 i$
35. $a^4 + 4a^3 bi - 6a^2b^2 - 4ab^3 i + b^4$
36. $a^5 + 5a^4 bi - 10a^3b^2 - 10a^2b^3i + 5ab^4 + b^5i$
37. $a^n + na^{n-1}bi - \binom{n}{2} a^{n-2}b^2 - \binom{n}{3} a^{n-3}b^3i + \cdots$
38. $x = -1/2 \pm i\sqrt{15}/2$
39. $x = 1/2 \pm i\sqrt{31}/2$
40. $x = -2 \pm i$
41. $x = -1.5 \pm i\sqrt{7}/2$
42. $x = -4 \pm 3i$
43. $x = -9.6 \pm 2.8i$
44. $x = -10 \pm 24i$
45. $x = -0.0833 \pm 1.6309i$
46. $x = -0.39 \pm 0.72i$

47. $z^* = 1 - i,$ $zz^* = 2$
48. $z^* = -16i$ $zz^* = 256$
49. $z^* = 32$ $zz^* = 32^2 = 1024$
50. $z^* = 10 + 5i$ $zz^* = 125$
51. $z^* = 16 - 11i$ $zz^* = 16^2 + 11^2 = 377$
52. $z^* = 33 - 2i$ $zz^* = 33^2 + 4 = 1093$
53. $zz^* = 101$
54. $zz^* = |z|^2$
55. A real number.
56. A pure imaginary number.
57. $(1/2)(1 - i)$
58. $15/26 - (10/26)i$
59. $6/5 + (9/10)i$
60. $6/13 + (21/26)i$
61. $[(ac + bd) + (cb - ad)i]/(c^2 + d^2)$
62. 3
63. $(50 - 75i)/25$
64. $1 - i$
65. $6 - 3i$
66. $8.4 + 2.8i$
67. $z = 6 - i$
68. $\sqrt{R^2 + L^2\omega^2}$
69. $\sqrt{R^2 + 1/(\omega^2 C^2)}$
70. $[\omega L - 1/\omega C]$

Glossary

abscissa Abscissa means X-coordinate.
absolute value The absolute value of a real number a is:

$$|a| = a \quad \text{if } a \geq 0; \qquad |a| = -a \quad \text{if } a < 0$$

The absolute value of a complex number $a + bi$ is $\sqrt{a^2 + b^2}$.
argument The argument of a function is the independent variable that is put into the function. See Chapter 8.
arithmetic series An arithmetic series is a sum of terms of the form:

$$a + (a + d) + (a + 2d) + (a + 3d) + \cdots + [a + (n - 1)d]$$

See Chapter 14.
asymptote An asymptote is a straight line that is a close approximation to a particular curve as the curve goes off to infinity in one direction. The curve comes very close to the asymptote line, but it never touches it.
axiom An axiom is a statement that is assumed to be true without proof.
axis (1) The x-axis in Cartesian coordinates is the line $y = 0$. The y axis is the line $x = 0$. (2) The axis of a figure is a line about which the figure is symmetrical.

base In the equation $x = b^y$, b is the base.
binomial A binomial is the sum of two terms.

cartesian coordinates A Cartesian coordinate system is a system in which each point on a plane is identified by an ordered pair of numbers representing its distances from two perpendicular lines, or in which each point in space is similarly identified by its distances from three perpendicular planes. See Chapter 9.

circle A circle is a set of points in a plane that are all the same distance from a given point.

circumference The circumference of a closed curve (such as a circle) is the total distance around the outer edge of the curve.

closure property An arithmetic operation obeys the closure property with respect to a given set of numbers if the result of the operation will always be in that set if the operands (the input numbers) are.

coefficient Coefficient is a technical term for something that multiplies something else, usually a fixed number multiplying a variable.

common logarithm A common logarithm is a logarithm to the base 10.

commutative property An operation has the commutative property if the order of the two quantities involved doesn't matter.

complex number A complex number is formed by adding a pure imaginary number to a real number. The general form of a complex number is $a + bi$, where a and b are real numbers and $i = \sqrt{-1}$. See Chapter 20.

conic sections The four curves circle, ellipse, parabola, and hyperbola are called conic sections because they can be formed by the intersection of a plane with a right circular cone.

conjugate The conjugate of a complex number is formed by reversing the sign of the imaginary part.

coordinates The coordinates of a point are a set of numbers that identify the location of that point. See Chapter 9.

counting numbers The counting numbers are the same as the natural numbers: 1, 2, 3, 4, 5, 6, 7, ... They're the numbers you use to count something.

Cramer's rule Cramer's rule is a method for solving a set of simultaneous linear equations using determinants. See Chapter 19.

denominator The denominator is the bottom part of a fraction.

dependent variable The dependent variable stands for any of the set of output numbers of a function. In the equation $y = f(x)$, y is the dependent variable and x is the independent variable.

determinant The determinant of a matrix is a number that determines some important properties of the matrix. See Chapter 19.

diameter The diameter of a circle is the length of a line segment joining two points on the circle and passing through the center.

difference The difference between two numbers is the result of subtracting one from the other.

discriminant. See **quadratic formula**.

distributive property The distributive property says that $a(b + c) = ab + ac$ for all a, b, and c.

dividend In the division problem $a/b = c$, a is called the dividend.

divisor In the division problem $a/b = c$, b is called the divisor.

eccentricity The eccentricity of a conic section is a number that indicates the shape of the conic section. See Chapter 12.

ellipse An ellipse is the set of all points in a plane such that the sum of the distances to two fixed points is a constant.

equation An equation is a statement that says two mathematical expressions have the same value.

even number An even number is a number that is divisible evenly by 2, such as 2, 4, 6, 8, 10, ...

exponent An exponent is a number that indicates the operation of repeated multiplication. See Chapter 5.

factor A factor is one of two or more expressions that can be multiplied together to get a given expression; they are said to be factors of that expression.

factorial The factorial of a positive integer is the product of all of the integers from 1 up to that number. The exclamation point ! is used to designate the factorial. For example, $4! = 4 \times 3 \times 2 \times 1 = 24$.

factoring Factoring is the process of splitting an expression into the product of two or more simpler expressions, called factors.

fraction A fraction is a number of the form a/b, which is defined by the equation $(a/b) \times b = a$. The fractional number a/b is the same as the answer to the division problem $a \div b$. The top of the fraction (in this case a) is called the numerator, and the bottom of the fraction (b) is called the denominator.

function A function is a rule that turns one number into another number. See Chapter 8.

geometric series A geometric series is a sum of terms of the form

$$a + ar + ar^2 + ar^3 + ar^4 + \cdots + ar^{n-1}$$

See Chapter 14.

graph The graph of an equation is the set of points (with values given by coordinates) that make the equation true. See Chapter 9.

hyperbola A hyperbola is the set of points in a plane such that the difference between the distances to two fixed points is a constant.

hypotenuse The hypotenuse is the side of a right triangle that is opposite the right angle.

i i is the basic unit for imaginary numbers. i is defined by the equation $i^2 = -1$.

identity An identity is an equation that is true for all possible values of the unknowns it contains.

imaginary number An imaginary number (or a pure imaginary number) is a number of the form bi, where b is a real number and $i = \sqrt{-1}$. See Chapter 20.

independent variable The independent variable stands for any of the set of input numbers to a function.

inequality An inequality is a statement of the form "a is greater than b," written $a > b$, or "a is less than b," written $a < b$.

integers The set of integers contains zero, the natural numbers, and the negatives of the natural numbers:

$$\ldots -6, -5, -4, -3, -2, -1, 0, 1, 2, 3, 4, 5, 6, \ldots$$

inverse function The inverse function of a function is the function that does exactly the opposite of the original function.

irrational number An irrational number is a real number that cannot be expressed as the ratio of two integers.

logarithm The equation $y = a^x$ can be written $x = \log_a y$, which means "x is the logarithm of y to the base a."

major axis The major axis of an ellipse is the line segment joining two points on the ellipse that passes through the two focus points. See Chapter 12.

matrix A matrix is a table of numbers arranged in rows and columns. See Chapter 19.

monomial A monomial is an algebraic expression that does not involve any additions or subtractions.

multinomial A multinomial is the sum of two or more monomials.

natural numbers The natural numbers are the numbers 1, 2, 3, 4, 5, 6, ...

negative number A negative number is a number that is less than zero. See Chapter 3.

numerator The numerator is the top part of a fraction.

odd number An odd number is a natural number that is not divisible by 2, such as 1, 3, 5, 7, 9, ...

ordinate The ordinate of a point is another name for y-coordinate.

origin The origin is the point (0, 0) of a cartesian coordinate system.

parabola A parabola is the set of all points that are equally distant from a fixed point (called the focus) and a fixed line (called the directrix).

pi The Greek letter π (pi) is used to represent the ratio between the circumference of a circle and its diameter:

$$\pi = \frac{\text{(circumference of circle)}}{\text{(diameter)}}$$

This ratio is the same for any circle. π is an irrational number with the decimal approximation 3.14159.

plane A plane is a flat surface (like a table top) that stretches off to infinity in all directions.

polynomial A polynomial in x is an algebraic expression of the form

$$a_nx^n + a_{n-1}x^{n-1} + \cdots + a_3x^3 + a_2x^2 + a_1x + a_0$$

where $a_0, a_1, \ldots, a_n$ are constants that are the coefficients of the polynomial. The degree of the polynomial is the highest power of the variable that appears (in this case n).

positive number A positive number is any number greater than zero.

postulate A postulate is a fundamental statement that is assumed to be true without proof.

power A power of a number indicates repeated multiplication. For example, the third power of 5 is $5^3 = 5 \times 5 \times 5 = 125$.

Pythagorean theorem The Pythagorean theorem relates the three sides of a right triangle: $c^2 = a^2 + b^2$, where c is the side opposite the right angle (called the hypotenuse) and a and b are the sides adjacent to the right angle.

quadratic equation A quadratic equation in x is an equation of the form

$$ax^2 + bx + c = 0$$

quadratic formula The quadratic formula states that the solutions to the equation $ax^2 + bx + c = 0$ are

$$x = \frac{-b \pm \sqrt{b^2 - 4ac}}{2a}$$

The quantity $b^2 - 4ac$ is called the **discriminant.**

radical The radical symbol $\sqrt{\ }$ is used to indicate a root of a number.
radius The radius of a circle is the distance from the center of the circle to a point on the circle.
ratio The ratio of two numbers a and b is a/b.
rational number A rational number is any number that can be expressed as the ratio of two integers.
real numbers The set of real numbers is the set of all numbers that can be represented by a point on a number line. The set of real numbers includes all rational numbers and all irrational numbers.
reciprocal The reciprocal of a number a is $1/a$.
root The process of taking a root of a number is the opposite of raising that number to a power. See Chapter 6.

scientific notation Scientific notation is a short way of writing very large or very small numbers. A number in scientific notation is expressed as a number between 1 and 10 multiplied by a power of 10. See Chapter 5.
semimajor axis The semimajor axis of an ellipse is equal to one-half of the longest distance across the ellipse (the major axis.)
semiminor axis The semiminor axis of an ellipse is equal to one-half the shortest distance across the ellipse.
simultaneous equations A system of simultaneous equations is a group of equations that must all be true at the same time.
slope The slope of a line is a number that measures how steep the line is. A horizontal line has a slope of zero. A vertical line has an infinite slope. See Chapter 9.
solution The solution of an equation is the value(s) of the variable(s) contained in that equation that make(s) the equation true.
square The square of a number is found by multiplying that number by itself.
square root The square root of a number a (written $\sqrt{a}$) is a number that, when multiplied by itself, gives a.
substitution property The substitution property states that, if $a = b$, then you can replace the expression a anywhere it appears by the expression b if you want to.
sum The sum is the result when two or more numbers are added.
synthetic division Synthetic division is a short way of dividing a polynomial by a binomial of the form $x - b$. See Chapter 13.

term A term is a part of a sum. For example, in the expression $ax^2 + bx + c$, ax^2 is the first term, bx is the second term, and c is the third term.
translation A translation occurs when we shift the axes of a cartesian coordinate system without rotation. See Chapter 12.

Common Logarithm Table

The table gives log (a + b)

a b:	.00	.01	.02	.03	.04	.05	.06	.07	.08	.09
1.0	.0000	.0043	.0086	.0128	.0170	.0212	.0253	.0294	.0334	.0374
1.1	.0414	.0453	.0492	.0531	.0569	.0607	.0645	.0682	.0719	.0755
1.2	.0792	.0828	.0864	.0899	.0934	.0969	.1004	.1038	.1072	.1106
1.3	.1139	.1173	.1206	.1239	.1271	.1303	.1335	.1367	.1399	.1430
1.4	.1461	.1492	.1523	.1553	.1584	.1614	.1644	.1673	.1703	.1732
1.5	.1761	.1790	.1818	.1847	.1875	.1903	.1931	.1959	.1987	.2014
1.6	.2041	.2068	.2095	.2122	.2148	.2175	.2201	.2227	.2253	.2279
1.7	.2304	.2330	.2355	.2380	.2405	.2430	.2455	.2480	.2504	.2529
1.8	.2553	.2577	.2601	.2625	.2648	.2672	.2695	.2718	.2742	.2765
1.9	.2788	.2810	.2833	.2856	.2878	.2900	.2923	.2945	.2967	.2989
2.0	.3010	.3032	.3054	.3075	.3096	.3118	.3139	.3160	.3181	.3201
2.1	.3222	.3243	.3263	.3284	.3304	.3324	.3345	.3365	.3385	.3404
2.2	.3424	.3444	.3464	.3483	.3502	.3522	.3541	.3560	.3579	.3598
2.3	.3617	.3636	.3655	.3674	.3692	.3711	.3729	.3747	.3766	.3784
2.4	.3802	.3820	.3838	.3856	.3874	.3892	.3909	.3927	.3945	.3962
2.5	.3979	.3997	.4014	.4031	.4048	.4065	.4082	.4099	.4116	.4133
2.6	.4150	.4166	.4183	.4200	.4216	.4232	.4249	.4265	.4281	.4298
2.7	.4314	.4330	.4346	.4362	.4378	.4393	.4409	.4425	.4440	.4456
2.8	.4472	.4487	.4502	.4518	.4533	.4548	.4564	.4579	.4594	.4609
2.9	.4624	.4639	.4654	.4669	.4683	.4698	.4713	.4728	.4742	.4757
3.0	.4771	.4786	.4800	.4814	.4829	.4843	.4857	.4871	.4886	.4900
3.1	.4914	.4928	.4942	.4955	.4969	.4983	.4997	.5011	.5024	.5038
3.2	.5052	.5065	.5079	.5092	.5105	.5119	.5132	.5145	.5159	.5172
3.3	.5185	.5198	.5211	.5224	.5237	.5250	.5263	.5276	.5289	.5302
3.4	.5315	.5328	.5340	.5353	.5366	.5378	.5391	.5403	.5416	.5428
3.5	.5441	.5453	.5465	.5478	.5490	.5502	.5515	.5527	.5539	.5551
3.6	.5563	.5575	.5587	.5599	.5611	.5623	.5635	.5647	.5658	.5670
3.7	.5682	.5694	.5705	.5717	.5729	.5740	.5752	.5763	.5775	.5786
3.8	.5798	.5809	.5821	.5832	.5843	.5855	.5866	.5877	.5888	.5899
3.9	.5911	.5922	.5933	.5944	.5955	.5966	.5977	.5988	.5999	.6010
4.0	.6021	.6031	.6042	.6053	.6064	.6075	.6085	.6096	.6107	.6117
4.1	.6128	.6138	.6149	.6160	.6170	.6180	.6191	.6201	.6212	.6222
4.2	.6232	.6243	.6253	.6263	.6274	.6284	.6294	.6304	.6314	.6325
4.3	.6335	.6345	.6355	.6365	.6375	.6385	.6395	.6405	.6415	.6425
4.4	.6435	.6444	.6454	.6464	.6474	.6484	.6493	.6503	.6513	.6522
4.5	.6532	.6542	.6551	.6561	.6571	.6580	.6590	.6599	.6609	.6618
4.6	.6628	.6637	.6646	.6656	.6665	.6675	.6684	.6693	.6702	.6712
4.7	.6721	.6730	.6739	.6749	.6758	.6767	.6776	.6785	.6794	.6803
4.8	.6812	.6821	.6830	.6839	.6848	.6857	.6866	.6875	.6884	.6893
4.9	.6902	.6911	.6920	.6928	.6937	.6946	.6955	.6964	.6972	.6981
5.0	.6990	.6998	.7007	.7016	.7024	.7033	.7042	.7050	.7059	.7067
5.1	.7076	.7084	.7093	.7101	.7110	.7118	.7126	.7135	.7143	.7152
5.2	.7160	.7168	.7177	.7185	.7193	.7202	.7210	.7218	.7226	.7235
5.3	.7243	.7251	.7259	.7267	.7275	.7284	.7292	.7300	.7308	.7316
5.4	.7324	.7332	.7340	.7348	.7356	.7364	.7372	.7380	.7388	.7396
5.5	.7404	.7412	.7419	.7427	.7435	.7443	.7451	.7459	.7466	.7474
5.6	.7482	.7490	.7497	.7505	.7513	.7520	.7528	.7536	.7543	.7551
5.7	.7559	.7566	.7574	.7582	.7589	.7597	.7604	.7612	.7619	.7627
5.8	.7634	.7642	.7649	.7657	.7664	.7672	.7679	.7686	.7694	.7701
5.9	.7709	.7716	.7723	.7731	.7738	.7745	.7752	.7760	.7767	.7774

a b:	.00	.01	.02	.03	.04	.05	.06	.07	.08	.09
6.0	.7782	.7789	.7796	.7803	.7810	.7818	.7825	.7832	.7839	.7846
6.1	.7853	.7860	.7868	.7875	.7882	.7889	.7896	.7903	.7910	.7917
6.2	.7924	.7931	.7938	.7945	.7952	.7959	.7966	.7973	.7980	.7987
6.3	.7993	.8000	.8007	.8014	.8021	.8028	.8035	.8041	.8048	.8055
6.4	.8062	.8069	.8075	.8082	.8089	.8096	.8102	.8109	.8116	.8122
6.5	.8129	.8136	.8142	.8149	.8156	.8162	.8169	.8176	.8182	.8189
6.6	.8195	.8202	.8209	.8215	.8222	.8228	.8235	.8241	.8248	.8254
6.7	.8261	.8267	.8274	.8280	.8287	.8293	.8299	.8306	.8312	.8319
6.8	.8325	.8331	.8338	.8344	.8351	.8357	.8363	.8370	.8376	.8382
6.9	.8388	.8395	.8401	.8407	.8414	.8420	.8426	.8432	.8439	.8445
7.0	.8451	.8457	.8463	.8470	.8476	.8482	.8488	.8494	.8500	.8506
7.1	.8513	.8519	.8525	.8531	.8537	.8543	.8549	.8555	.8561	.8567
7.2	.8573	.8579	.8585	.8591	.8597	.8603	.8609	.8615	.8621	.8627
7.3	.8633	.8639	.8645	.8651	.8657	.8663	.8669	.8675	.8681	.8686
7.4	.8692	.8698	.8704	.8710	.8716	.8722	.8727	.8733	.8739	.8745
7.5	.8751	.8756	.8762	.8768	.8774	.8779	.8785	.8791	.8797	.8802
7.6	.8808	.8814	.8820	.8825	.8831	.8837	.8842	.8848	.8854	.8859
7.7	.8865	.8871	.8876	.8882	.8887	.8893	.8899	.8904	.8910	.8915
7.8	.8921	.8927	.8932	.8938	.8943	.8949	.8954	.8960	.8965	.8971
7.9	.8976	.8982	.8987	.8993	.8998	.9004	.9009	.9015	.9020	.9025
8.0	.9031	.9036	.9042	.9047	.9053	.9058	.9063	.9069	.9074	.9079
8.1	.9085	.9090	.9096	.9101	.9106	.9112	.9117	.9122	.9128	.9133
8.2	.9138	.9143	.9149	.9154	.9159	.9165	.9170	.9175	.9180	.9186
8.3	.9191	.9196	.9201	.9206	.9212	.9217	.9222	.9227	.9232	.9238
8.4	.9243	.9248	.9253	.9258	.9263	.9269	.9274	.9279	.9284	.9289
8.5	.9294	.9299	.9304	.9309	.9315	.9320	.9325	.9330	.9335	.9340
8.6	.9345	.9350	.9355	.9360	.9365	.9370	.9375	.9380	.9385	.9390
8.7	.9395	.9400	.9405	.9410	.9415	.9420	.9425	.9430	.9435	.9440
8.8	.9445	.9450	.9455	.9460	.9465	.9469	.9474	.9479	.9484	.9489
8.9	.9494	.9499	.9504	.9509	.9513	.9518	.9523	.9528	.9533	.9538
9.0	.9542	.9547	.9552	.9557	.9562	.9566	.9571	.9576	.9581	.9586
9.1	.9590	.9595	.9600	.9605	.9609	.9614	.9619	.9624	.9628	.9633
9.2	.9638	.9643	.9647	.9652	.9657	.9661	.9666	.9671	.9675	.9680
9.3	.9685	.9689	.9694	.9699	.9703	.9708	.9713	.9717	.9722	.9727
9.4	.9731	.9736	.9741	.9745	.9750	.9754	.9759	.9764	.9768	.9773
9.5	.9777	.9782	.9786	.9791	.9795	.9800	.9805	.9809	.9814	.9818
9.6	.9823	.9827	.9832	.9836	.9841	.9845	.9850	.9854	.9859	.9863
9.7	.9868	.9872	.9877	.9881	.9886	.9890	.9894	.9899	.9903	.9908
9.8	.9912	.9917	.9921	.9926	.9930	.9934	.9939	.9943	.9948	.9952
9.9	.9956	.9961	.9965	.9969	.9974	.9978	.9983	.9987	.9991	.9996

Index

MOVE TO THE HEAD OF YOUR CLASS

THE EASY WAY!

Barron's presents THE EASY WAY SERIES—specially prepared by top educators, it maximizes effective learning, while minimizing the time and effort it takes to raise your grades, brush up on the basics, and build your confidence. Comprehensive and full of clear review examples, **THE EASY WAY SERIES** is your best bet for better grades, quickly!

0-8120-9409-3 Accounting the Easy Way, 3rd Ed.—$13.95, Can. $17.95
0-8120-9393-3 Algebra the Easy Way, 3rd Ed.—$12.95, Can. $16.95
0-8120-1943-1 American History the Easy Way, 2nd Ed.—$13.95, Can. $18.95
0-7641-0299-0 American Sign Language the Easy Way—$12.95, Can. $16.95
0-8120-9134-5 Anatomy and Physiology the Easy Way—$14.95, Can. $19.95
0-8120-9410-7 Arithmetic the Easy Way, 3rd Ed.—$13.95, Can. $18.95
0-8120-4286-7 Biology the Easy Way, 2nd Ed.—$12.95, Can. $16.95
0-7641-1079-9 Bookkeeping the Easy Way, 3rd Ed.—$12.95, Can. $17.50
0-8120-4760-5 Business Law the Easy Way—$12.95, Can. $17.50
0-7641-0314-8 Business Letters the Easy Way, 3rd Ed.—$11.95, Can. $15.95
0-8120-4627-7 Business Mathematics the Easy Way, 2nd Ed.—$13.95, Can. $18.95
0-8120-9141-8 Calculus the Easy Way, 3rd Ed.—$12.95, Can. $16.95
0-8120-9138-8 Chemistry the Easy Way, 3rd Ed.—$12.95, Can. $16.95
0-7641-0659-7 Chinese the Easy Way—$13.95, Can. $18.95
0-8120-4253-0 Computer Programming in Basic the Easy Way, 2nd Ed.—$9.95, Can. $13.95
0-8120-2800-7 Computer Programming in Fortran the Easy Way—$11.95, Can. $15.95
0-7641-0752-6 Computer Programming in Java the Easy Way—$18.95, Can. $25.50
0-8120-2799-X Computer Programming in Pascal the Easy Way—$14.95, Can. $19.95
0-8120-9144-2 Electronics the Easy Way, 3rd Ed.—$13.95, Can. $18.95
0-8120-9142-6 English the Easy Way, 3rd Ed.—$12.95, Can. $16.95
0-8120-9505-7 French the Easy Way, 3rd Ed.—$12.95, Can. $16.95
0-7641-0110-2 Geometry the Easy Way, 3rd Ed.—$12.95, Can. $16.95
0-8120-9145-0 German the Easy Way, 2nd Ed.—$13.95, Can. $18.95
0-8120-9146-9 Italian the Easy Way, 2nd Ed.—$12.95, Can. $16.95
0-8120-9627-4 Japanese the Easy Way—$13.95, Can. $18.95
0-8120-9139-6 Math the Easy Way, 3rd Ed.—$11.95, Can. $15.95
0-8120-9601-0 Microeconomics the Easy Way—$12.95, Can. $16.95
0-7641-0236-2 Physics the Easy Way, 3rd Ed.—$13.95, Can. $18.95
0-8120-9412-3 Spanish the Easy Way, 3rd Ed.—$12.95, Can. $16.95
0-8120-9852-8 Speed Reading the Easy Way—$12.95, Can. $16.95
0-8120-9143-4 Spelling the Easy Way, 3rd Ed.—$12.95, Can. $16.95
0-8120-9392-5 Statistics the Easy Way, 3rd Ed.—$12.95, Can. $16.95
0-8120-4389-8 Trigonometry the Easy Way, 2nd Ed.—$12.95, Can. $16.95
0-8120-9147-7 Typing the Easy Way, 3rd Ed.—$14.95, Can. $19.95
0-8120-9765-3 World History the Easy Way, Vol. One—$12.95, Can. $16.95
0-8120-9766-1 World History the Easy Way, Vol. Two—$12.95, Can. $16.95
0-8120-4615-3 Writing the Easy Way, 2nd Ed.—$13.95, Can. $18.95

Barron's Educational Series, Inc.
250 Wireless Boulevard • Hauppauge, New York 11788
In Canada: Georgetown Book Warehouse • 34 Armstrong Avenue, Georgetown, Ontario L7G 4R9
www.barronseduc.com

$ = U.S. Dollars Can. $ = Canadian Dollars

Prices subject to change without notice. Books may be purchased at your local bookstore, or by mail from Barron's. Enclose check or money order for total amount plus sales tax where applicable and 15% for postage and handling (minimum charge $4.95 U.S. and Canada). All books are paperback editions.

(#45) R4/99

4345

QA
152.2
.D69

1996